T0255058

LONDON MATHEMATICAL SOCIETY LECTURE NOTE SERIES

Managing Editor: Professor N.J. Hitchin, Mathematical Institute,
University of Oxford, 24–29 St Giles, Oxford OX1 3LB, United Kingdom

The titles below are available from booksellers, or from Cambridge University Press at www.cambridge.org

London Mathematical Society Lecture Note Series. 314

Spectral Generalizations of Line Graphs

On graphs with least eigenvalue −2

Dragoš Cvetković
University of Belgrade

Peter Rowlinson
University of Stirling

Slobodan Simić
University of Belgrade

CAMBRIDGE
UNIVERSITY PRESS

CAMBRIDGE
UNIVERSITY PRESS

University Printing House, Cambridge CB2 8BS, United Kingdom

One Liberty Plaza, 20th Floor, New York, NY 10006, USA

477 Williamstown Road, Port Melbourne, VIC 3207, Australia

314-321, 3rd Floor, Plot 3, Splendor Forum, Jasola District Centre, New Delhi - 110025, India

103 Penang Road, #05-06/07, Visioncrest Commercial, Singapore 238467

Cambridge University Press is part of the University of Cambridge.

It furthers the University's mission by disseminating knowledge in the pursuit of education, learning and research at the highest international levels of excellence.

www.cambridge.org
Information on this title: www.cambridge.org/9780521836630

© Dragoš Cvetković, Peter Rowlinson, Slobodan Simić 2004

This publication is in copyright. Subject to statutory exception and to the provisions of relevant collective licensing agreements, no reproduction of any part may take place without the written permission of Cambridge University Press.

First published 2004

A catalogue record for this publication is available from the British Library

Library of Congress Cataloging in Publication data
Cvetković, Dragoš M.
Spectral generalizations of line graphs : on graphs with least negative eigenvalue −2 /
Dragoš Cvetković, Peter Rowlinson, Slobodan Simić.
p. cm.
Includes bibliographical references and index.
ISBN 0-521-83663-8 (pbk.)
1. Graph theory. 2. Eigenvalues. I. Rowlinson, Peter. II. Simic, Slobodan
III. Title.
QA166.C837 2004
511´.5–dc22 2003065393

ISBN 978-0-521-83663-0 Paperback

Cambridge University Press has no responsibility for the persistence or accuracy of URLs for external or third-party internet websites referred to in this publication, and does not guarantee that any content on such websites is, or will remain, accurate or appropriate.

In memory of our late parents:

Jelka Cvetković	(1904–1993)
Mladen Cvetković	(1901–1979)
Irene Rowlinson	(1910–2001)
Arthur Rowlinson	(1908–1996)
Olga Simić	(1916–2002)
Kosta Simić	(1907–1978)

Contents

Preface

The eigenvalues discussed in this book are those of a $(0, 1)$-adjacency matrix of a finite undirected graph. Line graphs, familiar to graph-theorists for decades, have the property that their least eigenvalue is greater than or equal to -2. This property is shared with generalized line graphs, which can be viewed as line graphs of certain multigraphs. Apart from these classes of examples there are only finitely many further connected graphs with spectrum in the interval $[-2, \infty)$, and these are called exceptional graphs. This book deals with line graphs, generalized line graphs and exceptional graphs, in the context of spectral properties of graphs. Having worked in spectral graph theory for many years, the authors came to see the need for a single source of information on the principal results in this area. Work began early in 2000, and the principal motivation for writing the book at this juncture was the construction of the maximal exceptional graphs in 1999. The working title has become the subtitle on the grounds that 'Graphs with least eigenvalue -2' might appear unreasonably specialized to the casual observer. In fact, the subtitle is not wholly accurate in that it is necessary to treat also the graphs with least eigenvalue greater than -2.

The requirement that the spectrum of a graph lies in $[-2, \infty)$ is a natural one, and in principle not a restriction at all. The reason is to be found in the classical result of H. Whitney, who showed in 1932 that two connected graphs (with more than three vertices) are isomorphic if and only if their line graphs are isomorphic.

The titles of Chapters 2, 3 and 5, namely 'Forbidden subgraphs', 'Root systems' and 'Star complements' reflect three major techniques and three periods in the study of graphs with least eigenvalue -2. Of course, early results were often improved using later techniques, but on considering the interplay between techniques, the authors decided that a presentation broadly in chronological order was the most natural approach.

The forbidden subgraph technique (Chapter 2) was introduced by A. J. Hoffman and others in the 1960s. It is based on the fact that the property of having least eigenvalue greater than or equal to -2 is a hereditary property, that is, a property which the graph shares with all its induced subgraphs. For any hereditary property $\mathcal{P}$ we can consider graphs without property $\mathcal{P}$ which are minimal with respect to the induced subgraph relation: such graphs are the minimal forbidden subgraphs for graphs with property $\mathcal{P}$. For graphs with least eigenvalue greater than or equal to -2, the collection of minimal forbidden subgraphs is finite.

The subject of Chapter 3 is the root system technique introduced by P. J. Cameron, J. M. Goethals, J. J. Seidel and E. E. Shult [CaGSS] in 1976. Root systems were already known in the theory of Lie algebras and in other parts of mathematics, and it turned out that graphs with least eigenvalue -2 can be elegantly described by means of root systems. The description relies on the use of Gram matrices of certain sets of vectors to represent the graphs in question. Generalized line graphs (including line graphs) can be represented in the root system D_n for some n while the existence of the exceptional root system E_8 in 8-dimensional Euclidean space (containing extremely densely packed sets of vectors at 60 and 90 degrees) accounts for the existence of graphs with least eigenvalue -2 which are not generalized line graphs. Chapter 4 uses the tools introduced in Chapter 3 to investigate regular graphs; many spectral characterization theorems for regular line graphs are presented, among them some results from Chapter 2 in an improved form with shorter proofs.

The star complement technique was introduced into the study of graphs with least eigenvalue -2 by the authors of this book in 1998 [CvRS4]. One of the main results presented in Chapter 5 is a characterization of exceptional graphs by exceptional star complements, and this enables all of the maximal exceptional graphs to be constructed (Chapter 6).

Preliminary results in spectral graph theory are given in Chapter 1, while Chapter 7 contains miscellaneous results that do not fit readily into the earlier chapters. It is relatively straightforward to describe a means of constructing exceptional graphs, but the results of the construction make for a fairly elaborate picture. Accordingly the technical descriptions of the 187 regular exceptional graphs and the 473 maximal exceptional graphs are consigned to the Appendix. The authors are grateful to M. Lepović (University of Kragujevac, Serbia & Montenegro) for his assistance in completing the tables in the Appendix, which throughout the book are referred to as Tables A1 to A7. Table A2 contains a description of the 573 exceptional graphs with least eigenvalue greater than -2.

The book brings together many independent discoveries and overlapping results, and provides more than 250 references to the literature. The vast majority

of the material has not previously appeared in book form. The classification by
P. J. Cameron *et al* [CaGSS] using root systems has been summarized in various
forms in the monographs [BrCN], [CaLi] and [GoRo]. In this book an outline
appears in Section 3.5, following the presentation of a lesser known approach
due to M. Doob and D. Cvetković [CvDo2]. Further, we acknowledge a debt to
[BrCN, chapter 3] as the source of our proof of Theorem 4.1.5, and as a guide
to results on lattices.

Inevitably it has been necessary to limit the scope of the book. A more
ambitious work on graphs with least eigenvalue -2 could elaborate not only
on the connections with Lie algebras and lattices but also on the relation to
distance-regular graphs, association schemes, block designs, signed graphs,
Coxeter systems, Weyl groups and many other combinatorial or algebraic ob-
jects. We have merely drawn attention to such connections by short comments
and relevant references at the appropriate places. Many of these links to other
mathematical areas are described in the book [BrCN] and the expository paper
[CaST].

The authors are grateful for financial support from the United Kingdom
Engineering & Physical Sciences Research Council (EPSRC); the Serbian
Academy of Science & Arts; the Serbian Ministry for Science, Technology
and Development; the Universities of Belgrade and Stirling; and the University
of Montenegro (S.S. in the period 2000–2002).

Belgrade   D. Cvetković
Stirling   P. Rowlinson
Belgrade   S. Simić

August 2003

1

Introduction

In Section 1.1 we introduce notation and terminology which will be used throughout the book. In particular, we define *line graphs*, *generalized line graphs* and *exceptional graphs*, all of which have least eigenvalue greater than or equal to -2. Sections 1.2 and 1.3 contain some theorems related to graph spectra which will be used in other chapters. A short history of research on graphs with least eigenvalue -2 is given in Section 1.4.

1.1 Basic notions and results

Unless stated otherwise, the graphs we consider are finite undirected graphs without loops or multiple edges, and the eigenvalues we consider are those in the spectrum of a $(0, 1)$-adjacency matrix (as defined below). A comprehensive introduction to the theory of graph spectra is given in the monograph [CvDSa], along with some of the underlying results from matrix theory. Further results concerning the spectrum of an adjacency matrix can be found in [CvDGT] and [CvRS2]. Here we present only the basic notions which are needed frequently in other chapters. We recommend as a general reference on graph theory the book by Harary [Har]; and as general references on algebraic graph theory the texts by N. L. Biggs [Big] and C. Godsil and G. Royle [GoRo]. Some material related to graphs with least eigenvalue -2 can be found in the books [BrCN] and [CaLi].

If G is a graph with vertices $1, 2, \ldots, n$ then its *adjacency matrix* is the $n \times n$ matrix A $(= A(G))$ whose (i, j)-entry a_{ij} is 1 if the vertices i, j are adjacent (written $i \sim j$), and 0 otherwise. As an example, the adjacency matrix of a 4-cycle is illustrated in Fig. 1.1.

The characteristic polynomial $\det(xI - A)$ of the adjacency matrix A of G is called the *characteristic polynomial of G* and denoted by $P_G(x)$. The

1

$$A = \begin{pmatrix} 0 & 1 & 0 & 1 \\ 1 & 0 & 1 & 0 \\ 0 & 1 & 0 & 1 \\ 1 & 0 & 1 & 0 \end{pmatrix} \qquad G:$$

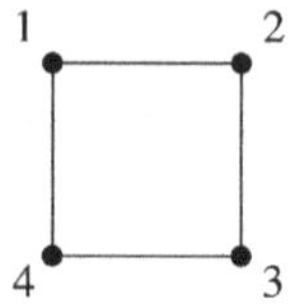

Figure 1.1: A labelled graph G and its adjacency matrix A.

eigenvalues of A (i.e. the zeros of $\det(xI - A)$) and the spectrum of A (which consists of the n eigenvalues) are called the *eigenvalues* and the *spectrum* of G, respectively: these notions are independent of vertex labelling because a re-ordering of vertices results in a similar adjacency matrix. The eigenvalues of G are usually denoted by $\lambda_1, \lambda_2, \ldots, \lambda_n$; they are real because A is symmetric. Unless we indicate otherwise, we shall assume that $\lambda_1 \geq \lambda_2 \geq \cdots \geq \lambda_n$ and use the notation $\lambda_i = \lambda_i(G)$ ($i = 1, 2, \ldots, n$). The least eigenvalue is also denoted by $\lambda(G)$. Clearly, isomorphic graphs have the same spectrum. In an *integral* graph, all eigenvalues are integers.

If G has distinct eigenvalues $\mu_1, \mu_2, \ldots, \mu_m$ with multiplicities $k_1, k_2, \ldots, k_m$ respectively, we shall frequently write $\mu_1^{k_1}, \mu_2^{k_2}, \ldots, \mu_m^{k_m}$ for the spectrum of G.

The eigenvalues of A are the numbers λ satisfying $A\mathbf{x} = \lambda\mathbf{x}$ for some non-zero vector $\mathbf{x} \in \mathbb{R}^n$. Each such vector $\mathbf{x}$ is called an *eigenvector* of the matrix A (or of the labelled graph G) belonging to the eigenvalue λ. The relation $A\mathbf{x} = \lambda\mathbf{x}$ can be interpreted in the following way: if $\mathbf{x} = (x_1, x_2, \ldots, x_n)^T$ then $\lambda x_u = \sum_{v \sim u} x_v$, where the summation is over all neighbours v of the vertex u. If λ is an eigenvalue of A then the set $\{\mathbf{x} \in \mathbb{R}^n : A\mathbf{x} = \lambda\mathbf{x}\}$ is a subspace of $\mathbb{R}^n$, called the *eigenspace* of λ and denoted by $\mathcal{E}(\lambda)$ or $\mathcal{E}_G(\lambda)$. Such eigenspaces are called eigenspaces of G. Of course, relabelling of the vertices in G will result in a permutation of coordinates in eigenvectors (and eigenspaces).

Example 1.1.1. For the eigenvalues λ of the graph G in Fig. 1.1 we have

$$P_G(\lambda) = \begin{vmatrix} \lambda & -1 & 0 & -1 \\ -1 & \lambda & -1 & 0 \\ 0 & -1 & \lambda & -1 \\ -1 & 0 & -1 & \lambda \end{vmatrix} = \lambda^4 - 4\lambda^2 = 0.$$

The eigenvalues in non-increasing order are $\lambda_1 = 2$, $\lambda_2 = 0$, $\lambda_3 = 0$, $\lambda_4 = -2$ with linearly independent eigenvectors $\mathbf{x}_1$, $\mathbf{x}_2$, $\mathbf{x}_3$, $\mathbf{x}_4$, where $\mathbf{x}_1 = (1, 1, 1, 1)^T$, $\mathbf{x}_2 = (1, 1, -1, -1)^T$, $\mathbf{x}_3 = (-1, 1, 1, -1)^T$, $\mathbf{x}_4 = (1, -1, 1, -1)^T$. Hence G is a graph with least eigenvalue -2. We have $\mathcal{E}(2) = \langle \mathbf{x}_1 \rangle$, $\mathcal{E}(0) = \langle \mathbf{x}_2, \mathbf{x}_3 \rangle$

and $\mathcal{E}(-2) = \langle \mathbf{x}_4 \rangle$, where angle brackets denote the subspace spanned by the enclosed vectors. □

Example 1.1.2. The eigenvalues of an n-cycle are $2\cos\frac{2\pi j}{n}$ $(j = 0, 1, \ldots, n-1)$ [Big, p. 17]. Thus the largest eigenvalue is 2 (with multiplicity 1) and the second largest is $2\cos\frac{2\pi}{n}$ (with multiplicity 2). The least eigenvalue is -2 (with multiplicity 1) if n is even, and $2\cos\frac{(n-1)\pi}{n}$ (with multiplicity 2) if n is odd. □

Example 1.1.3. The well-known Petersen graph has spectrum $3, 1^5, (-2)^4$; again we have a graph with least eigenvalue -2. □

The following remarks on matrices will serve to establish more notation.

Since A is a symmetric matrix with real entries there exists an orthogonal matrix U such that $U^T A U$ is a diagonal matrix, D say. Here $D = \text{diag}(\lambda_1, \lambda_2, \ldots, \lambda_n)$ (where $\lambda_1, \lambda_2, \ldots, \lambda_n$ are the eigenvalues of A in some order), and the columns of U are corresponding eigenvectors which form an orthonormal basis of $\mathbb{R}^n$. If this basis is constructed by stringing together orthonormal bases of the eigenspaces of A then $D = \mu_1 E_1 + \cdots + \mu_m E_m$ where $\mu_1, \ldots, \mu_m$ are the distinct eigenvalues of A and each E_i has block diagonal form $\text{diag}(O, \ldots, O, I, O, \ldots O)$ $(i = 1, \ldots, m)$. Then A has the *spectral decomposition*

(1.1) $$A = \mu_1 P_1 + \cdots + \mu_m P_m$$

where $P_i = U E_i U^T$ $(i = 1, \ldots, m)$. For fixed i, if $\mathcal{E}(\mu_i)$ has $\{\mathbf{x}_1, \ldots, \mathbf{x}_d\}$ as an orthonormal basis then

(1.2) $$P_i = \mathbf{x}_1 \mathbf{x}_1^T + \cdots + \mathbf{x}_d \mathbf{x}_d^T$$

and P_i represents the orthogonal projection of $\mathbb{R}^n$ onto $\mathcal{E}(\mu_i)$ with respect to the standard orthonormal basis of $\mathbb{R}^n$. Moreover $P_i^2 = P_i = P_i^T$ $(i = 1, \ldots, m)$ and $P_i P_j = O$ $(i \neq j)$.

Let $\{\mathbf{e}_1, \ldots, \mathbf{e}_n\}$ be the standard orthonormal basis of $\mathbb{R}^n$. The mn numbers $\alpha_{ij} = \|P_i \mathbf{e}_j\|$ are called the *angles* of G; they are the cosines of the (acute) angles between axes and eigenspaces. We shall assume throughout that $\mu_1 > \cdots > \mu_m$. If also we order the columns of the matrix (α_{ij}) lexicographically then this matrix is a graph invariant, called the *angle matrix* of G.

We shall also need the observation that for any polynomial f, we have

$$f(A) = f(\mu_1)P_1 + \cdots + f(\mu_m)P_m.$$

In particular, P_i is a polynomial in A for each i; explicitly, $P_i = f_i(A)$ where

$$f_i(x) = \frac{\prod_{s \neq i}(x - \mu_s)}{\prod_{s \neq i}(\mu_i - \mu_s)}.$$

Next we present certain notation, definitions and results from graph theory.

As usual, K_n, C_n and P_n denote respectively the *complete graph*, the *cycle* and the *path* on n vertices. A connected graph with n vertices is said to be *unicyclic* if it has n edges, for then it contains a unique cycle. If this cycle has odd length, then the graph is said to be *odd-unicyclic*. A connected graph with n vertices and $n + 1$ edges is called a *bicyclic* graph. A complete subgraph of a graph G is called a *clique* of G, while a *coclique* is an induced subgraph without edges. Further, $K_{m,n}$ denotes the *complete bipartite* graph on $m + n$ vertices. A graph of the form $K_{1,n}$ is called an *n-claw* or a *star*. (The term 'star' is used in different contexts in Sections 3.1 and 5.1.) More generally, $K_{n_1,n_2,\dots,n_k}$ denotes the *complete k-partite graph* with parts (colour classes) of size $n_1, n_2, \dots, n_k$.

Vertices (or edges) are said to be *independent* if they are pairwise non-adjacent. Any set of independent edges in a graph G is called a *matching* of G. A matching of G is *perfect* if each vertex of G is the endvertex of an edge from the matching. The *cocktail party graph* $CP(n)$ is the unique regular graph with $2n$ vertices of degree $2n - 2$; it is obtained from K_{2n} by deleting a perfect matching. The degree of a vertex v is denoted by $\deg(v)$. An edge that contains a vertex of degree 1 is called a *pendant* edge.

A regular graph of degree r is called *r-regular*. A *strongly regular* graph, with parameters (n, r, e, f), is an r-regular graph of order n, other than K_n or its complement, such that any two adjacent vertices have e common neighbours and any two non-adjacent vertices have f common neighbours. The concept of a strongly regular graph was introduced in 1963 by R. C. Bose [Bos], and there is now an extensive literature on graphs of this type; see, for example, [BrLi].

A graph is called *semi-regular bipartite*, with parameters (n_1, n_2, r_1, r_2), if it is bipartite (i.e. 2-colourable) and vertices in the same colour class have the same degree (n_1 vertices of degree r_1 and n_2 vertices of degree r_2, where $n_1 r_1 = n_2 r_2$).

The *complement* of a graph G is denoted by $\overline{G}$, while mG denotes the graph consisting of m disjoint copies of G. The *subdivision graph* $S(G)$ is obtained from G by inserting a vertex of degree 2 in each edge of G. The *total graph* of G is denoted by $T(G)$: its vertices are the vertices and edges of G, and these are adjacent in $T(G)$ if and only if they are adjacent or incident in G. We write $V(G)$ for the vertex set of G, and $E(G)$ for the edge set of G. We call $|V(G)|$ the *order* of G, and we say that G is *empty* if $V(G) = \emptyset$. If G, H are graphs, with $V(G) = \{v_1, \dots, v_n\}$, then the *corona* $G \circ H$ is obtained from G by adding n

disjoint copies of H and joining v_i by an edge to each vertex in the i-th copy of H ($i = 1, \ldots, n$). The *union* of disjoint copies of the graphs G and H is denoted by $G \mathbin{\dot\cup} H$. The *join* $G \bigtriangledown H$ of (disjoint) graphs G and H is the graph obtained from $G \mathbin{\dot\cup} H$ by joining each vertex of G to each vertex of H. The graph $K_1 \bigtriangledown H$ is called the *cone* over H, while $K_2 \bigtriangledown H$ is called the *double cone* over H.

If uv is an edge of G we write $G - uv$ for the graph obtained from G by deleting uv. For $v \in V(G)$, $G - v$ denotes the graph obtained from G by deleting the vertex v and all edges incident with v. More generally, for $U \subseteq V(G)$, $G - U$ is the subgraph of G induced by $V(G) \setminus U$. If each vertex of $G - U$ is adjacent to a vertex of U then U is called a *dominating set* in G.

Definition 1.1.4. *The* line graph $L(H)$ *of a graph H is the graph whose vertices are the edges of H, with two vertices in $L(H)$ adjacent whenever the corresponding edges in H have exactly one vertex in common.*

If $G = L(H)$ for some graph H, then H is called a *root graph* of G. If $E(H) = \emptyset$ then G is the empty graph. Accordingly, we take a line graph to mean a graph of the form $L(H)$, where $E(H)$ is non-empty; note that we may assume if necessary that H has no isolated vertices. If H is connected, then the same is true of $L(H)$. If H is disconnected, then each non-trivial component of H gives rise to a connected component of $L(H)$.

We mention a simple, but useful, observation.

Proposition 1.1.5. *If $L(H)$ is a regular connected graph, then H is either regular or semi-regular bipartite.*

The proof is straightforward, and therefore omitted.

The *incidence matrix* of the graph H is a matrix B whose rows and columns are indexed by the vertices and edges of H, respectively. The (v, e)-entry of B is

$$b_{ve} = \begin{cases} 0 \text{ if } v \text{ is not incident with } e, \\ 1 \text{ if } v \text{ is incident with } e. \end{cases}$$

Thus the columns of B are the characteristic vectors of the edges of H as subsets of $V(H)$. Now we find easily that

(1.3) $$B^T B = A(G) + 2I,$$

where $G = L(H)$. Thus $A(G) + 2I$ is a positive semi-definite matrix and hence all of its eigenvalues are non-negative. Consequently the least eigenvalue of G is not less than -2; this is the most remarkable spectral property of line graphs. We shall see in Section 2.2 that the least eigenvalue is strictly greater than -2 if and only if the rank of the matrix B is $|V(H)|$.

The class of graphs with spectrum in the interval $[-2, \infty)$ also contains the cocktail party graphs. Are there any others? This natural question was posed by A. J. Hoffman [Hof8], who constructed a wide class of examples in which line graphs and cocktail party graphs are combined as follows.

Definition 1.1.6. *Let H be a graph with vertex set $\{v_1, \ldots, v_n\}$, and let $a_1, \ldots, a_n$ be non-negative integers. The generalized line graph $L(H; a_1, \ldots, a_n)$ consists of disjoint copies of $L(H)$ and $CP(a_1), \ldots, CP(a_n)$ along with all edges joining a vertex $\{v_i, v_j\}$ of $L(H)$ with each vertex in $CP(a_i)$ and $CP(a_j)$.*

We give immediately an alternative definition prompted by the observation that Definition 1.1.4 extends to the line graph of a loopless multigraph $\hat{H}$: vertices in $L(\hat{H})$ are adjacent if and only if the corresponding edges in $\hat{H}$ have exactly one vertex in common. (In particular, multiple edges between two given vertices of $\hat{H}$ are non-adjacent in $L(\hat{H})$.) If $G = L(\hat{H})$ then we call $\hat{H}$ a *root multigraph* of G. We say that a *petal* is added to a graph when we add a pendant edge and then duplicate this edge to form a pendant 2-cycle.

Definition 1.1.6′. *Let H be a graph with vertex set $\{v_1, \ldots, v_n\}$, and let $a_1, \ldots, a_n$ be non-negative integers. The generalized line graph $G = L(H; a_1, \ldots, a_n)$ is the graph $L(\hat{H})$, where $\hat{H}$ is the multigraph $H(a_1, \ldots, a_n)$ obtained from H by adding a_i petals at vertex v_i $(i = 1, \ldots, n)$.*

This construction of a generalized line graph is illustrated in Fig. 1.2. Note that the cocktail party graph $CP(n)$ is the generalized line graph $L(K_1; n)$.

The incidence matrix $C = (c_{ve})$ of $\hat{H} = H(a_1, \ldots, a_n)$ is defined as for H with the following exception: if e and f are the edges between v and w in a

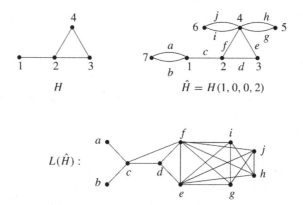

Figure 1.2: Construction of a generalized line graph

petal at v then $\{c_{we}, c_{wf}\} = \{-1, 1\}$. (Note that all other entries in row w are zero.) For example, the incidence matrix of the multigraph $\hat{H}$ from Fig. 1.2 is:

$$\begin{pmatrix} 1 & 1 & 1 & 0 & 0 & 0 & 0 & 0 & 0 & 0 \\ 0 & 0 & 1 & 1 & 0 & 1 & 0 & 0 & 0 & 0 \\ 0 & 0 & 0 & 1 & 1 & 0 & 0 & 0 & 0 & 0 \\ 0 & 0 & 0 & 0 & 1 & 1 & 1 & 1 & 1 & 1 \\ 0 & 0 & 0 & 0 & 0 & 0 & -1 & 1 & 0 & 0 \\ 0 & 0 & 0 & 0 & 0 & 0 & 0 & 0 & -1 & 1 \\ -1 & 1 & 0 & 0 & 0 & 0 & 0 & 0 & 0 & 0 \end{pmatrix}.$$

Here the rows are indexed by $1, 2, \ldots, 7$ and the columns are indexed by $a, b, \ldots, j$.

Remark 1.1.7. It is straightforward to show that the generalized line graphs are the only graphs with least eigenvalue greater than or equal to -2 which arise when we allow -1s in an incidence matrix. In other words, if R is a $(-1, 0, 1)$-matrix with the same zero - non-zero pattern as the incidence matrix C of some multigraph, then $R^T R - 2I$ is the adjacency matrix of a graph G if and only if G is a generalized line graph. In this situation, R need not coincide with C; for instance if $\deg(v) = 1$ and e is the edge incident with v then the (v, e)-entry of R can be either 1 or -1. ☐

The following observation will be required in the next chapter.

Proposition 1.1.8. *Let C be the incidence matrix of a root multigraph $\hat{H}$ for which $|V(\hat{H})| = |E(\hat{H})| = n$. If $\mathrm{rank}(C) = n$ then $\det(C) = \pm 2^k$, where k is the number of components of $\hat{H}$.*

Proof. Suppose that $\mathrm{rank}(C) = n$. Note first that if $\hat{H}$ is not connected then C has block-diagonal form with each diagonal block a (square) invertible incidence matrix. Accordingly it suffices to prove the result in the case $k = 1$, and we assume that $\hat{H}$ is connected. Then $\hat{H}$ is either a unicyclic graph or a tree with a single petal attached. It is immediate that $\det(C) = \pm\det(C')$, where C' is the incidence matrix of any graph obtained from $\hat{H}$ by deleting an endvertex. Repeated use of this property shows that $\det(C) = \pm\det(R)$ where R is the incidence matrix of an odd cycle or a single petal. In the latter case, $\det(R) = \pm \begin{vmatrix} 1 & -1 \\ 1 & 1 \end{vmatrix} = \pm 2$. In the former case, we have $R^T R = A(C_m) + 2I$ for some odd integer $m \geq 3$ and so we have (cf. Example 1.1.2):

$$\det(R)^2 = \prod_{j=0}^{m-1} 2(1 + \cos\frac{2\pi j}{m}) = \prod_{j=0}^{m-1} 4\cos^2\frac{\pi j}{m}.$$

Now $\prod_{j=0}^{m-1} 2\cos\frac{\pi j}{m} = \pm 2$, for example by considering an appropriate Chebyshev polynomial (cf. [CvDSa, p. 73]). The result follows.   □

Proposition 1.1.9. *A regular connected generalized line graph is either a line graph or a cocktail party graph.*

Proof. Suppose that G is a connected generalized line graph $L(H; a_1, a_2, \ldots, a_n)$ which is neither a line graph nor a cocktail party graph. Then H contains an edge ij with $a_i > 0$. Now the degree of ij as a vertex of G is greater than the degree of an edge in any of the a_i petals at i. This completes the proof.   □

Next we turn to a frequently used graph transformation known as *switching*, or *Seidel switching*. Many elementary properties of this transformation can be found in [Sei4]. Given a subset U of vertices of the graph G, the graph G_U obtained from G by switching with respect to U differs from G as follows: for $u \in U, v \notin U$ the vertices u, v are adjacent in G_U if and only if they are non-adjacent in G. Thus if G has adjacency matrix $A(G) = \begin{pmatrix} A_U & B^T \\ B & C \end{pmatrix}$, where A_U is the adjacency matrix of the subgraph induced by U, then G_U has adjacency matrix $A(G_U) = \begin{pmatrix} A_U & \overline{B}^T \\ \overline{B} & C \end{pmatrix}$, where $\overline{B}$ is obtained from B by interchanging 0s and 1s. When G is regular, this formulation makes it straightforward to find a necessary and sufficient condition on U for G_U to be regular of the same degree:

Proposition 1.1.10. *Suppose that G is regular with n vertices and degree r. Then G_U is regular of degree r if and only if U induces a regular subgraph of degree k, where $|U| = n - 2(r - k)$.*

Note that switching with respect to the subset U of $V(G)$ is the same as switching with respect to its complement in $V(G)$. Switching is described easily in terms of the *Seidel matrix* S of G defined as follows: the (i, j)-entry of S is 0 if $i = j$, -1 if i is adjacent to j, and 1 otherwise. Thus $S = J - I - 2A$, where J is the all-1 matrix and A is the adjacency matrix of G. The Seidel matrix of G_U is $D^{-1}SD$ where D is the (involutory) diagonal matrix whose i-th diagonal entry is 1 if $i \in U$, -1 if $i \notin U$. Now it is easy to see that switching with respect to U and then with respect to V is the same as switching with respect to $(U \setminus V) \cup (V \setminus U)$; it follows that switching determines an equivalence relation on graphs.

Note that switching-equivalent graphs have similar Seidel matrices and hence the same Seidel spectrum. For regular graphs, the matrices J, A and $J - 2I - A$ are simultaneously diagonalizable and so we deduce the following.

Proposition 1.1.11. *If G and G_U are regular of the same degree, then G and G_U are cospectral.*

Example 1.1.12. Let S_1, S_2, S_3 be sets of vertices of $L(K_8)$ which induce subgraphs isomorphic to $4K_1$, $C_5 \dot{\cup} C_3$ and C_8, respectively. The graphs Ch_1, Ch_2, Ch_3 obtained from $L(K_8)$ by switching with respect to S_1, S_2, S_3 respectively are called the *Chang graphs*. The graphs $L(K_8), Ch_1, Ch_2, Ch_3$ are regular of degree 12, cospectral and pairwise non-isomorphic (see, for example, [BrCN, p. 105], and also [Sei4]). These graphs are strongly regular with parameters $(28, 12, 6, 4)$.

If we switch $L(K_8)$ with respect to the set of neighbours of a vertex v, we obtain a graph H in which v is an isolated vertex. If we delete v from H we obtain a 16-regular graph of order 27 which is called the *Schläfli* graph; it is strongly regular with parameters $(27, 16, 10, 8)$. □

Like the Petersen graph, the Chang graphs and the Schläfli graph are not generalized line graphs. Connected graphs with least eigenvalue greater than or equal to -2 which are not generalized line graphs are called *exceptional* graphs.

Theorem 1.1.13 [Sei4]. *The result of switching $L(K_8)$ with respect to any set of $4k$ vertices $(k \in I\!N)$ that induce the line graph of a regular graph is again $L(K_8)$ or one of the Chang graphs. The Chang graphs can be obtained from $L(K_8)$ by switching only in this way.*

Example 1.1.14. If we switch the graph $L(K_{4,4})$ with respect to four independent vertices, then we obtain a 6-regular exceptional graph of order 16 called the *Shrikhande* graph; it is strongly regular with parameters $(16, 6, 2, 2)$. If we switch $L(K_{4,4})$ with respect to the vertices of an induced subgraph $L(K_{4,2})$ then we obtain a 10-regular exceptional graph of order 16 called the *Clebsch* graph; it is strongly regular with parameters $(16, 10, 6, 6)$.

These graphs are represented in Fig. 1.3. In Fig. 1.3(a), the vertices of $L(K_{4,4})$ are shown as the points of intersection of four horizontal and four vertical lines, two vertices being adjacent in $L(K_{4,4})$ if and only if the corresponding points

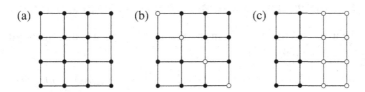

Figure 1.3: Construction of the graphs in Example 1.1.14.

are collinear. In Figs. 1.3(b) and 1.3(c), the white vertices are those in switching sets which yield the Shrikhande and Clebsch graphs, respectively.   □

The following results on graph switching will be useful in the sequel.

Proposition 1.1.15 [Sei4]. *Let G be a graph of even order, with p vertices of odd degree and q vertices of even degree. If G' is switching equivalent to G then either G' has p vertices of odd degree and q vertices of even degree, or G' has q vertices of odd degree and p vertices of even degree.*

Proof. Let $G' = G_U$. It is straightforward to check that if $|U|$ is even then parities of degrees are preserved when switching with respect to U; if $|U|$ is odd, then parities are reversed.   □

Proposition 1.1.16 (cf. [Sei2]). *For any graph G and any vertex v of G there exists a unique switching-equivalent graph G' which has v as an isolated vertex.*

Proof. The graph G' is necessarily obtained from G by switching with respect to the set of neighbours of v.   □

In Example 1.1.12, the Schläfli graph is constructed by means of such a switching.

Proposition 1.1.17 [BuCS1]. *Let I(G) be the collection of graphs obtained by isolating in turn the vertices of the graph G. The graphs G_1 and G_2 are switching-equivalent if and only if $I(G_1) = I(G_2)$.*

Proof. If $I(G_1)$ and $I(G_2)$ contain a common graph, then the graphs G_1 and G_2 are switching-equivalent. Now suppose that G_1 and G_2 are switching-equivalent. Label the vertices of G_1 and G_2 by $v_1, \ldots, v_n$ in an arbitrary way. There exists a graph G_1' switching-equivalent to G_1 such that G_1' and G_2 are isomorphic. Let φ be any isomorphism from G_1' to G_2, and consider the vertices v_i, $\varphi(v_i)$. Of course, by isolating v_i in G_1' and $\varphi(v_i)$ in G_2 we obtain the same graphs; but, by Proposition 1.1.16, the graph obtained by isolating v_i in G_1 is the same as the graph obtained by isolating v_i in G_1'. Therefore, $I(G_1) = I(G_2)$.   □

1.2 Some general theorems from spectral graph theory

We present several theorems from the theory of graph spectra which will be used later in the book.

Theorem 1.2.1. *Let $\lambda_1 = r, \lambda_2, \ldots, \lambda_n$ $(\lambda_1 \geq \lambda_2 \geq \cdots \geq \lambda_n)$ be the spectrum of a graph G. The graph G is regular (of degree r) if and only if $\sum_{i=1}^n \lambda_i^2 = n\lambda_1$.*

In the case of regularity the number of components is equal to the multiplicity of λ_1.

In view of this theorem the structural property of being a regular connected graph can be established from the spectrum.

Theorem 1.2.2. *Let G be a connected graph. Then the following statements are equivalent:*

(i) *G is bipartite;*
(ii) *if λ is an eigenvalue of G then $-\lambda$ is also an eigenvalue of G with the same multiplicity;*
(iii) *if r is the largest eigenvalue of G then $-r$ is an eigenvalue of G.*

See [CvDSa], p. 94 and p. 87, for additional information on these theorems.

In situations where the spectrum of a graph does not provide sufficient structural information, a natural way of extending spectral techniques is to bring into consideration the eigenvectors of an adjacency matrix.

A graph is completely determined by eigenvalues and eigenvectors in the following way. Let A be the adjacency matrix of a graph G with vertices $1, 2, \ldots, n$ and eigenvalues $\lambda_1, \lambda_2, \ldots, \lambda_n$. If $\mathbf{x}_1, \mathbf{x}_2, \ldots, \mathbf{x}_n$ are linearly independent eigenvectors of A corresponding to $\lambda_1, \lambda_2, \ldots, \lambda_n$ respectively, if $V = (\mathbf{x}_1|\mathbf{x}_2|\cdots|\mathbf{x}_n)$ and if $D = \mathrm{diag}(\lambda_1, \lambda_2, \ldots, \lambda_n)$, then

$$A = VDV^{-1}$$

(cf. Equation (1.1)). Since G is determined by A, we have proved:

Theorem 1.2.3. *Any graph is determined by its eigenvalues and a basis of corresponding eigenvectors.*

Sometimes valuable information about a graph can be obtained from its eigenvectors alone, as in the following (straightforward) result:

Theorem 1.2.4. *A graph is regular if and only if its adjacency matrix has an eigenvector all of whose components are equal to 1.*

The largest eigenvalue ($\mu_1 = \lambda_1$) of a graph G is called the *index* of G; since adjacency matrices are non-negative there is a corresponding eigenvector whose entries are all non-negative (see, for example, [Gan], vol II, p. 66).

A matrix A is called *reducible* if there is a permutation matrix P such that the matrix $P^{-1}AP$ is of the form $\begin{pmatrix} X & O \\ Y & Z \end{pmatrix}$, where X and Z are square matrices. Otherwise, A is called *irreducible*. Note that if A is symmetric and reducible then $Y = O$ in the above definition. In particular, the adjacency matrix of a graph G is irreducible if and only if G is connected. Now a symmetric non-negative matrix is irreducible if and only if its largest eigenvalue is simple and there exists a corresponding eigenvector whose entries are all positive (see, for example, [Gan], vol. II, pp. 53–54, 79). Accordingly we have the following result.

Theorem 1.2.5. *A graph is connected if and only if its index is a simple eigenvalue with a positive eigenvector.*

If G is a regular graph of degree r then G has index r and the all-1 vector $\mathbf{j}$ is a corresponding eigenvector. It follows from the orthogonality of eigenspaces that if $(x_1, x_2, \ldots, x_n)^T$ is any eigenvector corresponding to an eigenvalue other than r, then $\sum_{i=1}^{n} x_i = 0$.

Next we discuss the relation between eigenvalues and walks in a graph. By a *walk of length k* in a graph we mean any sequence of (not necessarily different) vertices $v_0, v_1, \ldots, v_k$ such that for each $i = 1, 2, \ldots, k$ there is an edge from v_{i-1} to v_i. The walk is *closed* if $v_k = v_0$. The following result has a straightforward proof by induction on k.

Theorem 1.2.6. *If A is the adjacency matrix of a graph, then the (i, j)-entry $a_{ij}^{(k)}$ of the matrix A^k is equal to the number of walks of length k that originate at vertex i and terminate at vertex j.*

We shall also require the following result.

Theorem 1.2.7 [Hof3]. *For a graph G with adjacency matrix A, there exists a polynomial $P(x)$ such that $P(A) = J$ if and only if G is regular and connected. In this case we have*

$$P(x) = \frac{n(x - \mu_2) \cdots (x - \mu_m)}{(r - \mu_2) \cdots (r - \mu_m)}$$

where n is the number of vertices and $r, \mu_2, \ldots, \mu_m$ are the distinct eigenvalues of G in decreasing order.

Proof. If $P(A) = J$ then for any pair of distinct vertices i, j there exists a positive integer k such that the (i, j)-entry of A^k is non-zero. By Theorem 1.2.6 there exists an i-j walk of length k, and so G is connected. Regularity follows from the fact that A commutes with J.

Conversely, if G is r-regular and connected then $AJ = JA$ and r is a simple eigenvalue of A. Since $P(A)$ commutes with J, $P(A)$ and J are simultaneously diagonalizable. Both $P(A)$ and J have spectrum $n, 0, 0, \ldots, 0$, with the all-1 vector as an eigenvector corresponding to n. Hence $P(A) = J$. $\square$

Theorem 1.2.7 provides a good means of investigating the structure of regular graphs by means of spectra. The polynomial $P(x)$ is called the *Hoffman polynomial* of G.

It follows from Theorem 1.2.6 that the number of closed walks of length k is equal to the k-th spectral moment, since $\sum_{j=1}^{n} a_{jj}^{(k)} = \mathrm{tr}(A^k) = \sum_{j=1}^{n} \lambda_j^k$. From the spectral decomposition of A we have

$$(1.4) \qquad A^k = \mu_1^k P_1 + \mu_2^k P_2 + \cdots + \mu_m^k P_m$$

and so $a_{jj}^{(k)} = \sum_{i=1}^{m} \mu_i^k \alpha_{ij}^2$, where the α_{ij} are the angles of G. In particular, the vertex degrees $a_{jj}^{(2)}$ are determined by the spectrum and angles. More generally, by considering the numbers of walks in appropriate subgraphs we obtain the following result.

Theorem 1.2.8 [CvRo1]. *The spectrum and angles of a graph G determine (i) the degree of each vertex, (ii) the number of 3-cycles through each vertex, (iii) the number of 4-cycles in G, and (iv) the number of 5-cycles in G.*

It also follows from (1.4) that the number N_k of all walks of length k in G is given by

$$(1.5) \qquad N_k = \sum_{u,v} a_{uv}^{(k)} = \mathbf{j}^T A \mathbf{j} = \sum_{i=1}^{n} \mu_i^k \|P_i \mathbf{j}\|^2,$$

where $\mathbf{j}$ denotes the all-1 vector. The numbers $\beta_i = \|P_i \mathbf{j}\|/\sqrt{n}$ $(i = 1, \ldots, m)$ are called the *main angles* of G: they are the cosines of the (acute) angles between eigenspaces and $\mathbf{j}$. Note that $\sum_{i=1}^{m} \beta_i^2 = 1$. The eigenvalue μ_i is said to be a *main* eigenvalue if $P_i \mathbf{j} \neq \mathbf{0}$. By Theorem 1.2.5, the largest eigenvalue μ_1 is always a main eigenvalue. In view of (1.5) we have the following result.

Theorem 1.2.9. *The total number N_k of walks of length k in a graph G is given by*

$$(1.6) \qquad N_k = n \Sigma' \mu_i^k \beta_i^2,$$

where the sum Σ' is taken over all main eigenvalues μ_i.

The main eigenvalues of G are said to constitute the *main part* of the spectrum of G.

Theorem 1.2.10 [HaSc2]. *For a graph G, the following statements are equivalent:*

(i) $\mathcal{M}$ *is the main part of the spectrum;*

(ii) $\mathcal{M}$ *is the minimum set of eigenvalues the span of whose eigenvectors includes $\mathbf{j}$;*

(iii) $\mathcal{M}$ *is the set of those eigenvalues which have an eigenvector not orthogonal to* **j**.

Remark 1.2.11. If μ is a non-main eigenvalue of G then the eigenspace of μ in G is contained in the eigenspace of $-\mu - 1$ in $\overline{G}$. $\qquad\square$

The next three propositions (proved in [CvRS2, Section 4.5]) show that, given the eigenvalues of G, knowledge of the main angles of G is equivalent to knowledge of the spectrum of the Seidel matrix of G, or the spectrum of $\overline{G}$, or the spectrum of the cone over G.

Proposition 1.2.12. *For any graph G of order n, the characteristic polynomial $S_G(x)$ of the Seidel adjacency matrix of G is given by*

$$(1.7) \quad S_G(x) = (-2)^n P_G(-\tfrac{1}{2}(x+1)) \left(1 - n \sum_{i=1}^{m} \frac{\beta_i^2}{x+1+2\mu_i}\right).$$

Proposition 1.2.13. *For any graph G of order n, the complement $\overline{G}$ of G has characteristic polynomial*

$$(1.8) \quad P_{\overline{G}}(x) = (-1)^n P_G(-x-1) \left(1 - n \sum_{i=1}^{m} \frac{\beta_i^2}{x+1+\mu_i}\right).$$

Note that G is regular if and only if the main part of the spectrum consists of the index alone, and so we have the following result.

Corollary 1.2.14. *If G is an r-regular graph of order n then its complement $\overline{G}$ has characteristic polynomial*

$$(1.9) \quad P_{\overline{G}}(x) = (-1)^n P_G(-x-1)(x-n+r+1)/(x+1+r).$$

Proposition 1.2.15. *The cone over G has characteristic polynomial*

$$(1.10) \quad P_{K_1 \nabla G}(x) = P_G(x) \left(x - \sum_{i=1}^{m} \frac{n\beta_i^2}{x-\mu_i}\right).$$

Next we discuss the characteristic polynomials of certain line graphs and generalized line graphs. If G is a regular graph, then the characteristic polynomial of $L(G)$ can be expressed in terms of the characteristic polynomial of G, as follows.

Theorem 1.2.16 [Sac3]. *If G is a regular graph of degree r, with n vertices and m $(= \frac{1}{2}nr)$ edges, then the following relation holds:*

$$P_{L(G)}(x) = (x + 2)^{m-n} P_G(x - r + 2).$$

The proof of Theorem 1.2.16 follows by considering the common eigenvalues of $BB^T = A(G) + rI$ and $B^T B = A(L(G)) + 2I$, where B is the incidence matrix of G.

The next theorem shows that a relation between $P_G(x)$ and $P_{L(G)}(x)$ can be established for certain non-regular graphs.

Theorem 1.2.17 [Cve2]. *Let G be a semi-regular bipartite graph with n_1 independent vertices of degree r_1 and n_2 independent vertices of degree r_2, where $n_1 \geq n_2$. Then*

$$P_{L(G)}(x) = (x + 2)^{\beta} \sqrt{\left(-\frac{\alpha_1}{\alpha_2}\right)^{n_1 - n_2} P_G(\sqrt{\alpha_1 \alpha_2}) P_G(-\sqrt{\alpha_1 \alpha_2})},$$

where $\alpha_i = x - r_i + 2$ $(i = 1, 2)$ and $\beta = n_1 r_1 - n_1 - n_2$.

Proposition 1.2.18 [BuCS1]. *If G is a semi-regular bipartite graph with parameters (n_1, n_2, r_1, r_2) $(n_1 > n_2)$ and if $\lambda_1, \lambda_2, \ldots, \lambda_{n_2}$ are the first n_2 largest eigenvalues of G, then*

$$P_{L(G)}(x) = (x - r_1 - r_2 + 2)(x - r_1 + 2)^{n_1 - n_2}(x + 2)^{n_1 r_1 - n_1 - n_2 + 1}$$

$$\times \prod_{i=2}^{n_2} ((x - r_1 + 2)(x - r_2 + 2) - \lambda_i^2).$$

Proof. It is easy to see that $\lambda_1 = \sqrt{r_1 r_2}$ and that the spectrum of G contains at least $n_1 - n_2$ eigenvalues equal to 0. Since the spectrum of a bipartite graph is symmetric about zero, we can obtain Proposition 1.2.18 from Theorem 1.2.17 by straightforward calculation. $\qquad\square$

Turning now to the characteristic polynomials of certain generalized line graphs, we give a result which, in one special case, yields the whole spectrum.

Theorem 1.2.19 (cf. [CvDGT, p.52]). *Let G be a graph having vertex degrees $d_1, d_2, \ldots, d_n$. If $a_1, a_2, \ldots, a_n$ are non-negative integers such that $d_i + 2a_i = d$, $i = 1, 2, \ldots, n$, then*

$$P_{L(G; a_1, a_2, \ldots, a_n)}(x) = x^{\sum_{i=1}^{n} a_i} (x + 2)^{m-n+\sum_{i=1}^{n} a_i} P_G(x - d + 2).$$

Proof. An incidence matrix of $L(G; a_1, \ldots, a_n)$ has the form

$$C = \begin{pmatrix} B & L_1 & L_2 & \ldots & L_n \\ 0 & M_1 & 0 & \ldots & 0 \\ 0 & 0 & M_2 & \ldots & 0 \\ \vdots & \vdots & \vdots & \ddots & \vdots \\ 0 & 0 & 0 & \ldots & M_n \end{pmatrix}$$

where B is the incidence matrix of G; L_i is an $n \times 2a_i$ matrix in which all entries of the i-th row are 1, and all other entries are 0; and M_i is an $a_i \times 2a_i$ matrix of the form $(I_{a_i} \mid -I_{a_i})$, I_m being the identity matrix of order m. It is well known that for any matrix X, the matrices XX^T and $X^T X$ have the same non-zero eigenvalues. The theorem now follows from the fact that $C^T C = A + 2I$ where A is the adjacency matrix of $L(G; a_1, a_2, \ldots, a_n)$. $\square$

Next, we make some remarks concerning strongly regular graphs. From Theorem 1.2.6 we deduce immediately that a non-complete regular graph G of degree $r > 0$ is strongly regular if and only if there exist non-negative integers e and f such that the adjacency matrix A of G satisfies the following relation:

$$(1.11) \qquad A^2 = (e - f)A + fJ + (r - f)I.$$

The following result is due to S. S. Shrikhande and D. Bhagwandas.

Theorem 1.2.20 [ShBh]. *A connected r-regular graph G is strongly regular if and only if it has exactly three distinct eigenvalues $\mu_1 = r$, μ_2, μ_3. If G is strongly regular, with parameters (n, r, e, f), then*

$$e = r + \mu_2\mu_3 + \mu_2 + \mu_3 \quad and \quad f = r + \mu_2\mu_3.$$

Proof. If G is strongly regular then it follows from (1.11) that the minimal polynomial of an adjacency matrix has degree 3, and so G has exactly three distinct eigenvalues.

Now suppose that G has exactly three distinct eigenvalues $\mu_1 = r$, μ_2, μ_3. Then $r > 0$, G is not complete, and by Theorem 1.2.7,

$$(1.12) \qquad aA^2 + bA + cI = J \quad (a \neq 0),$$

where the polynomial $ax^2 + bx + c$ has roots μ_2, μ_3. Equating diagonal elements in (1.12), we find that $c = 1 - ar$. Further, the number of walks of length two between distinct vertices i and j is $\dfrac{1-b}{a}$ if $i \sim j$, and $\dfrac{1}{a}$ if

$i \not\sim j$. Hence G is strongly regular. Comparing (1.11) and (1.12), we obtain $e = r + \mu_2 + \mu_2 + \mu_2\mu_3$ and $f = r + \mu_2\mu_3$.

This completes the proof. □

Before we introduce the concept of a graph divisor we present some relevant theorems from matrix theory. The first applies not just to real symmetric matrices but to any Hermitian matrix. (Recall that the matrix A with complex entries a_{ij} is called *Hermitian* if $A^T = A^*$, i.e. $a_{ji} = \bar{a}_{ij}$ for all i, j.)

Theorem 1.2.21. *Let A be a Hermitian matrix with eigenvalues $\lambda_1 \geq \lambda_2 \geq \cdots \geq \lambda_n$ and let B be one of its principal submatrices. If the eigenvalues of B are $\nu_1 \geq \nu_2 \geq \cdots \geq \nu_m$ then $\lambda_{n-m+i} \leq \nu_i \leq \lambda_i$ ($i = 1, \ldots, m$).*

A proof can be found in [MaMi, p. 119]. The inequalities of Theorem 1.2.21 are known as *Cauchy's inequalities* and the whole theorem is known as the *Interlacing Theorem*. It is used frequently as a spectral technique in graph theory. The next result is attributed to C. C. Sims (see [HeHi]).

Theorem 1.2.22. *Let A be a real symmetric matrix with eigenvalues $\lambda_1 \geq \lambda_2 \geq \cdots \geq \lambda_n$. Given a partition $\{1, 2, \ldots, n\} = \Delta_1 \dot\cup \Delta_2 \dot\cup \cdots \dot\cup \Delta_m$ with $|\Delta_i| = n_i > 0$, consider the corresponding blocking $A = (A_{ij})$, where A_{ij} is an $n_i \times n_j$ block. Let e_{ij} be the sum of the entries in A_{ij} and set $B = (e_{ij}/n_i)$ (Note that e_{ij}/n_i is the average row sum in A_{ij}.) Then the spectrum of B is contained in the interval $[\lambda_n, \lambda_1]$.*

W. Haemers [Hae] has shown that the interlacing properties also hold for the matrices A and B of this theorem. If we assume that in each block A_{ij} from Theorem 1.2.22 all row sums are equal then we can say more:

Theorem 1.2.23 [Hay, PeSa]. *Let A be any matrix partitioned into blocks as in Theorem 1.2.22. Suppose that the block A_{ij} has constant row sums b_{ij}, and let $B = (b_{ij})$. Then the spectrum of B is contained in the spectrum of A (taking into account the multiplicities of the eigenvalues).*

The content of Theorem 1.2.23 justifies the introduction of the following definition. Note that in an adjacency matrix of a multi-digraph, the (i, j)-entry is the number of arcs from vertex i to vertex j.

Definition 1.2.24. *Given an $s \times s$ matrix $B = (b_{ij})$, suppose that the vertex set of a graph G is partitioned into (non-empty) subsets $X_1, X_2, \ldots, X_s$ such that for any $i, j = 1, 2, \ldots, s$, each vertex from X_i is adjacent to exactly b_{ij} vertices of X_j. The multi-digraph H with adjacency matrix B is called a* front divisor *of G, or briefly, a* divisor *of G.*

The concept of a divisor of a graph was introduced by Sachs [Sac2, Sac3]. The existence of a divisor means that the graph has a certain structure; indeed, a divisor can be interpreted as a homomorphic image of the graph. On the other hand, by Theorem 1.2.23, the characteristic polynomial of a divisor divides the characteristic polynomial of the graph (i.e. the spectrum of a divisor is contained in the spectrum of the graph). In this way the notion of a divisor can be seen as a link between spectral and structural properties of a graph.

Using Theorem 1.2.22 we first prove a general theorem which will be used in Chapter 3 to obtain spectral bounds for graphs represented in the root system E_8.

Theorem 1.2.25. *Let G be a regular graph of degree d and order n with eigenvalues $\lambda_1 = d, \lambda_2, \ldots, \lambda_n$. Let G_1 be an induced subgraph of G of order n_1 with mean degree d_1. Then*

$$(1.13) \qquad \frac{n_1(d - \lambda_n)}{n} + \lambda_n \leq d_1 \leq \frac{n_1(d - \lambda_2)}{n} + \lambda_2.$$

Proof. We partition $V(G)$ into $V(G_1)$ and its complement, and consider the corresponding blocking of the adjacency matrix of G. The average values of row sums of blocks form the matrix

$$\begin{pmatrix} d_1 & d - d_1 \\ \frac{(d - d_1)n_1}{n - n_1} & d - \frac{(d - d_1)n_1}{n - n_1} \end{pmatrix}.$$

The eigenvalues of this matrix are d and $d_1 - (d - d_1)n_1/(n - n_1)$. By Theorem 1.2.22 we have $\lambda_n \leq d_1 - (d - d_1)n_1/(n - n_1)$ and the left-hand inequality in (1.13) is proved.

In order to prove the right-hand inequality, we consider the complements $\overline{G}, \overline{G_1}$. The graph $\overline{G}$ is a regular graph on n vertices of degree $n - 1 - d$, and by Corollary 1.2.14 its least eigenvalue is $-\lambda_2 - 1$. The graph $\overline{G_1}$ is an induced subgraph of $\overline{G}$ with n_1 vertices and mean degree $n - 1 - d_1$. Applying the left-hand inequality of (1.13) to $\overline{G}$ and $\overline{G_1}$ we obtain the right-hand inequality. This completes the proof. □

We now prove a theorem which relates the divisor concept to switching in graphs.

Theorem 1.2.26 [Cve6]. *If G is a regular graph of order n which can be switched into a regular graph of degree r^*, then $r^* - \frac{n}{2}$ is an eigenvalue of G.*

Proof. If G is r-regular then the switching sets form a divisor with adjacency matrix

$$\begin{pmatrix} r - \frac{1}{2}(n - t - r^* + r) & \frac{1}{2}(n - t - r^* + r) \\ \frac{1}{2}(t - r^* + r) & r - \frac{1}{2}(t - r^* + r) \end{pmatrix},$$

where t is the size of a switching set ($1 \leq t < n$). This matrix has eigenvalues r and $r^* - \frac{n}{2}$, and the theorem follows. $\qquad\square$

Since graph eigenvalues are algebraic integers, we deduce the following.

Corollary 1.2.27. *A regular graph of odd order cannot be switched to another regular graph.*

Corollary 1.2.28. *Let G be a regular graph of degree r and order n, with q as its least eigenvalue. If G can be switched to another regular graph of the same degree then $r - \frac{n}{2} \geq q$, i.e. $n \leq 2r - 2q$. Since $q \geq -r$, we have $r - \frac{n}{2} \geq -r$, i.e. $r \geq n/4$.*

Example 1.2.29. There is no cospectral pair of non-isomorphic cubic graphs with fewer than 14 vertices. Accordingly it follows from Corollary 1.2.28 that the existence of cospectral cubic graphs cannot be explained by switching. $\qquad\square$

Example 1.2.30. If $L(K_s)$ ($s > 1$) can be switched to another regular graph of the same degree then by Theorem 1.2.26, $2s - 4 - \frac{s(s-1)}{4} \geq -2$, whence $s \leq 8$. When $s = 8$ there are three different graphs switching equivalent to $L(K_8)$ which are cospectral with $L(K_8)$, but not isomorphic to $L(K_8)$. These three graphs are the Chang graphs of Example 1.1.12. $\qquad\square$

Example 1.2.30 can be generalized to all regular line graphs as follows. Let $L(G)$ be regular with G connected; there are two cases.

(1) Suppose first that G is regular of degree r with n vertices. Then $L(G)$ is of degree $2r - 2$ and has $nr/2$ vertices. If $L(G)$ can be switched into another regular graph of the same degree then $2r - 2 - nr/4$ is an eigenvalue of $L(G)$. Clearly, $2r - 2 - nr/4 \geq -2$, which implies $n \leq 8$.

(2) Let G be a semi-regular bipartite graph with parameters (n_1, d_1, n_2, d_2) (so that $n_1 d_1 = n_2 d_2$). Then $L(G)$ has $n_1 d_1$ vertices and degree $d_1 + d_2 - 2$. Therefore, we have

$$d_1 + d_2 - 2 - n_1 d_1/2 \geq -2, \qquad n_1 d_1 \leq 2(d_1 + d_2),$$

$$n_1 \leq 2(1 + d_2/d_1) = 2(1 + n_1/n_2), \qquad 1/n_1 + 1/n_2 \geq 1/2.$$

Suppose that $n_1 \leq n_2$. Then $n_1 \leq 4$. If $n_1 = 1$, then $L(G) = K_{n_2}$ and no exceptional graph arises. Similarly if $n_1 = 2$, then $L(G) = L(K_{2,n_2})$ and again no exceptional graph arises. Further, if $n_1 = 3$ then $n_2 \in \{4, 5, 6\}$ and if $n_1 = 4$ then $n_2 = 4$. Hence $n_1 + n_2 \leq 9$.

In this way the exceptional graphs which are cospectral with regular line graphs, and obtained from them by switching, can be determined by quite elementary means.

Next, we consider a very general graph operation called NEPS (*non-complete extended p-sum*) of graphs.

Definition 1.2.31. *Let $\mathcal{B}$ be a set of non-zero binary n-tuples, i.e. $\mathcal{B} \subseteq \{0, 1\}^n \setminus \{(0, \ldots, 0)\}$. The NEPS of graphs $G_1, \ldots, G_n$ with basis $\mathcal{B}$ is the graph with vertex set $V(G_1) \times \cdots \times V(G_n)$, in which two vertices, say $(x_1, \ldots, x_n)$ and $(y_1, \ldots, y_n)$, are adjacent if and only if there exists an n-tuple $(\beta_1, \ldots, \beta_n) \in \mathcal{B}$ such that $x_i = y_i$ whenever $\beta_i = 0$, and x_i is adjacent to y_i (in G_i) whenever $\beta_i = 1$.*

Clearly the NEPS construction generates many binary graph operations in which the vertex set of the resulting graph is the Cartesian product of the vertex sets of the graphs on which the operation is performed (see [CvDSa, pp. 65–66], and the references cited in [CvDSa]).

We now mention some special cases in which a graph is the NEPS of graphs $G_1, \ldots, G_n$ with basis $\mathcal{B}$. In particular, for $n = 2$ we have the following familiar operations:

(i) the *sum* $G_1 + G_2$, when $\mathcal{B} = \{(0, 1), (1, 0)\}$;
(ii) the *product* $G_1 \times G_2$, when $\mathcal{B} = \{(1, 1)\}$;
(iii) the *strong product* $G_1 * G_2$, when $\mathcal{B} = \{(0, 1), (1, 0), (1, 1)\}$.

The *p-sum* of graphs is a NEPS in which the basis consists of all n-tuples with exactly p entries equal to 1. The *J-sum* of graphs, where J is a subset of $\{1, \ldots, n\}$, is a NEPS with a basis consisting of all n-tuples in which the number of 1s belongs to J.

The notion of NEPS arises in a natural way when studying spectral properties of graphs obtained by binary operations of the type mentioned above. The next theorem appears also in [CvDSa, pp. 68–69].

Theorem 1.2.32. *If $\lambda_{i1}, \ldots, \lambda_{in_i}$ is the spectrum of G_i $(i = 1, \ldots, n)$, then the spectrum of the NEPS of $G_1, \ldots, G_n$ with basis $\mathcal{B}$ consists of all possible values $\Lambda_{i_1,\ldots,i_n}$ where*

$$(1.14) \quad \Lambda_{i_1,\ldots,i_n} = \sum_{\beta \in \mathcal{B}} \lambda_{1i_1}^{\beta_1} \cdots \lambda_{ni_n}^{\beta_n} \quad (i_k = 1, \ldots, n_k; \ k = 1, \ldots, n).$$

Thus if $\lambda_1, \ldots, \lambda_n$ and $\mu_1, \ldots, \mu_m$ are the eigenvalues of G and H, respectively, then:

$\lambda_i + \mu_j$ $(i = 1, \ldots, n; \ j = 1, \ldots, m)$ are the eigenvalues of $G + H$;

$\lambda_i \mu_j$ $(i = 1, \ldots, n; \ j = 1, \ldots, m)$ are the eigenvalues of $G \times H$;

$\lambda_i + \mu_j + \lambda_i \mu_j$ $(i = 1, \ldots, n; \ j = 1, \ldots, m)$ are the eigenvalues of $G * H$.

Example 1.2.33. We have $L(K_{m,n}) = K_m + K_n$. Since K_n has spectrum $n - 1, (-1)^{n-1}$ we obtain $m + n - 2, (n - 2)^{m-1}, (m - 2)^{n-1}, (-2)^{(m-1)(n-1)}$ for the spectrum of $L(K_{m,n})$.

1.3 Elementary spectral characterizations

In this section some elementary instances of the following problem are considered:

Given the spectrum, or some spectral characteristics of a graph, determine all graphs from a given class of graphs having the given spectrum, or the given spectral characteristics.

We shall see that in some cases the solution of such a problem can provide a characterization of a graph up to isomorphism. In other cases we can deduce structural details (cf. [CvDSa, Chapter 3]), and we have seen in Section 1.2 several examples of spectral properties equivalent to structural properties. In yet further cases, graphs are characterized by a mixture of spectral and structural properties.

We say that a graph G is *characterized by its spectrum* if the only graphs cospectral with G are those isomorphic to G. Note first that this condition is satisfied by graphs which are characterized by invariants (such as the number of vertices and edges) which can be determined from the spectrum. Examples include the complete graphs and graphs with one edge, together with their complements. Given the spectrum of a graph G we can always establish whether or not G is regular (see Theorem 1.2.1). It follows that if G or $\overline{G}$ is regular of degree 1 then G is characterized by its spectrum. Thus the cocktail party graphs are characterized by their spectra. By inspecting eigenvalues of cycles (Example 1.1.2) we can also show the following.

Theorem 1.3.1 [Cve2]. *Any regular graph of degree 2 is characterized by its spectrum.*

Remark 1.3.2. From the spectrum of a regular graph G we can find the spectrum of $\overline{G}$ (see Corollary 1.2.14), and so it follows from Theorems 1.2.1 and 1.3.1 that any graph of order n which is regular of degree $n - 3$ is characterized by its spectrum. $\qquad \square$

It is straightforward to show that a graph of the form $m K_n$ is characterized by its spectrum, a fact established in complementary form in [Fin]:

Theorem 1.3.3. *For each positive integer n, the complete multipartite graph* $K_{n,n,...,n}$ *is characterized by its spectrum.*

The next result, however, does not admit a transition to the complement.

Theorem 1.3.4 [Cve2]. *The spectrum of the graph G consists of the natural numbers* $n_1 - 1, \ldots, n_k - 1$, *together with s numbers equal to* 0 *and* $n_1 + \cdots + n_k - k$ *numbers equal to* -1, *if and only if G has as its components s isolated vertices together with k complete graphs on* $n_1, \ldots, n_k$ *vertices, respectively.*

The following elementary result will also be useful.

Theorem 1.3.5 [Doo3]. *Let G be a bipartite graph with eigenvalues* $\mu_1 > \mu_2 > \mu_3$ *of multiplicities* m_1, m_2, m_3 *respectively. Then* $\mu_3 = -\mu_1, \mu_2 = 0, m_1 = m_3$ *and G is the disjoint union of* m_1 *complete bipartite graphs* K_{r_i,s_i}, *where* $r_i s_i = \mu_1^2$ $(i = 1, \ldots, m_1)$, *and* $m_2 - \sum_{i=1}^{m_1}(r_i + s_i - 2)$ *isolated vertices.*

In the 1960s A. J. Hoffman investigated the extent to which regular connected graphs are determined by their distinct eigenvalues. One of his results is the following; a proof can be found in [CvDSa, Section 6.2].

Theorem 1.3.6 [Hof4]. *Let G be a regular connected graph of order n with at most 4 distinct eigenvalues. The following are equivalent:*

(i) *the graph H is cospectral with G,*
(ii) *the graph H is a connected regular graph of order n with the same set of distinct eigenvalues as G.*

1.4 A history of research on graphs with least eigenvalue -2

The origins of the topic lie in the elementary observation that line graphs have least eigenvalue greater than or equal to -2 (see Section 1.1). It was a natural problem to search for all the graphs with such a remarkable property, and interest deepened when it became apparent that generalized line graphs are not the only graphs to share the property.

Exceptional graphs first arose in the spectral characterizations of some classes of line graphs by A. J. Hoffman [Hof2] and others in the 1960s (cf. [CvRS2, pp. 12–14]). In particular, the three Chang graphs on 28 vertices, the Schläfli graph on 27 vertices, the Shrikhande and Clebsch graphs on

16 vertices and the ubiquitous Petersen graph were among the first exceptional graphs to be identified. The results from that era which proved to be of particular importance include the spectral characterization of the line graphs $L(K_n)$ $(n > 8)$ (cf. [Hof4],[Hof1],[Hof2],[Shr1],[Cha1],[Cha2],[Conn]), and a complete enumeration of strongly regular graphs with least eigenvalue −2 by J. J. Seidel [Sei3]. The techniques used were the method of forbidden subgraphs and switching. However, a characterization theorem for graphs with least eigenvalue −2 by A. J. Hoffman and D. K. Ray-Chaudhuri [HoRa3] remained unpublished because the authors were not satisfied with the length of proof based on the forbidden subgraph technique.

In 1970, L. W. Beineke [Bei1] characterized line graphs by means of forbidden induced subgraphs (nine in number). Although this result does not mention eigenvalues explicitly, it is relevant to the study of graphs with least eigenvalue −2.

In 1976 the key paper [CaGSS] introduced root systems into the study of graphs with least eigenvalue −2. The main result is that an exceptional graph can be represented in the exceptional root system E_8. In particular, it is proved in this way that an exceptional graph has at most 36 vertices and each vertex has degree at most 28.

The regular exceptional graphs, 187 in number, were found in [BuCS1], [BuCS2] in 1976 by a mixture of mathematical reasoning and a computer search, while the problem of a suitable description of all exceptional graphs remained open. Enumeration of the regular exceptional graphs made it possible to improve the existing characterization theorems for several classes of line graphs. In the papers [CvDo2] and [CvRa1] some effort was made to eliminate computer searches from the proofs of various characterization theorems for regular line graphs. The book [BrCN] also contains some simplified proofs and a computer-free proof of an intermediate result (Theorem 4.1.5).

In 1979 M. Doob and D. Cvetković [DoCv] characterized graphs with least eigenvalue greater than −2, and described the 573 exceptional ones (20 on six vertices, 110 on seven and 443 on eight vertices).

Generalized line graphs were introduced by A. J. Hoffman [Hof6] in 1970, and studied extensively by D. Cvetković, M. Doob and S. Simić [CvDS1, CvDS2] in 1980. Generalized line graphs were characterized by a collection of 31 forbidden induced subgraphs in [CvDS1, CvDS2], and independently by S. B. Rao, N. M. Singhi and K. S. Vijayan in [RaSV2]. In 1984, M. Doob and D. Cvetković [CvDo2] provided an alternative means of constructing the root systems used to describe the graphs with least eigenvalue −2 (see Section 3.3).

A family of minimal forbidden subgraphs for graphs with least eigenvalue -2 was found in [KuRS] and [BuNe]. These forbidden graphs have at most 10 vertices and there are 1812 of them (see Section 2.4).

The subject attracted less attention from researchers for some years, the book [BrCN] and the paper [BuNe] being most important publications for almost two decades. Much information on the problems mentioned above can be found in the books [BrCN], [CaLi], [CvDSa], [CvDGT], [CvRS2], and in the expository paper [BuNe].

Now we turn to recent developments. In 1998 the paper [CvRS4] introduced the star complement technique into the study of graphs with the least eigenvalue -2. The technique has its origins in the Schur complement of a principal submatrix (see [Pra, p. 17]); its application in a graph theory context was noted independently by M. N. Ellingham [Ell] and P. Rowlinson [Row1] in 1993. It was proved in [CvRS4] that a graph is exceptional if and only if it has an exceptional star complement. Based on this observation, the maximal exceptional graphs, 473 in number, were first found by computer in 1999 and then derived theoretically in [CvLRS2, CvRS6]. Independently, A. Munemasa and M. Kitazume also found the maximal exceptional graphs by computer in 1999, by constructing maximal sets of vectors at $60°$ or $90°$ in the root system E_8 (see Chapter 3). At the time of writing their work remains unpublished, but it confirms the results found by the star complement technique.

2

Forbidden subgraphs

In this chapter we extend some structural characterizations of line graphs to generalized line graphs, with an emphasis on the technique of forbidden subgraphs. We describe a collection of minimal forbidden subgraphs for graphs whose smallest eigenvalue is at least -2, and we note some implications concerning the characterization of certain graphs by their spectra.

2.1 Line graphs

In this section we discuss characterizations of line graphs and the extent to which a root graph is determined by its line graph. We give three characterizations of line graphs, two of which will be extended to generalized line graphs in Section 2.3. The first, due to J. Krausz, is in terms of an edge-covering by cliques (complete subgraphs).

Theorem 2.1.1 [Kra]. *A graph is a line graph if and only if its edge set can be partitioned into non-trivial cliques such that:*

(i) *two cliques have at most one vertex in common;*
(ii) *each vertex is in at most two cliques.*

The proof of Krausz's theorem is not difficult (see, for example, [Har]): in a line graph $L(G)$, a non-trivial clique $K(v)$ arises from each vertex v of degree at least 2 in G, and the collection of all such cliques satisfies (i) and (ii). For the converse, given a collection $\mathcal{C}$ of non-trivial cliques satisfying (i) and (ii), we add to $\mathcal{C}$ a trivial clique for every vertex in just one clique of $\mathcal{C}$, and construct a root graph as the intersection graph on the enlarged collection of cliques. In this way we establish a one-to-one correspondence between root graphs and systems of cliques satisfying conditions (i) and (ii). Such systems are said to be *complete*.

Note that, in a complete system covering G, if C and D are cliques such that $C \cap D = \{v\}$ then the neighbours of v are precisely the neighbours of v in $C \cup D$, and any edges between $C - v$ and $D - v$ are independent. In any graph, two cliques satisfying these conditions are said to form a pair of *bridged cliques*; and if, further, G contains an edge between $C - v$ and $D - v$ then C and D form a pair of *linked cliques*.

The second characterization of line graphs is based on a property of triangles: a triangle H in a graph G is said to be an *odd triangle* if there exists a vertex in G adjacent to an odd number of vertices from H; otherwise, H is said to be an *even triangle*. The characterization that follows is due to A. C. M. van Rooij and H. S. Wilf; a proof can be found in [Har, Chapter 8].

Theorem 2.1.2 [RoWi]. *A graph is a line graph if and only if it does not have $K_{1,3}$ as an induced subgraph and, whenever two odd triangles have a common edge, their remaining two vertices are adjacent.*

The third characterization is based on the fact that the property of being a line graph is hereditary: a graph property $\mathcal{P}$ is *hereditary* if it is inherited by every induced subgraph of a graph with property $\mathcal{P}$. Thus if H is an induced subgraph of G and H does not have the hereditary property $\mathcal{P}$ then G does not have the property $\mathcal{P}$: in this situation we say that H is a *forbidden subgraph* for the property $\mathcal{P}$; and H is a *minimal forbidden subgraph* if, further, all vertex-deleted subgraphs of H have property $\mathcal{P}$. Graphs having the hereditary property $\mathcal{P}$ can be characterized by a collection of minimal forbidden subgraphs, and the crucial question is whether there exists such a collection which is finite.

Let G be a graph with eigenvalues $\lambda_1(G) \geq \lambda_2(G) \geq \cdots \geq \lambda_n(G)$. For any real a and any integer $i \in \{1, 2, \ldots, n\}$, each of the properties $\lambda_i(G) \leq a$, $\lambda_i(G) < a$ is hereditary. Analogously, if we denote the i-th smallest eigenvalue by $\lambda^i(G)$, then the properties $\lambda^i(G) \geq a$, $\lambda^i(G) > a$ are also hereditary. (In each case, the property is satisfied vacuously by an induced subgraph of order less than i.) These observations follow from the Interlacing Theorem (Theorem 1.2.21).

In this chapter the following hereditary properties of graphs G will be of special interest:

$\lambda(G) \geq -2$;
G is a line graph;
G is a generalized line graph.

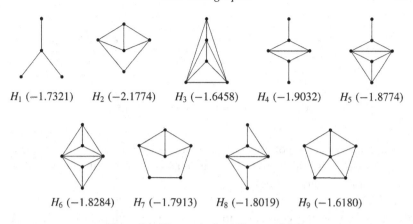

Figure 2.1: The graphs from Theorem 2.1.3.

L. W. Beineke proved that line graphs are characterized by a collection of only nine forbidden subgraphs.

Theorem 2.1.3 [Bei1]. *A graph is a line graph if and only if it has no induced subgraph isomorphic to one of the graphs $H_1, \ldots, H_9$ from Fig. 2.1.*

The numbers in parentheses in Fig. 2.1 denote the least eigenvalues of the corresponding graphs. A proof of Theorem 2.1.3 can be found in [Har, Chapter 8].

Remark 2.1.4. It follows from Theorem 2.1.3 that there exists a polynomial time algorithm for checking whether a given graph is a line graph. One needs only to compare the subgraphs of the input graph with the nine graphs from Fig. 2.1. More sophisticated algorithms can be found in the literature [Leh, Rou]. □

It was pointed out by L. Šoltés that just the first seven graphs from Fig. 2.1 can be used to characterize line graphs as follows.

Theorem 2.1.5 [Sol]. *A connected graph is a line graph if and only if it has no induced subgraphs isomorphic to one of the graphs $H_1, \ldots, H_7$ and is not one of the graphs of Fig. 2.2.*

We conclude this section with a well-known result of H. Whitney which answers the question: when is a root graph determined by its line graph? A proof appears in [Har, Chapter 8].

Theorem 2.1.6 [Whi]. *If G and H are connected graphs, then $L(G) = L(H)$ if and only if $G = H$ or $\{G, H\} = \{K_{1,3}, K_3\}$.*

Figure 2.2: The graphs from Theorem 2.1.5

The exceptional case of Theorem 2.1.6 arises from two different complete systems of cliques which cover K_3: one consists of K_3 itself, and the other consists of three cliques of order 2.

In Section 2.3 we obtain analogues of the results of this section for generalized line graphs. It is convenient to discuss first the multiplicity of -2 as an eigenvalue of such graphs.

2.2 The eigenspace of -2 for generalized line graphs

As noted in Section 1.1, if G is a generalized line graph, then its least eigenvalue is greater than or equal to -2. To investigate the multiplicity (possibly zero) of -2 as an eigenvalue of G, consider the relation

$$(2.1) \qquad\qquad C^T C = A(G) + 2I,$$

where C is a vertex-edge incidence matrix of a root multigraph $\hat{H}$ of G.

Lemma 2.2.1. *If G is a graph satisfying* (2.1), *then -2 is its least eigenvalue if and only if $C\mathbf{x} = \mathbf{0}$ for some non-zero vector $\mathbf{x}$.*

Proof. Clearly, if $C\mathbf{x} = \mathbf{0}$ and $\mathbf{x} \neq \mathbf{0}$ then $\mathbf{x}$ is an eigenvector of G corresponding to -2. Conversely, if $A(G)\mathbf{x} = -2\mathbf{x}$ then $C^T C\mathbf{x} = \mathbf{0}$, and so $(C\mathbf{x})^T(C\mathbf{x}) = 0$. The lemma follows.   □

We can also make the following observation.

Proposition 2.2.2. *A non-zero vector $\mathbf{x}$ is an eigenvector for $L(\hat{H})$ corresponding to -2 if and only if $C\mathbf{x} = \mathbf{0}$, for any incidence matrix C of $\hat{H}$.*

From Lemma 2.2.1 we see that the multiplicity of -2 as an eigenvalue of a generalized line graph is easily found if the rank of C is known. In the case of line graphs we can therefore use the following result of H. Sachs, which in its original form allows loops but not multiple edges (see also [Nuff]).

Lemma 2.2.3 [Sac3]. *Let B be an incidence matrix of a connected graph H with n vertices ($n > 1$). Then*

$$(2.2) \qquad \mathrm{rank}(B) = \begin{cases} n-1 & \text{if } H \text{ is bipartite,} \\ n & \text{if } H \text{ is non-bipartite.} \end{cases}$$

Proof. Let $B = (b_{ij})$, with rows $B_1, \ldots, B_n$, and assume that the rows are linearly dependent, say

$$(2.3) \qquad c_1 B_1 + \cdots + c_n B_n = \mathbf{0} \text{ and } (c_1, \ldots, c_n) \neq (0, \ldots, 0).$$

If two vertices v_s and v_t are joined by the edge e_j then $b_{sj} = b_{tj} = 1$, while $b_{kj} = 0$ for all $k \neq s, t$. Consequently, from (2.3) we obtain $c_s = -c_t$.

It follows that for any path $i_1 i_2 \ldots i_k$ beginning at a vertex for which $c_{i_1} = c \neq 0$, the coefficients $c_{i_1}, c_{i_2}, \ldots, c_{i_k}$ are alternately c and $-c$. Since H is connected we deduce that H is bipartite and that $\dim\{\mathbf{x} \in \mathbb{R}^n : \mathbf{x}^T B = \mathbf{0}\} = 1$. The result follows. $\square$

Let $m(\lambda, G)$ denote the multiplicity of λ as an eigenvalue of G. From Lemmas 2.2.1 and 2.2.3 we have the following result.

Theorem 2.2.4. *Let H be a connected graph with n vertices and m edges. Then*

$$(2.4) \qquad m(-2, L(H)) = \begin{cases} m-n+1 & \text{if } H \text{ is bipartite,} \\ m-n & \text{if } H \text{ is non-bipartite.} \end{cases}$$

The following results of M. Doob [Doo2] can now be deduced as immediate consequences (see also [Doo8]). Recall that the smallest eigenvalue of a graph G is denoted by $\lambda(G)$.

Corollary 2.2.5. *Let H be a connected graph. Then $\lambda(L(H)) > -2$ if and only if H is a tree or an odd-unicyclic graph.*

Corollary 2.2.6. *Let H be a (connected) graph with diameter d. Then*

$$-2 \leq \lambda(L(H)) \leq -2\cos\frac{\pi}{d+1},$$

and these bounds are best possible.

Proof. It remains to consider the second inequality. Since the diameter of $L(H)$ is not less than $d-1$, $L(H)$ has a path P_d as an induced subgraph. By the Interlacing Theorem we have $\lambda(L(H)) \leq \lambda(P_d) = -2\cos\frac{\pi}{d+1}$, and equality holds when $H = P_{d+1}$. $\square$

To obtain an analogue of (2.4) for generalized line graphs, we can proceed in the same way as above.

Lemma 2.2.7. *Suppose that C is an incidence matrix of a connected root multigraph $H(a_1, a_2, \ldots, a_n)$ for which $(a_1, a_2, \ldots, a_n) \neq (0, 0, \ldots, 0)$. Then*

$$(2.5) \qquad \qquad \text{rank}(C) = n + \sum_{i=1}^{n} a_i.$$

Proof. Let $C = (c_{ij})$, with rows $C_1, \ldots, C_r$, and suppose that $c_1 C_1 + \cdots + c_r C_r = \mathbf{0}$. Our multigraph contains vertices h and i joined by two edges, say e_j and e_k. From Section 1.2 we know that, without loss of generality, $c_{hj} = c_{hk} = 1$, $c_{ij} = -c_{ik} = 1$ and $c_{hl} = c_{il} = 0$ for all $l \neq j, k$. It follows that $c_h = c_i = 0$. Tracing paths from h as in the proof of Lemma 2.2.3, we find that $c_1 = c_2 = \cdots = c_r = 0$. The lemma follows. $\square$

Now from Lemmas 2.2.1 and 2.2.7 we obtain the following analogue of Theorem 2.2.4.

Theorem 2.2.8. *Suppose that H is a connected graph with n vertices and m edges. If $(a_1, a_2, \ldots, a_n) \neq (0, 0, \ldots, 0)$, then*

$$(2.6) \qquad m(-2, L(H; a_1, a_2, \ldots, a_n)) = m - n + \sum_{i=1}^{n} a_i.$$

Theorem 2.2.8 was first proved in [CvDS2] (see also [CvDS1]). We shall see in the next section that it is useful in the construction of minimal graphs which are not generalized line graphs.

We conclude this section with some remarks on the relation between a line graph $L(H)$ and the cycle space of H, defined as follows. Let $\overline{V}(H)$ be the vector space over $GF(2)$ with basis $V(H)$, and let $\overline{E}(H)$ be the vector space over $GF(2)$ with basis $E(H)$. The *boundary operator* ∂ is the linear map $\overline{E}(H) \rightarrow \overline{V}(H)$ defined by $\partial(uv) = u + v$. An element of $\overline{E}(H)$ with boundary 0 is called a *cycle vector*: it is the sum of vertices in edge-disjoint cycles. The subspace of $\overline{E}(H)$ consisting of all cycle vectors is called the *cycle space* of H. Cycles $Z_1, \ldots, Z_r$ are said to be *independent* if the corresponding cycle vectors $\sum_{v \in V(Z_1)} v, \ldots, \sum_{v \in V(Z_r)} v$ are linearly independent.

Given a vector $\mathbf{x} = (x_e)_{e \in E(H)}$ in $\overline{E}(H)$, and a vertex u of H, let $c_u = \sum_{e \in E(H), u \in e} x_e$. It is straightforward to show that $A(L(H))\mathbf{x} = -2\mathbf{x}$ if and only if $c_i + c_j = 0$ whenever i and j are adjacent. M. Doob [Doo8] exploited this property in a matroid context to show that the eigenspace of -2 is spanned by vectors constructed from even cycles in H by labelling their edges alternately 1 and -1, and labelling other edges 0. This leads to the following result concerning the dimension of the eigenspace.

Theorem 2.2.9 [Doo8]. *The multiplicity of the eigenvalue* -2 *in the line graph* $L(H)$ *is equal to the maximal number of independent even cycles in* H.

We return to this topic in Section 5.2, where a basis for the eigenspace of -2 is constructed for both line graphs and generalized line graphs. The construction uses even cycles as above, but does not require cycle spaces.

2.3 Generalized line graphs

Here we extend the results of Section 2.1 to generalized line graphs. Accordingly we discuss characterizations of generalized line graphs and the extent to which a generalized line graph determines its root graph. Theorems 2.3.1, 2.3.3 and 2.3.4 appear in [Sim3]; here we shall follow the methods used in [CvDS2] (see also [CvDS1]).

Recall first that a cocktail party graph is isomorphic to a clique with a perfect matching removed. For our further needs we now define a *generalized cocktail party graph* (GCP) as a graph isomorphic to a clique with some independent edges removed. We denote by $GCP(n, m)$ the graph obtained from the complete graph K_n by the deletion of m independent edges $(0 \leq m \leq \lfloor \frac{1}{2}n \rfloor)$. Thus every vertex of $GCP(n, m)$ is of degree $n - 1$ or $n - 2$. A vertex of degree $n - 1$ is said to be of *type a* while the others are of *type b*. Two non-adjacent vertices of type *b* are called *partners*. Note that $GCP(1, 0)$ is the trivial graph, while $GCP(2, 1)$ is the only GCP that is not connected.

Theorem 2.3.1. *A graph G without isolated vertices is a generalized line graph if and only if its edges can be partitioned into non-trivial connected GCPs such that*

  (i) *two GCPs have at most one common vertex;*
 (ii) *each vertex is in at most two GCPs;*
(iii) *if two GCPs have a common vertex, then it is of type a in both of them.*

Proof. If $G = L(H; a_1, \ldots, a_n)$ then each vertex i of H for which $\deg(i) > 1$ or $a_i > 0$ determines a non-trivial connected generalized cocktail party graph in G, and the collection of all such GCPs satisfies conditions (i)–(iii). Conversely, if the edges of a graph G are partitioned by a family $\mathcal{C}$ of GCPs satisfying (i)–(iii), then we add to $\mathcal{C}$ a trivial GCP for each vertex lying in just one GCP of $\mathcal{C}$. Now take H to be the graph with vertices $1, 2, \ldots, n$ corresponding to GCPs in the extended family, with two vertices adjacent if and only if the corresponding GCPs have a common vertex. For each $i \in \{1, 2, \ldots, n\}$, let a_i be one-half the number of vertices of type b in the i-th GCP. Then $G = L(H; a_1, \ldots, a_n)$. $\square$

We call a partition of edges of G into GCPs a *cover*. A cover of G is said to be *proper* when it satisfies the conditions (i)–(iii) of Theorem 2.3.1.

We know from Theorem 2.1.6 that, except for one pair of root graphs (K_3 and $K_{1,3}$), if two connected line graphs are isomorphic then, apart from isolated vertices, the root graphs are isomorphic too. We show next that a similar result holds for generalized line graphs: except for the six pairs of *connected* graphs shown in Fig. 2.3, if two connected generalized line graphs are isomorphic then, apart from isolated vertices, the root graphs are isomorphic. If a generalized line graph is disconnected without isolated vertices then it suffices to examine connected components, while if isolated vertices are allowed, the remaining pair displayed in Fig. 2.3 is the only new exception that arises.

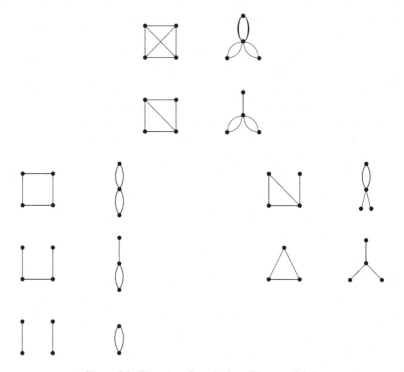

Figure 2.3: The pairs of graphs from Theorem 2.3.4.

Proposition 2.3.2. *Let G be a connected generalized line graph with more than six vertices. Then two vertices of G have the same set of neighbours if and only if they are partners in a (unique) GCP of a proper cover.*

Proof. Let $\hat{H}$ be a root multigraph of G, and note that vertices of type a in G are single edges in $\hat{H}$, while vertices of type b arise from petals. The sufficiency of the stated condition follows from Theorem 2.3.1. Accordingly, suppose that u and v are vertices with the same neigbours that do not satisfy the conclusion of the proposition. Note that u and v are non-adjacent. There are three cases to consider.

Case 1. u and v are both of type a. Here there are two independent edges in $\hat{H}$ such that any further edge adjacent to one of them is also adjacent to the other. Since double edges have only one vertex in common with an edge of type a, $\hat{H}$ can consist only of two independent edges together with some of the four possible edges joining their endvertices. Thus G is one of the five graphs P_3, C_4, $GCP(4, 1)$, $GCP(5, 2)$ and $GCP(6, 3)$.

Case 2. u is of type a and v is of type b. In this case, $\hat{H}$ contains a pair of edges in a petal, a third edge non-adjacent to these, and some of the possible edges joining an end vertex of the petal with an end vertex of the third edge. Thus G is either $K_{1,3}$ or $K_2 \nabla 3K_1$.

Case 3. u and v are both of type b. Now $\hat{H}$ consists of two petals joined by a single edge, and so G is $K_{1,4}$.

Since none of the eight graphs arising above has more than six vertices, the proof is complete. □

We can now settle the uniqueness question for proper covers of generalized line graphs.

Theorem 2.3.3 [CvDS2]. *If G is a connected generalized line graph with more than six vertices, then there exists one and only one partition of the edges of G into non-trivial connected GCPs satisfying the three conditions of Theorem 2.3.1.*

Proof. Suppose, by way of contradiction, that there exist two different partitions of $E(G)$ satisfying the given conditions. By Proposition 2.3.2, the GCPs other than cliques are the same in each partition. For each such GCP, add an edge between each pair of partners. Then we obtain a line graph with two different clique covers, contradicting Theorem 2.1.6. □

More generally, a proper cover of a generalized line graph is determined completely by one of its GCPs. Indeed, starting from some fixed GCP, we can recognize the types of its vertices. Next, if we pick from it any vertex of a-type which has neighbours external to the GCP, then this vertex along with the external neighbours forms a new GCP, and so forth. In a finite number of steps all GCPs in the cover can be reconstructed. This is the basis for an algorithm which determines whether a given graph is a generalized line graph and outputs the possible root multigraphs; details appear in [Sim4].

Theorem 2.3.4. *Except for the pairs in Fig. 2.3, if two connected generalized line graphs are isomorphic then their root multigraphs are also isomorphic.*

Proof. We know from Theorem 2.3.3 that it suffices to consider generalized line graphs which have at most six vertices, and are such that the types of vertices are not uniquely determined. The possibilities for such graphs are listed in the proof of Proposition 2.3.2. For each of the graphs in cases (2) and (3), there is just one isomorphism class of proper covers and so the corresponding root graphs are isomorphic. (In case (2), a vertex of least degree can be of type a or b.) Accordingly, we consider case (1). Note first that the octahedron $GCP(6, 3)$ has two isomorphism classes of proper covers: in one case, a cover consists of four triangles (and all vertices are of type a) and in the other, the cover consists of the octahedron itself (and all vertices are of type b). The corresponding root multigraphs are respectively K_4 and $D_{1,3}$, where $D_{1,3}$ is the graph obtained from $K_{1,3}$ by duplicating edges. The remaining graphs in case (1) are induced subgraphs of the octahedron, and each is the line graph of two root multigraphs obtained from K_4 and $D_{1,3}$ by deleting edges. All seven pairs obtained in this way are shown in Fig. 2.3. $\qquad\square$

Theorems 2.3.3 and 2.3.4 are contained implicitly in a paper of P. J. Cameron [Cam].

We saw in Theorem 2.1.3 that there are exactly 9 graphs that are not line graphs yet have the property that every proper induced subgraph is a line graph. Since the property of being a generalized line graph is hereditary, we can also look for minimal graphs that are not generalized line graphs. We use the abbreviation MNGLG for such a graph (a 'minimal non-generalized line graph'). We shall see that there are exactly 31 MNGLGs. They are displayed in Fig. 2.4, where the least eigenvalue of each graph is shown in parentheses. Note that the graphs $G^{(1)}, \ldots, G^{(11)}$ have least eigenvalue less than -2 and are ordered lexicographically by non-decreasing spectral moments. The graphs $G^{(12)}, \ldots, G^{(31)}$ have least eigenvalue greater than -2 and are ordered similarly.

In order to understand the origin of these 31 MNGLGs, we discuss some of their spectral properties. We first note that such a graph G must be connected (for if not, each component would be a generalized line graph and then G would be also). We consider first those minimal non-generalized line graphs G with $\lambda(G) < -2$.

Lemma 2.3.5. *Let G be a minimal non-generalized line graph with eigenvalues $\lambda_1 \geq \lambda_2 \geq \cdots \geq \lambda_n$. If $\lambda_n < -2$ then $\lambda_{n-1} > -2$.*

Proof. We know by interlacing that $\lambda_{n-1} \geq -2$, for otherwise a vertex-deleted subgraph of G would have least eigenvalue less than -2. Suppose by way

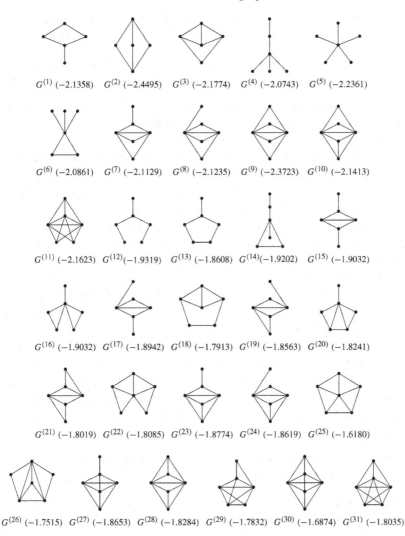

$G^{(1)}$ (−2.1358) $G^{(2)}$ (−2.4495) $G^{(3)}$ (−2.1774) $G^{(4)}$ (−2.0743) $G^{(5)}$ (−2.2361)

$G^{(6)}$ (−2.0861) $G^{(7)}$ (−2.1129) $G^{(8)}$ (−2.1235) $G^{(9)}$ (−2.3723) $G^{(10)}$ (−2.1413)

$G^{(11)}$ (−2.1623) $G^{(12)}$(−1.9319) $G^{(13)}$ (−1.8608) $G^{(14)}$(−1.9202) $G^{(15)}$ (−1.9032)

$G^{(16)}$ (−1.9032) $G^{(17)}$ (−1.8942) $G^{(18)}$ (−1.7913) $G^{(19)}$ (−1.8563) $G^{(20)}$ (−1.8241)

$G^{(21)}$ (−1.8019) $G^{(22)}$ (−1.8085) $G^{(23)}$ (−1.8774) $G^{(24)}$ (−1.8619) $G^{(25)}$ (−1.6180)

$G^{(26)}$ (−1.7515) $G^{(27)}$ (−1.8653) $G^{(28)}$ (−1.8284) $G^{(29)}$ (−1.7832) $G^{(30)}$ (−1.6874) $G^{(31)}$ (−1.8035)

Figure 2.4: The graphs from Theorem 2.3.18.

of contradiction that $\lambda_{n-1} = -2$, and let $P_G(x) = (x+2)^k Q(x)$, where $k > 0$ and $Q(-2) \neq 0$. By interlacing, each vertex-deleted subgraph $G - i$ has -2 as an eigenvalue of multiplicity at least k. Since $P'_G(x) = \Sigma_{i=1}^{n} P_{G-i}(x)$, $(x+2)^k$ divides $P'_G(x)$, a contradiction. $\qquad\square$

Remark 2.3.6. In the next section we shall consider minimal forbidden subgraphs G for graphs H with the property $\lambda(H) \geq -2$. For such graphs G, Lemma 2.3.5 and its proof hold without any changes. $\qquad\square$

The only graphs with six or fewer vertices that satisfy the hypotheses of Lemma 2.3.5 are the graphs $G^{(i)}$ ($1 \leq i \leq 11$) in Fig. 2.4 (cf. [CvPe]).

Lemma 2.3.7. *If G is a minimal non-generalized line graph for which $\lambda(G) \geq -2$ then $\lambda(G) > -2$.*

Proof. Suppose by way of contradiction that $\lambda(G) = -2$. Then the adjacency matrix A of G is such that $\text{rank}(A + 2I) < n$, where n is the order of G. Hence $A + 2I = Q^T Q$ where Q is a $d \times n$ matrix of rank d for some $d < n$. Now Q has d linearly independent columns which form an invertible matrix M such that $M^T M = A' + 2I$, where A' is the adjacency matrix of a generalized line graph $L(\hat{H})$. We may suppose that $\hat{H}$ is chosen to have the smallest possible number of components, say k, and that $M = (\mathbf{q}_1 | \mathbf{q}_2 | \cdots | \mathbf{q}_d)$, where $\mathbf{q}_1, \mathbf{q}_2, \ldots, \mathbf{q}_d$ are the first d columns of Q. If R is the incidence matrix of the root multigraph $\hat{H}$ then $UM = R$ for some orthogonal matrix U. Replacing Q with UQ, we may assume that $M = R$, and then all non-zero entries of $\mathbf{q}_1, \ldots, \mathbf{q}_d$ are ± 1. Now let $\mathbf{c}$ be any one of the last $n - d$ columns of Q, say $\mathbf{c} = \Sigma_{i=1}^d c_i \mathbf{q}_i$. Then $c_i = \det(R_i)/\det(R)$, where R_i is obtained from R by replacing $\mathbf{q}_i$ with $\mathbf{c}$ (Cramer's Rule). By Proposition 1.1.8, $\det(R) = \pm 2^k$; and if $\det(R_i) \neq 0$ then $\det(R_i) = \pm 2^{k_i}$ (for some $k_i \geq k$) because the columns of R_i determine another generalized line graph with least eigenvalue greater than -2. Hence $c_i \in \{0, \pm 2^{k_i - k}\}$ and so the entries of $\mathbf{c}$ are integers. Since $\mathbf{c}^T \mathbf{c} = 2$, the only non-zero entries of $\mathbf{c}$ are two entries ± 1. Since $Q^T Q = A + 2I$, $\mathbf{c}$ represents an edge which may be added to $\hat{H}$. If H^* is the multigraph obtained from $\hat{H}$ by adding all such edges then Q is the incidence matrix of H^* and $G = L(H^*)$, a contradiction. $\square$

In view of Lemmas 2.3.5 and 2.3.7, we can use the Interlacing Theorem once more to deduce the following result.

Proposition 2.3.8. *Let G be a minimal non-generalized line graph, and let i be any vertex of G. Then $\lambda(G - i) \geq -2$ and the multiplicity of -2 as an eigenvalue of $G - i$ is at most 1.*

Remark 2.3.9. In view of Proposition 2.3.8, we can deduce from Theorems 2.2.4 and 2.2.8 the form of each vertex-deleted subgraph of a minimal non-generalized line graph G. If we choose a vertex j such that $G - j$ is connected, then $G - j = L(\hat{H})$, where the root multigraph $\hat{H}$ has one of the forms: a tree with at most two petals, a unicyclic graph with at most one petal, or a bicyclic graph with an odd cycle. Now it is possible to prove that the 31 graphs in Fig. 2.4 are all the minimal non-generalized line graphs by starting from the generalized line graphs described above and adding a new vertex. However, the arguments are intricate and we present instead a proof from [CvRS7] that a minimal non-generalized line graph has at most 6 vertices. It is then straightforward to check that the 31 graphs in Fig. 2.4 are precisely the MNGLGs with at most 6 vertices; for this purpose one can use the graph tables from [CvPe]. □

The following additional abbreviations will be useful in the sequel: GLG denotes a generalized line graph, while NGLG denotes a graph that is not a generalized line graph. Also, NLG denotes a graph that is not a line graph, while MNLG denotes a graph that is minimal among NLGs. All of the graphs $G^{(1)}, \ldots, G^{(31)}$ of Fig. 2.4 are MNGLGs; in particular, the graphs $G^{(j)}$ ($j = 3, 15, 18, 21, 23, 25, 28$) are MNLGs.

Consideration of 6 of the 7 pairs in Fig. 2.3 shows that a connected graph can be both a line graph and a GLG of the form $L(H; a_1, \ldots, a_n)$ with not all a_i zero (cf. Theorem 2.3.4).

Remark 2.3.10. If G is an MNGLG, then

(a) G is connected;
(b) for each $v \in V(G)$, the graph $G - v$ is a GLG;
(c) either all graphs $G - v$ ($v \in V(G)$) are line graphs (case 1), or at least one of them is a *GLG* but not a line graph (case 2);
(d) in case 1 from (c), G is an MNLG and hence is one of the seven graphs $G^{(j)}$ ($j = 3, 15, 18, 21, 23, 25, 28$) from Fig. 2.4 (since the graphs H_1 ($= K_{1,3}$) and H_3 ($= K_5 - e$) from Fig. 2.1 are GLGs);
(e) in case 2 from (c), at least one vertex-deleted subgraph $G - v$ has as an induced subgraph one of the graphs H_1 or H_3 from Fig. 2.1 (since the remaining graphs from Fig. 2.1 are not GLGs). □

Lemma 2.3.11. *If G is an MNGLG with at least seven vertices, then there exists a vertex v of G such that $G - v = L(\hat{H})$ where $\hat{H}$ is a connected root graph with at least one petal.*

Proof. Since H has at least 7 vertices, it follows from parts (b) and (d) of Remark 2.3.10 that there exists a vertex v of G such that $G - v = L(\hat{H})$ where $\hat{H}$ is a root graph with at least one petal. We show that v can be selected in such a way that $\hat{H}$ is connected.

Assume first that $G - v$ is disconnected, i.e. that v is a cutvertex for G. Note that $G - v$ has a component K such that the components other than K form a GLG but not a line graph. Now it suffices to find a vertex u in K such that $G - u$ is connected. If K contains just one vertex we are done; otherwise, if K contains at least two vertices, then at least two are not cutvertices of K, while at least one is not a cutvertex of G. This completes the proof. □

Now we introduce an edge-colouring technique. Suppose that G is a rooted graph, with v as a root, such that $G - v$ is a GLG, say $G - v = L(\hat{H})$. Then we represent G by the coloured root graph $\mathcal{H} = (\hat{H}; c)$ obtained from $\hat{H}$ by specifying an edge-colouring c as follows: the edges of $\hat{H}$ that correspond in G to the vertices adjacent to v are coloured red, while the remaining edges are coloured blue. Conversely, given $\hat{H}$ and a (red-blue) edge-colouring c, we obtain a unique rooted graph G from $(\hat{H}; c)$ by adding to $L(\hat{H})$ a vertex v whose neighbours are the vertices corresponding to the red edges. In due course, we shall consider in turn the three possible edge-colourings of the petal in the root graph of $G - v$, but first we need a few further notions.

Let $\hat{H}$ be a root graph endowed with a red-blue edge-colouring c. Let G ($=$ $L_r(\hat{H}; c)$) be the rooted graph obtained by extending $L(\hat{H})$ by a new vertex r (the root of G) which is adjacent to the vertices of $L(\hat{H})$ corresponding to red edges of $(\hat{H}; c)$. We say that two edges in the root graph $\hat{H}$ are *adjacent* if the corresponding vertices in $L(\hat{H})$ are adjacent, while two edges are *incident* in $\hat{H}$ if they have a vertex in common. (Thus the edges of a petal are incident but non-adjacent.)

An *extension* of a coloured root graph $(\hat{H}; c)$ is obtained by adding an (un-coloured) edge r to $\hat{H}$ in such a way that $L(\hat{H} + r) = L_r(\hat{H}; c)$. This means that $\hat{H} + r$ is a root graph in which r is adjacent to all red edges of $(\hat{H}; c)$, and not adjacent to any blue edge. Note that the edge r can be a pendant edge (in which case we add a new vertex as well); also, r can be in a petal (in which case it duplicates a blue pendant edge in the coloured root graph). A *reduction* of $\hat{H}$ is a root graph $\hat{H} \setminus e$ obtained from $\hat{H}$ by deleting an edge e and any isolated vertex in $\hat{H} - e$.

Clearly, if a coloured root graph $(\hat{H}; c)$ has an extension, then $L_r(\hat{H}; c)$ ($=$ $L(\hat{H} + r)$) is a GLG. The converse need not to be true: if $(\hat{H}; c)$ has no extension, the graph $L_r(\hat{H}; c)$ may still be a GLG, as the following example shows.

Example 2.3.12. Let the coloured root graph $\mathcal{H}_1$ ($= (\hat{H}_1; c_1)$) consist of three petals at a single vertex (cf. Fig. 2.3), with just one edge in each petal being red. It has no extension, but $L_r(\mathcal{H}_1)$ is a GLG. The explanation is found in the existence of the coloured root graph $\mathcal{H}_2$ ($= (\hat{H}_2; c_2)$), which consists of K_4 with just three red edges meeting at a fixed vertex. We have $\hat{H}_1 \neq \hat{H}_2$, yet $L_r(\mathcal{H}_1) = L_r(\mathcal{H}_2)$ (cf. Theorem 2.3.4). On the other hand, $(\hat{H}_2; c_2)$ does have an extension (by a pendant edge attached at the vertex common to the red edges). □

The above effect appears only in 'small graphs': by Theorem 2.3.4 we know that in connected root graphs with at least seven edges it cannot arise.

Proposition 2.3.13. *Let $\mathcal{H}$ be a connected coloured root graph with at least seven edges. Then $L_r(\mathcal{H})$ is a GLG if and only if $\mathcal{H}$ has an extension.*

Proof. We have already noted that if $\mathcal{H}$ has an extension, then $L_r(\mathcal{H})$ is a GLG. Suppose next that $L_r(\mathcal{H})$ is a GLG, say $L_r(\mathcal{H}) = L(\hat{F})$, where $\hat{F}$ has no isolated vertices. Let f be the edge of $\hat{F}$ corresponding to r, so that $L(\hat{H}) = L(\hat{F} \setminus f)$. Since $\hat{H}$ is connected, and since $\hat{H}$ and $\hat{F} \setminus f$ have at least 7 edges, we know from Theorem 2.3.4 that $\hat{H} = \hat{F} \setminus f$. It follows that $\mathcal{H}$ has an extension (by the edge f). □

We show first that an MNGLG has at most eight vertices; the proof is taken from [CvRS7]. An alternative proof appears in [CvDS2]; see also [RaSV1] and [Vij1].

Theorem 2.3.14. *If G is a connected graph with at least nine vertices then G is not an MNGLG.*

Proof. Suppose, by way of contradiction, that G is an MNGLG. Then, by Lemma 2.3.11, there exists a vertex r of G such that $G - r = L(\hat{H})$, where $\hat{H}$ is a connected root graph with at least one petal. Let $\mathcal{H} = (\hat{H}; c)$ where c is the red-blue edge-colouring of $\hat{H}$ in which the red edges correspond to the vertices of $L(\hat{H})$ adjacent to r. By Proposition 2.3.13, $\mathcal{H}$ has no extension. To prove the theorem it is sufficient to show that $\hat{H}$ contains an edge e such that the reduction $\hat{H} \setminus e$ is connected and has no extension with respect to the colouring c (or, more precisely, the restriction of c to $\hat{H} \setminus e$). For $\hat{H} \setminus e$ contains at least seven edges and has no extension, and so by Proposition 2.3.2, $L_r(\hat{H} \setminus e; c)$ is an NGLG. Since $L_r(\hat{H} \setminus e; c)$ is an induced subgraph of G ($= L_r(\hat{H}; c)$), this contradicts the assumption that G is an MNGLG.

The existence of the edge e will be shown in different ways depending on the colouring of the edges p, p' in the petal whose existence is guaranteed by Lemma 2.3.11.

(i) *both p and p' are blue.* In this case we can take e to be either p or p', for then if $(\hat{H} \setminus e; c)$ has an extension, so too does $(\hat{H}; c)$.

(ii) *one of p, p' is blue, the other is red.* Consider the edges of the reduction $(\hat{H} - p) \setminus p'$. Since this subgraph is connected and has at least six edges, there exists at least one edge which belongs to some (other) petal, or belongs to some cycle, or is a pendant edge. Now we take e to be one such edge. (Note that, in each variant, there are at least two choices for e.)

(iii) *both p and p' are red.* Let v be the vertex of the petal incident with edges other than p and p', say with $q_1, \ldots, q_s$. If one of these edges, say q_1, is blue then we can choose $e \neq q_1$ as in case (ii) (having in mind that in case (ii) there are at least two choices for e). Accordingly, we assume that $q_1, \ldots, q_s$ are all red. Now consider the connected graph $\hat{H} - p'$: we may suppose that $(\hat{H} - p'; c)$ has an extension (by an edge f, say), for otherwise we are done. If f is incident with v then we can simply replace p' to obtain an extension of $(\hat{H}; c)$, a contradiction. Thus f must be incident with u, where $p = uv$; in addition, f is incident with all edges $q_1, \ldots, q_s$, and so $s = 1$. Let $q = q_1 = vw$. Then all edges incident with w are red, while all other edges except p and p' are blue. Denote these new red edges by $r_1, \ldots, r_t$. Consider now the restriction of c to any graph $\hat{H} \setminus e'$ (with $e' \neq p, p', q$), such that $\hat{H} \setminus e'$ is connected; again we may assume that $(\hat{H} \setminus e'; c)$ has an extension (by an edge f', say). Then f' must be incident with v, and also with the endvertices (other than w) of all the edges $r_1, \ldots, r_t$. It follows that $t \leq 2$. (Note that e' can be one of the edges $r_1, \ldots, r_t$.) Thus there are at least three blue edges. In this situation, we choose e as a (red or blue) pendant edge, if any, or as a blue edge in a petal, or as a blue edge in a cycle. Then $\hat{H} \setminus e$ is connected and $(\hat{H} \setminus e; c)$ has no extension.

This completes the proof. $\qquad\qquad\qquad\qquad\qquad\qquad\qquad\qquad\qquad\qquad\qquad\square$

Theorem 2.3.14 represents the essential step towards our goal, since all MNGLGs can now be found by inspecting all connected graphs with up to eight vertices. However we can improve Proposition 2.3.13 by adapting the proof to take account of the effects illustrated in Example 2.3.12. We obtain the following.

Proposition 2.3.15. *Let $(\hat{H}; c)$ be a connected coloured root graph $\mathcal{H}$ with five or six edges and at least one petal, such that $\hat{H}$ is not equal to one of the multigraphs of Fig. 2.3. Then $L_r(\mathcal{H})$ is a GLG if and only if $\mathcal{H}$ has an extension.*

Now we can prove the following result.

Theorem 2.3.16. *If G is a connected graph with seven or eight vertices then G is not an MNGLG.*

Proof. Suppose, by way of contradiction, that G $(= L_r(\hat{H}; c))$ is an MNGLG, where $\hat{H}$ is a connected root graph with six or seven edges. Arguing as in the proof of Theorem 2.3.14, we can find a connected reduction $\hat{H} \setminus e$ such that either $\hat{H} \setminus e$ has no extension (with respect to c) or $\hat{H} \setminus e$ is one of the root graphs excluded in Proposition 2.3.15. In the former case we can complete the proof as in Theorem 2.3.14. In the latter case, $\hat{H}$ contains as a subgraph one of the excluded multigraphs of Fig. 2.3, and so either (i) G has seven vertices and $\hat{H} \setminus e$ is the multigraph with two petals and a pendant edge at a single vertex, or (ii) G has eight vertices and $\hat{H} \setminus e$ is the multigraph with three petals at a single vertex. The second possibility cannot arise because, as we noted in Remark 2.3.9, no connected vertex-deleted subgraph of a minimal non-generalized line graph has three petals. Now $(\hat{H}; c)$ can be obtained from $\hat{H} \setminus e$ by adding an edge and then colouring the edges. To complete the proof we have to show that no extension of $\hat{H} \setminus e$ yields an MNGLG. First note that the edges in each petal are all coloured in the same way: both blue, or one blue and the other red, or both red; for otherwise G has as an induced subgraph one of the graphs $G^{(1)}, G^{(2)}, G^{(3)}$ from Fig. 2.4. Now $\hat{H} \setminus e$ can be extended in three ways: by adding a pendant edge at the vertex of degree 5, by duplicating the pendant edge to form a petal, or by adding a pendant edge at a vertex of degree 1; thus we obtain at most 24 ($= 9 + 3 + 12$) coloured root graphs. We can now check, using the tables from [CvDGT], that none of these coloured root graphs gives rise to an MNGLG. This completes the proof. $\square$

If we combine Theorems 2.3.14 and 2.3.16, and inspect the graphs with 6 vertices, we obtain the following result.

Corollary 2.3.17. *The MNGLGs are the* 31 *graphs shown in Fig. 2.4*

The 31 graphs which arise in the above corollary were originally identified independently by B. D. McKay (private communication) and the authors of [CvDS1, CvDS2]. In the former case, a computer was used to search for minimal non-generalized line graphs with nine or fewer vertices; in the latter case, use was made of tables of graphs and their spectra which would subsequently appear in [CvDSa] and [CvPe]. We state the resulting characterization as follows.

Theorem 2.3.18. *A graph G is a generalized line graph if and only if it does not contain any of the* 31 *graphs in Fig. 2.4 as an induced subgraph.*

This theorem was first announced by D. Cvetković, M. Doob and S. Simić in [CvDS1]; their proof appeared in [CvDS2]. Meanwhile the result was proved independently by S. B. Rao, N. M. Singhi and K. S. Vijayan [RaSV1]; these authors also extended to generalized line graphs the characterization of line graphs by odd triangles given in Theorem 2.1.2.

Theorem 2.3.18 is an analogue of the characterization of line graphs by forbidden subgraphs in Theorem 2.1.3. However, it contains implicitly the following characterization of exceptional graphs, in that the twenty graphs $G^{(i)}$ ($12 \leq i \leq 31$) are minimal exceptional graphs.

Theorem 2.3.19. *A connected graph G is exceptional if and only if its least eigenvalue is greater than or equal to -2 and it contains as an induced subgraph one of the graphs $G^{(i)}$ ($12 \leq i \leq 31$) in Fig. 2.4.*

Proof. Suppose that G is an exceptional graph. Then by definition its least eigenvalue is greater than or equal to -2. Since G is not a generalized line graph, it contains one of the graphs $G^{(i)}$ ($i = 1, 2, \ldots, 31$) as an induced subgraph. This cannot be one of the graphs $G^{(i)}$ (i = 1,2,...,11) since then, by the Interlacing Theorem, G would have least eigenvalue less than -2. Hence G contains as an induced subgraph one of the graphs $G^{(i)}$ ($i = 12, 13, \ldots, 31$).

The proof in other direction is immediate.   □

It is sometimes convenient to denote the graphs $G^{(i)}$ ($12 \leq i \leq 31$) by $F_1, F_2, \ldots, F_{20}$, as indicated in Table A2.

Next we describe the graphs whose least eigenvalue is greater than -2. These graphs were determined by M. Doob and D. Cvetković, whose result is as follows.

Theorem 2.3.20 [DoCv]. *If H is a connected graph with least eigenvalue greater than -2 then one of the following holds:*

(i) $H = L(T; 1, 0, \ldots, 0)$ *where T is a tree;*
(ii) $H = L(K)$ *where K is a tree or an odd-unicyclic graph;*
(iii) *H is one of the 20 graphs $F_1, F_2, \ldots, F_{20}$ on 6 vertices given in Table A2;*
(iv) *H is one of 110 graphs on 7 vertices given in Table A2;*
(v) *H is one of 443 graphs on 8 vertices given in Table A2.*

Sketch proof. If H is a generalized line graph then it can be identified as a graph from Theorem 2.2.4 or 2.2.8 in which the multiplicity of the eigenvalue -2 is equal to zero. We then see immediately that H is a graph of type (i) or (ii).

If H is not a generalized line graph, then by Theorem 2.3.19 it is either one of the graphs (iii) or can be obtained as a connected extension of a graph from (iii). A computer search then yields the 110 graphs (iv) on 7 vertices and the 443 graphs (v) on 8 vertices as given in Table A2. Finally we find that every connected one-vertex extension of any of the 443 graphs of type (v) has least eigenvalue smaller than or equal to -2.   □

The twenty 6-vertex graphs of type (iii) are identified in [CvPe]. The 110 graphs of type (iv) are identified in [CvLRS1] by means of the list of 7-vertex graphs in [CvDGT].

Remarks 2.3.21. The 573 exceptional graphs (iii)–(v) were first found with the aid of computer by F. C. Bussemaker at the request of the authors of [DoCv]. The method was different from that presented in the proof of Theorem 2.3.20 since generalized line graphs had not yet been characterized by forbidden subgraphs. However, the use of root systems in the theory of graphs with least eigenvalue -2 (cf. Chapters 3 and 4) demonstrated the existence of exceptional graphs of orders

6, 7 and 8 with least eigenvalue greater than -2. These were found by computer as connected extensions of forbidden subgraphs for line graphs (cf. Theorem 2.1.3). The above proof of Theorem 2.3.20 is independent of root systems. The 573 exceptional graphs from Theorem 2.3.20 and are mentioned in [DoCv] and [Cve7]; their adjacency matrices and smallest eigenvalues are given in the microfiche appendix to the paper[BuNe]. Table A2 in this book was produced independently by M. Lepović. □

Corollary 2.3.22. *If G is a connected regular graph with least eigenvalue greater than -2 then G is either a complete graph or an odd cycle.*

The well-known inequality $\lambda(G) \le -1$, for any non-trivial connected graph G, has been refined in [Hon]: if we order connected graphs on n vertices ($n > 2$) by decreasing least eigenvalues then the first graph is K_n and the second is K_{n-1} with a pendant edge attached. We mention in passing that, according to [Cons], the last graph in this ordering is $K_{p,p}$ with $\lambda(K_{p,p}) = -p$ for $n = 2p$, and $K_{p,p+1}$ with $\lambda(K_{p,p+1}) = -\sqrt{p(p+1)}$ for $n = 2p + 1$.

Recall that x is a limit point of a set S of reals if any open interval containing x contains an element of S different from x. Limit points of graph eigenvalues have been studied in several papers; in particular, the papers [Hof12] and [Doo13] contain results on the limit points of least eigenvalues greater than -2.

Let Λ be the set of least eigenvalues of graphs. Let T be a tree with at least two edges, and for such an edge e let $\hat{A}(T, e)$ be the matrix obtained from $A(L(T))$ by replacing 0 with -1 in the diagonal position corresponding to e. We shall say that the pair (T, e) is *proper* provided that $\lambda(\hat{A}(T, e)) < \lambda(L(T))$. (It was conjectured in [Hof12] that every (T, e) is proper, but so far there is no proof.)

Proposition 2.3.23 [Hof12]. *If (T, e) is proper, then $\lambda(\hat{A}(T, e))$ is a limit point of Λ. Conversely, if $\lambda > -2$ is a limit point of Λ, then $\lambda = \lambda(\hat{A}(T, e))$ for some proper pair (T, e).*

It is shown in [Yong] that, for $n \ge 4$, if G is not a complete graph on n vertices, then

$$\lambda(G) < -\frac{1 + \sqrt{1 + 4\frac{n-3}{n-1}}}{2}.$$

When n tends to infinity, this upper bound tends to $\tau = -\frac{1+\sqrt{5}}{2} \approx -1.61803$. It is straightforward to check that if K_n^+ denotes the graph obtained from K_n by adding a pendant edge, then the sequence $(\lambda(K_n^+))$ is decreasing and converges to τ.

The value τ is the largest limit point of Λ, and the second largest limit point is $-\sqrt{3}$. In Proposition 2.3.23, the limit point τ is obtained when $T = K_{1,2}$, and the next limit point $-\sqrt{3}$ is obtained when $T = K_{1,3}$. All graphs with least eigenvalue greater than τ, and all graphs with least eigenvalue greater than $-\sqrt{3}$, have been determined in [CvSt].

It is well known that for a graph H with at least five vertices, the automorphism groups of H and $L(H)$ are isomorphic [Whi]. We conclude this section with a theorem concerning the automorphism group of a generalized line graph; this result follows from Theorem 2.3.3. The wreath product of permutation groups which appears here is defined in [Hall, Section 5.9].

Theorem 2.3.24. *Let* $G = L(H; a_1, a_2, \ldots, a_n)$, *where* G *and* H *are connected graphs. Let* Γ *be the group of automorphisms of* H *which preserve the vertex labellings. Then either the automorphism group of* G *is* $\left(\prod_{i=1}^{n} S_2 \, Wr \, S_{a_i}\right) Wr \, \Gamma$ *or* G *is one of the following graphs:* K_1, C_4, $GCP(5, 2)$, $GCP(6, 3)$, $K_{1,3}$, $K_2 \nabla 3K_1$ *and* $K_{1,4}$.

Corollary 2.3.25. *Unless* G *is one of the exceptions in Theorem 2.3.24, the automorphism group of the root graph* $\hat{H}$ *is a homomorphic image of the automorphism group of the generalized line graph* $L(\hat{H})$, *with distinct automorphisms of* $\hat{H}$ *inducing distinct automorphisms in* $L(\hat{H})$

In [Cam] P. J. Cameron proves this Corollary by a different method, using the structure of the root system F_4.

2.4 Some other classes of graphs

Let $\mathcal{G}(-2)$ be the class of graphs G for which $\lambda(G) \geq -2$. In this section we explain why a graph H which is minimal with respect to the property $\lambda(H) < -2$ has at most ten vertices. It follows that there exists a finite collection of minimal forbidden subgraphs for $\mathcal{G}(-2)$, and a computer search establishes that there are 1812 such forbidden subgraphs. The Rayleigh quotient method illustrated below is one of several techniques developed by A. J. Hoffman, D. K. Ray-Chaudhuri and others in the 1960s which can be used in conjunction with a computer to avoid a brute force approach. The regular graphs in $\mathcal{G}(-2)$ are discussed in Chapter 4, but for cubic graphs we are able to use the forbidden subgraph technique to construct all the exceptional graphs which arise in that case.

The first observation concerning Rayleigh quotients follows immediately from the fact that a real symmetric matrix is orthogonally diagonalizable.

Lemma 2.4.1. *Let G be a graph of order n with adjacency matrix A. If* $\mathbf{x}$ *is a non-zero vector in* $I\!R^n$ *such that* $\mathbf{x}^T A\mathbf{x}/\mathbf{x}^T\mathbf{x} < \rho$ $(\rho \in I\!R)$ *then* $\lambda(G) < \rho$.

The second observation will be needed in Section 2.6.

Lemma 2.4.2. *Let H be an induced subgraph of the graph G with* $\lambda(H) = \lambda(G) = \lambda$, *and let* $\mathbf{x}'$ *be an eigenvector of H corresponding to* λ, *say* $\mathbf{x}' = (x_i)_{i \in V(H)}$.

   (i) *If* $\mathbf{x}$ *is the vector obtained from* $\mathbf{x}'$ *by adding components* $x_i = 0$ *for each* $i \in V(G) \setminus V(H)$ *then* $\mathbf{x}$ *is an eigenvector of G corresponding to* λ.
  (ii) *If* $j \in V(G) \setminus V(H)$ *then* $\Sigma\{x_i : i \in V(H), i \sim j\} = 0$.
 (iii) *If* λ *is a non-main eigenvalue of G then it also a non-main eigenvalue of H.*

Proof. If A is the adjacency matrix of G then $\mathbf{x}^T A\mathbf{x} = \lambda\mathbf{x}^T\mathbf{x}$. Since λ is the least eigenvalue of A, $\mathbf{x}$ is an eigenvector of G corresponding to λ. The second statement follows by equating the j-th entries of $A\mathbf{x}$ and $\lambda\mathbf{x}$. The last statement follows immediately from the relation $\mathbf{j}^T\mathbf{x} = 0$. $\square$

We can now exhibit some classes of graphs which cannot appear as induced subgraphs of any graph in $\mathcal{G}(-2)$. We shall make use of these particular graphs in the next section. In Fig. 2.5, each diagram consists of a graph, with some edges shown as broken lines, together with the components of an associated vector $\mathbf{x} = (x_1, x_2, \dots, x_n)^T$. Each diagram illustrates the class of spanning subgraphs whose edges are represented by the full lines and any subset of the set of broken lines. It is easy to verify that if G is such a spanning subgraph, with adjacency matrix A, then the Rayleigh quotient $\mathbf{x}^T A\mathbf{x}/\mathbf{x}^T\mathbf{x}$ is less than -2. (For instance, if $G \in \mathcal{G}_1$ then the maximum value of $\mathbf{x}^T(A(G) + 2I)\mathbf{x}$ is $-\frac{1}{2}$.) By Lemma 2.4.1, we have $\lambda(G) < -2$, and so we can deduce the following result.

Theorem 2.4.3. *If G is a graph for which* $\lambda(G) \geq -2$, *then G does not contain as an induced subgraph a graph from any of the classes shown in Fig. 2.5.*

Now we consider minimal forbidden subgraphs for the graphs G with the (hereditary) property $\lambda(G) \geq -2$. We require the following lemma, which is a special case of Theorem 5.1.6.

Lemma 2.4.4. *Let* λ *be a simple eigenvalue of the connected graph G, and let K be a connected induced subgraph of G which does not have* λ *as an eigenvalue. Then there exists a vertex* $k \in V(G) \setminus V(K)$ *such that* $G - k$ *is also a connected graph without* λ *as an eigenvalue.*

Proof. The vertices of G may be labelled $1, 2, \dots, n$ in such a way that (i) each vertex is adjacent to one of its predecessors, (ii) the vertices of K are $1, 2, \dots, r$,

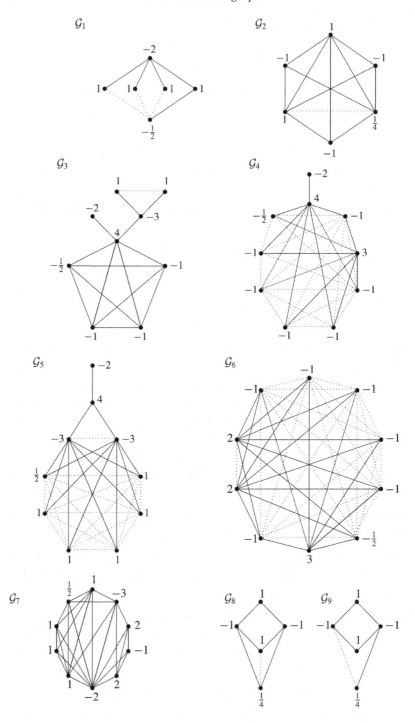

Figure 2.5: Some classes of forbidden subgraphs.

where $r < n$. If A is the adjacency matrix of G then one column of $A - \lambda I$, say the k-th column $\mathbf{c}_k$, is a linear combination of the preceding columns. Note that $k > r$ because λ is not an eigenvalue of K.

Given $i > k$, let h be least such that $h \sim i$. Observe that $h \neq k$, for otherwise $\mathbf{c}_k$ has i-th entry 1, while all preceding columns have i-th entry 0. It follows that $G - k$ is connected.

To see that $G - k$ does not have λ as an eigenvalue, note first that the columns of $A - \lambda I$ other than $\mathbf{c}_k$ form an $n \times (n-1)$ matrix M of rank $n - 1$. From the symmetry of $A - \lambda I$ we know that the k-th row of M is a linear combination of the remaining rows of M. On deleting this row, we obtain a non-singular matrix $A' - \lambda I$, where A' is the adjacency matrix of $G - k$. □

Theorem 2.4.5. *If the graph H is minimal with respect to the property $\lambda(H) < -2$ then H has at most 10 vertices.*

Proof. By minimality, H is connected. If every vertex-deleted subgraph of H is a generalized line graph then H is a minimal non-generalized line graph and hence one of the graphs $G^{(i)}$ ($i = 1, 2, \ldots, 11$) of Fig. 2.4. Thus H has order at most 6 in this case.

Now suppose that some vertex-deleted subgraph $H - j$ is not a generalized line graph. We may assume that $H - j$ is connected, for otherwise we may replace j with a vertex j' in a suitable component of $H - j$ such that j' is not a cutvertex of H. Since $H - j$ is exceptional we know (by Theorem 2.3.18) that $H - j$ has an induced subgraph F isomorphic to one of the graphs $G^{(12)}, \ldots G^{(31)}$ of Fig. 2.4. As we noted in Remark 2.3.6, the argument of Lemma 2.3.5 shows that $\lambda_{n-1}(H) > -2$, and so the multiplicity of -2 as an eigenvalue of $H - j$ is at most 1. If -2 is not an eigenvalue of $H - j$ then $H - j$ has order at most 8, by Theorem 2.3.20. If -2 is a simple eigenvalue of $H - j$ then by Lemma 2.4.4, there exists a vertex k of $H - j$ such that $H - j - k$ is connected with F as an induced subgraph. Now $H - j - k$ has order at most 8, and the theorem follows. □

Many of the graphs of order 10 represented in Fig. 2.5 are minimal forbidden graphs, and so the bound in Theorem 2.4.5 is best possible. There are 1812 minimal forbidden subgraphs altogether, and the following table gives the number f_n on n vertices ($n = 5, 6, \ldots, 10$).

n	5	6	7	8	9	10
f_n	3	8	14	67	315	1405

The 1812 graphs in question were found as the result of a computer search by F. C. Bussemaker and A. Neumaier: they are listed in a microfiche appendix to the paper [BuNe]. It had previously been shown in [RaSV1] that the set of forbidden subgraphs is finite. This result was improved in [KuRS], where it was shown that a forbidden subgraph has at most 10 vertices. The rather long argument from this paper was simplified in [Vij1]. The authors of [BuNe] derived further properties of these graphs, and used (among other things) a generalization of Lemma 2.3.7 to facilitate a fast computer search. The simple proof of Theorem 2.4.5 given here is based on ideas developed in [CvRS4], where it is shown how to treat exceptional graphs without reference to the root systems described in Chapter 3.

The finiteness of the above collection of minimal forbidden subgraphs implies the existence of 'eigenvalue gaps' at ± 2 in the following sense. No graph has least eigenvalue lying between -2 and the largest value of $\lambda(H)$ arising among the 1812 graphs H. This largest value is $-\rho$, where $\rho = 2.006594\ldots$, the largest solution of the equation $(\lambda^3 - \lambda)^2(\lambda^2 - 3)(\lambda^2 - 4) = 1$; moreover $\lambda(H) = -\rho$ if and only if H is obtained from P_9 by adding a pendant edge to a vertex at distance 2 from an endvertex [BuNe]. It follows that no graph has a largest eigenvalue λ_1 in the interval $(2, \rho)$, since the only connected graphs with $\lambda_1 \in (2, \sqrt{2 + \sqrt{5}}]$ are trees (see [BuNe, CvDG]).

Following M. Doob [Doo12], we say that the real number r has the *induced subgraph property*, if every graph G with $\lambda(G) < r$ has an induced subgraph H such that $\lambda(H) = r$. At first sight it might seem surprising that any such r exists, but in fact -2 has this property. In [Doo12] it was proved that the set of all real numbers with this property is precisely $\{0, -1, -\sqrt{2}, -2\}$. Here we shall restrict ourselves to a proof that -2 has the induced subgraph property.

Theorem 2.4.6. *Let G be a graph with $\lambda(G) < -2$. Then there exists an induced subgraph H of G such that $\lambda(H) = -2$.*

Proof. Clearly, we can restrict ourselves to connected graphs. We may also assume that G has at least ten vertices because, as reported in [Doo12], an exhaustive computer search by B. McKay shows that the result is true for all graphs of order nine or less.

Since $\lambda(G) < -2$, G is not a generalized line graph, and so it has as an induced subgraph one of 31 MNGLGs from Fig. 2.4 (see Theorem 2.3.18). If G contains one of the graphs $G^{(1)}, \ldots, G^{(11)}$, then the conclusion follows by direct inspection: three of these graphs contain C_4, another three contain $K_{1,4}$, and the remaining five contain $K_2 \nabla 3K_1$ as an induced subgraph. Otherwise, G has an induced subgraph F isomorphic to one of the remaining twenty MNGLGs (the graphs $G^{(12)}, \ldots, G^{(31)}$). Now G has a connected induced subgraph H of order 9 which contains F, and by Theorem 2.3.20, $\lambda(H) \leq -2$. If $\lambda(H) = -2$

then we are done, while if $\lambda(H) < -2$ the conclusion follows from the result for graphs of order 9. □

Remark 2.4.7. One can investigate the analogous property in respect of the largest eigenvalue: the problem is to find the real numbers r such that, for any graph with index greater than r, there exists an induced subgraph with index equal to r. It turns out that the numbers in question are 0, 1 and 2; one proof for $r = 2$ is based on Theorem 2.4.6, and another makes use of the Smith graphs (see Section 3.4). □

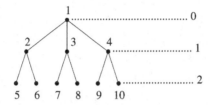

Figure 2.6: A subgraph of an exceptional graph.

In Chapter 4 we shall investigate the class of regular exceptional graphs. The 68 such graphs which are cospectral with a line graph are listed in Table A4, and all but one of them has the 3-claw $K_{1,3}$ as an induced subgraph. This motivates the study of exceptional regular graphs containing an induced 3-claw.

Suppose that G is such a graph with $\lambda(G) \geq -2$. Then G has none of the graphs $G^{(i)}$ ($1 \leq i \leq 11$) from Fig. 2.4 as an induced subgraph. In particular, the absence of $G^{(1)}$ and $G^{(2)}$ ensures that G has as a subgraph the graph H illustrated in Fig. 2.6; moreover the subgraph induced by $V(H)$ is obtained from H by adding edges between the six vertices 5,6,7,8,9,10 at distance 2 from vertex 1. Let H' be the subgraph induced by these six vertices. Examples of H' for several exceptional graphs G, easily analysed in the form they are constructed in Table A4, show that H' need not be regular. However, we are now in a position to construct the cubic exceptional graphs, first found in [BuCS1] using a computer-produced table of cubic graphs. The construction stems from [Cve9].

A cubic exceptional graph must contain, as an induced subgraph, one of Beineke's nine forbidden subgraphs $H_1, \ldots, H_9$ (see Fig. 2.1). Among cubic graphs, only H_1, H_2, H_4 and H_7 can occur. Let us find first the cubic exceptional graphs G containing $H_1 = K_{1,3}$ as an induced subgraph. The two graphs in Fig. 2.7 have least eigenvalue less than -2, and so neither appears as an induced subgraph of G. (In fact, the graphs T_1 and T_2 of Fig. 2.7 are among the 1812 forbidden subgraphs mentioned above; they appear in Table A1.2 as the first

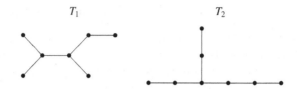

Figure 2.7: Two forbidden subgraphs.

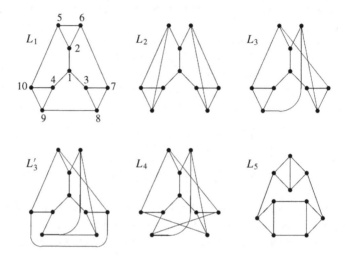

Figure 2.8: A construction of cubic exceptional graphs.

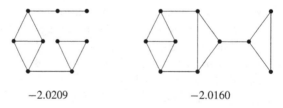

-2.0209 $\qquad$ -2.0160

Figure 2.9: Some forbidden subgraphs.

graph on 7 vertices and the first graph on 8 vertices, respectively.) It follows that there can be no vertex at distance 3 from vertex 1 in Fig. 2.6; in other words, G has diameter 2. If $5 \sim 6$, $7 \sim 8$ and $9 \sim 10$ then G is the graph L_1 of Fig. 2.8. Now suppose without loss of generality that $5 \not\sim 6$. To avoid T_1, we must join each of 7, 8, 9, 10 to vertex 5 or 6. We obtain the graphs L_2, L_3, L_3', L_4 of Fig. 2.8: the graphs L_3, L_3' are isomorphic, while L_4 is the Petersen graph.

A cubic graph with H_2 as an induced subgraph necessarily has $K_{1,3} \, (= H_1)$ as an induced subgraph, and so it remains to find the cubic exceptional graphs which have H_4 or H_7, but not H_1, as an induced subgraph. If G is such a graph then either G has one of graphs from Fig. 2.9 as an induced subgraph or G is the graph L_5 from Fig. 2.8. Since the graphs from Fig. 2.9 have least eigenvalue less than -2, the only new graph which can arise is L_5.

Theorem 2.4.8. *A cubic graph is exceptional if and only if it is one of the graphs L_i $(1 \leq i \leq 5)$ from Fig. 2.8.*

2.5 General characterizations

The characterizations of generalized line graphs given in Sections 2.1 and 2.3 are formulated without reference to the spectrum. In the remainder of this chapter we investigate the extent to which such graphs are characterized by spectral properties, and so the results can be considered as a continuation of the elementary characterization theorems from Section 1.3. The theorems given here date from the 1960s and are presented in their original form with outline proofs to illustrate the forbidden subgraph technique. Readers will find improved versions in Section 3 of Chapter 4, where complete proofs are given using the root system approach developed in Chapter 3.

We first examine the extent to which graphs with least eigenvalue greater than or equal to -2 are characterized by their spectra. Many appealing results on this topic were proved by A. J. Hoffman, D. K. Ray-Chaudhuri, M. Doob *et al*, independently of root sytems. Ray-Chaudhuri [Ray] showed that certain line graphs admit a characterization in terms of the least eigenvalue and some local (non-spectral) invariants. His result is the following, where as usual $\delta(G)$ denote the smallest degree of a vertex in the graph G.

Theorem 2.5.1. *Let G be a graph such that the following conditions hold:*

(i) $\lambda(G) = -2$;

(ii) $\delta(G) > 43$;

(iii) *for each edge uv, there are at least two vertices $\neq v$ adjacent to u but not to v, and at least two vertices $\neq u$ adjacent to v but not to u.*

Then G is a line graph, i.e. $G = L(H)$ for some graph H.

We give only the details of the proof sufficient to explain the provenance of condition (ii). If G is a graph which satisfies conditions (i) and (iii) then G has no induced $K_{1,4}$, for otherwise G has an induced subgraph from the class $\mathcal{G}_1$ depicted in Fig. 2.5. Now suppose that G has an induced $K_{1,3}$, with central vertex 0 and endvertices 1, 2, 3. Consider the following sets of vertices:

V_i is the set of vertices of G adjacent to 0 and i $(i = 1, 2, 3)$;

V_{ij} is the set of vertices of G adjacent to 0, i and j $(i \neq j)$;

V_{123} is the set of vertices of G adjacent to 0, 1, 2 and 3;
$\hat{V}_i$ is the set of vertices of G adjacent to i but not adjacent to 0.

Condition (iii) tells us that $|\hat{V}_i| \geq 2\,(i = 1, 2, 3)$. If say $|V_1| > 2$ then we have sufficient vertices to construct an induced subgraph from $\mathcal{G}_2$. If say $|V_{12}| > 12$ then we can construct an induced subgraph from some $\mathcal{G}_i$ ($i = 4, 6, 8, 9$). (In the absence of graphs from $\mathcal{G}_8$ or $\mathcal{G}_9$, a vertex in $\hat{V}_2$ has a 7-clique among its neighbours or among its non-neighbours.) If $|V_{123}| > 1$ then 0, 1, 2, 3 together with two vertices from V_{123} induce a subgraph from $\mathcal{G}_2$. We deduce that

$$\deg(0) \leq (|V_1| + |V_2| + |V_3|) + (|V_{12}| + |V_{13}| + |V_{23}|) + |V_{123}| \leq 43.$$

The argument outlined above replaces six lemmas in [Ray]. To prove Theorem 2.5.1 we may now assume that G has no induced $K_{1,3}$. This makes it possible, in the absence of graphs from $\mathcal{G}_i$ ($i = 4, 5, 6, 7, 8$) to cover G with a complete system of cliques. Where a vertex u lies in two bridged cliques C, D these are realized as follows. First choose a vertex v adjacent to u such that the number of common neighbours of u and v is minimal. Let T be the set of common neighbours, and let U consist of the vertices adjacent to u but not to v. If there exists $q \in T$ adjacent to every vertex in U then $V(C) = (T \setminus \{q\}) \cup \{u, v\}$, $V(D) = U \cup \{u, q\}$; otherwise, $V(C) = T \cup \{u, v\}$, $V(D) = U \cup \{u\}$. This paragraph conceals a multitude of details which the reader can find in a sequence of lemmas in [Ray].

Remark 2.5.2. Theorem 2.5.1 does not hold without any restriction on δ, the minimum vertex degree. For example, A. J. Hoffman constructed a graph on 28 vertices with $\delta = 12$ which satisfies conditions (i) and (iii) of Theorem 2.5.1 and is not a line graph. However, D. K. Ray-Chaudhuri conjectured that it is possible to reduce the upper bound for δ in Theorem 2.5.1, and we shall see in Chapter 3 that the more powerful root system approach enables us to replace condition (ii) with the inequality $\delta > 28$.   □

Remark 2.5.3. We can state a converse of Theorem 2.5.1 as follows: if G is the line graph $L(H)$, and if $\delta(H) > 3$, then G satisfies conditions (i) and (iii). Recall that (i) is proved in Section 1.2, while the condition $\delta(H) > 3$ ensures that (iii) follows from the definition of line graphs.   □

In the remainder of this section, we note the consequences of Theorem 2.5.1 for regular graphs. In the next section we say more about graphs which are cospectral with connected regular line graphs.

Theorem 2.5.4. *Let G be a regular graph which satisfies conditions (i)–(iii) of Theorem 2.5.1. If G contains a pair of linked cliques, then G is the line graph of a regular graph.*

Proof. By Theorem 2.5.1, $G = L(H)$ for some graph H. Since G is regular it follows that H is either regular or a semi-regular bipartite graph. Since G contains a pair of linked cliques, it contains $K_4 - e$ as an induced subgraph. Since $K_4 - e$ is not the line graph of a bipartite graph, H must be regular. $\square$

As before, let $\mathcal{G}(-2)$ be the class of graphs G for which $\lambda(G) \geq -2$ The following important theorem asserts that, apart from the cocktail party graphs, any connected regular graph in $\mathcal{G}(-2)$ of suitably large degree is a line graph.

Theorem 2.5.5. *Let G be a connected regular graph with degree greater than 16 and least eigenvalue not less than* -2. *Then G is either*

(a) *a line graph, or*
(b) *a cocktail party graph.*

This theorem is due to A. J. Hoffman and D. K. Ray-Chaudhuri but the corresponding paper [HoRa3] remained unpublished because the authors were not satisfied with the length of their proof based on the forbidden subgraph technique. The theorem will be proved in Chapter 4, in several improved forms, using root systems.

The next theorem of D. K. Ray-Chaudhuri [Ray] is a direct consequence of Theorem 2.5.5. Recall that the parameters (n_1, n_2, r_1, r_2) of a bipartite semi-regular graph G have the following meaning: in a 2-colouring of G, the colour classes consist of n_1 vertices of degree r_1 and n_2 vertices of degree r_2.

Theorem 2.5.6. *Let H be a connected regular graph of degree* $r > 9$ *and order n. If the graphs G and $L(H)$ are cospectral, then $G = L(H')$ where either*

(a) *H' is regular with degree r, or*
(b) *H' is semi-regular bipartite with parameters (n_1, n_2, r_1, r_2) where*
    $$r_1 + r_2 = 2r \text{ and } n_1 r_1 = n_2 r_2 = nr/2.$$

Proof. Since H is r-regular, $L(H)$ is $(2r - 2)$-regular, and $\lambda(L(H)) \geq -2$. If G is any graph cospectral with $L(H)$, then G is connected (cf. Theorem 1.2.1). Since $r > 9$, Theorem 2.5.5 is applicable to G, and so it is either a line graph or a cocktail party graph. The latter case is impossible since cocktail party graphs are characterized by their spectra (see Theorem 1.3.3). Hence $G = L(H')$ for some H'. By Proposition 1.1.5, H' is either regular or semi-regular bipartite with parameters as claimed. This completes the proof. $\square$

Note that conditions (a) and (b) in Theorem 2.5.6 are not mutually exclusive, and that in view of Theorem 1.2.16 we have the following corollary.

Corollary 2.5.7. *If H and H' are regular and L(H) and L(H') are cospectral, then H and H' are cospectral. In particular, H is bipartite if and only if H' is bipartite.*

We conclude this section with the following result.

Proposition 2.5.8 [BuCS1]. *If G is a connected bipartite semi-regular graph, then the parameters of G are determined from the spectrum of L(G).*

Proof. Let n_1, n_2, r_1, r_2 be the parameters of G. Then $L(G)$ has $n_1 r_1 (= n_2 r_2)$ vertices, is regular of degree $r_1 + r_2 - 2$ and has -2 as an eigenvalue of multiplicity $n_1 r_1 - n_1 - n_2 + 1$. Since all of these numbers can be determined from the spectrum of $L(G)$, we have a system of equations from which the parameters n_1, n_2, r_1, r_2 can be determined uniquely. This proves the proposition. □

2.6 Spectral characterizations of regular line graphs

Here we discuss, in four subsections, graphs which are cospectral with

(1) line graphs of complete graphs (triangular graphs);
(2) line graphs of complete bipartite graphs;
(3) the line graphs $L(K_{n,n,n})$;
(4) line graphs of block designs.

All of these graphs are regular and connected, and we know from Theorem 1.2.1 that any graph which shares a spectrum with one of them is itself regular and connected. In a majority of cases the graphs considered have at most four distinct eigenvalues. In view of Theorem 1.3.6, spectral characterizations of such graphs can take two equivalent forms: (i) the spectrum of a graph G is specified, (ii) the distinct eigenvalues of G are specified and G is assumed to be a connected regular graph. A. J. Hoffman *et al* used the second form; in this book we generally use the first.

1: $L(K_n)$ – triangular graphs
It follows easily from Theorem 1.2.16 that the eigenvalues of $L(K_n)$ are

(2.7) $\mu_1 = 2(n-2)$ with multiplicity 1;

(2.8) $\mu_2 = n-4$ with multiplicity $n-1$;

(2.9) $\mu_3 = -2$ with multiplicity $\frac{1}{2}n(n-3)$.

Moreover, $L(K_n)$ is a strongly regular graph:

(2.10) it is regular of degree $r = 2n - 4$;

(2.11) any two adjacent vertices have just $e = n - 2$ common neighbours;

(2.12) any two non-adjacent vertices have just $f = 4$ common neighbours.

Recall from Theorem 1.2.20 that these three parameters are related to the eigenvalues as follows: $r = \mu_1$, $e = \mu_1 + \mu_2\mu_3 + \mu_2 + \mu_3$ and $f = \mu_1 + \mu_2\mu_3$.

We now consider the following question: for which values of n is $L(K_n)$ determined by its spectrum or, equivalently (cf. Theorem 1.2.20), by the above parameters? This question was first posed (and answered) in the context of association schemes: the case $n \geq 9$ was settled by W.S. Connor [Conn], and the case $n \leq 6$ by S.S. Shrikhande [Shr1]. The two remaining cases were settled independently by A. J. Hoffman [Hof2, Hof1], and L. C. Chang [Cha1, Cha2].

Theorem 2.6.1. *Unless $n = 8$, $L(K_n)$ is characterized by its spectrum.*

Proof. It is straightforward to check from the spectra of graphs of order at most 6 (cf. [CvPe]) that the result holds for $n \leq 4$. Let G be a graph whose spectrum is given by (2.7)–(2.9), with $n > 4$.

Claim 1: $K_{1,3}$ is not an induced subgraph G.

Suppose that H is an induced subgraph $K_{1,3}$ with central vertex 0 and end-vertices 1, 2, 3. Let H_v be the subgraph induced by the five vertices 0, 1, 2, 3, v. Then (i) $H_v \neq K_{2,3}$ and (ii) H_v is not a 4-cycle with a pendant edge attached; for otherwise H_v has least eigenvalue less than -2. Also, (iii) $H_v \neq K_{1,3}\nabla K_1$ for otherwise H_v has $(-1, 1, 1, 1, -1)$ as an eigenvector corresponding to -2, and this contradicts Lemma 2.4.2.

It follows from (i) and (iii) that 0 is the only vertex adjacent to 1, 2 and 3. Hence, by (2.12), there are exactly nine additional vertices of G: each is adjacent to exactly two of the vertices 1, 2, 3. Now it follows from (ii) that each of the nine is adjacent to 0. Thus the degree of 0 is at least twelve. From (2.10) we have $12 \leq 2n - 4$, and so we assume henceforth that $n \geq 9$.

The vertices 0 and 1 have $n - 2$ common neighbours, and we have so far encountered six of them: three are adjacent to 2 (but not to 3), and three are adjacent to 3 (but not to 2). Let X_1 be the set of the remaining $n - 8$ neighbours of 0 and 1. Considering the common neigbours of 0 and 2, and the common neignbours of 0 and 3, we define X_2 and X_3 analogously. Clearly, the sets X_1, X_2, X_3 are pairwise disjoint. Now $2(n - 2) = \deg(0) \geq 12 + 3(n - 8)$, whence $n \leq 8$, contrary to assumption. This proves the claim.

Claim 2: The subgraph induced by the four common neighbours of two non-adjacent vertices is a 4-cycle.

Consider two non-adjacent vertices of G, say 1 and 2, with common neighbours 3, 4, 5, 6 which induce the subgraph H. Consider the $2n - 4$ vertices adjacent to 3. If v ($\neq 1, 2$) is adjacent to 3, then (in the absence of $K_{1,3}$) v is adjacent to 1 or 2. Let t be the number of neighbours of 3 which are adjacent to both 1 and 2. From (2.10) and (2.11), we obtain $(2n - 4) - 2 = (n - 2) + (n - 2) - t$, whence $t = 2$. Thus vertex 3 has degree 2 in H. Considering vertices 4, 5 and 6 in place of 3, we see that H is regular of degree 2. Thus $H = C_4$, proving the claim.

Let L be the subgraph induced by the vertices 1, 2, 3, 4, 5, 6 above. Note that $L = \overline{K}_2 \nabla K_{2,2} = K_{2,2,2}$ and that we may take the non-adjacent pairs to be $\{1, 2\}, \{3, 4\}$ and $\{5, 6\}$. Consider next the possible graphs L_{uv} induced by $V(L)$ and two further vertices u and v which are such that u and v are adjacent to 1 and 3, but not adjacent to 2 or 4.

Claim 3: Under the above assumptions, L_{uv} is forbidden in G if

(i) *u is adjacent to 5 and not adjacent to 6, while v is adjacent to 6 and not adjacent to 5, or*

(ii) *u and v are not adjacent, while both are adjacent to 5 but not adjacent to 6.*

Suppose, by way of contradiction, that G has an induced subgraph L_{uv}. Assume that (i) holds, and suppose that u and v are adjacent. Then all six edges between $\{6, u\}$ and $\{1, 3, v\}$ are present. Since 6 and u are non-adjacent, while 1, 3 and v are pairwise adjacent, this contradicts Claim 2. If u and v are non-adjacent, then the vertices 1, 4, u, v induce $K_{1,3}$, contradicting Claim 1. Next assume that (ii) holds. Considering the edges between $\{u, v\}$ and $\{1, 3, 5\}$, and the adjacencies among vertices 1, 3 and 5, we again obtain a contradiction. This proves the claim.

Claim 4: The $2n - 4$ neighbours of any vertex can be partitioned into two cliques of order $n - 2$.

With L as above, let W be the set consisting of the $n - 4$ common neigbours of 1 and 3 other than 5 and 6. If $w \in W$ then $w \notin V(L)$, w is adjacent to neither 2 nor 4, and by Claim 1, w is adjacent to one of the vertices 5 and 6. By (2.12), w is not adjacent to both 5 and 6. Now by Claim 3(i), the vertices in W are all adjacent to 5, or all adjacent to 6, and without loss of generality, we suppose the former. By Claim 3(ii), the vertices in W are pairwise adjacent, and together with vertices 3 and 5 induce a clique of order $n - 2$. By the same reasoning applied to vertices 1 and 4, we obtain a second clique of order $n - 2$. This clique is disjoint from the first, for if it contains vertex 5 instead of 6 then we may apply (2.11) to vertices 1 and 5 to deduce that $n \leq 4$. This proves the claim.

Finally, if the cliques from Claim 4 are extended by the vertex 1 then by (2.11) we obtain a pair of bridged cliques of order $n - 1$. Such a pair arises from each vertex of G and so the result follows from Theorem 2.1.1. $\square$

If $n = 8$ then the three Chang graphs are the only exceptions (see Example 1.1.12). A. J. Hoffman constructed one of them in [Hof2] and proposed a procedure for generating all of them in [Hof1]. The exceptions were enumerated first by L. C. Chang [Cha2], and independently by E. Seiden [Seid]. J. J. Seidel [Sei4] observed that each Chang graph can be obtained from $L(K_8)$ by switching in an essentially unique way (see Theorem 1.1.13).

2: The graphs $L(K_{m,n})$
We first recall from Example 1.2.33 that when $m > 1$ and $n > 1$, the eigenvalues of $L(K_{m,n})$ are given by

$$(2.13) \qquad m + n - 2, \ (m - 2)^{n-1}, (n - 2)^{m-1}, \ (-2)^{(m-1)(n-1)}.$$

Note that when $m = n$ we have as the spectrum of $L(K_{n,n})$ $(n > 1)$:

$$(2.14) \qquad 2n - 2, \ (n - 2)^{2(n-1)}, \ (-2)^{(n-1)^2}.$$

Since $L(K_{n,n})$ has just three distinct eigenvalues, it is a strongly regular graph with parameters

$$(2.15) \qquad r = 2n - 2, \ e = n - 2, \ f = 2.$$

The following result is due to S. S. Shrikhande.

Theorem 2.6.2 [Shr2]. *Unless $n = 4$, $L(K_{n,n})$ is characterized by its spectrum.*

Sketch proof. It is easy to show that the theorem holds for $n < 4$. The cases in which $5 \leq n \leq 9$ are more involved, and details may be found in [Shr2]. For $n > 9$ we can proceed as follows. First, any graph G cospectral with $L(K_{n,n})$ is a regular connected graph. Next, by Theorem 2.5.5, G is a line graph (since $n > 9$), and by Theorem 2.5.6, $G = L(H)$ where H is either a regular graph or a semi-regular bipartite graph. By Corollary 2.5.7, H and $K_{n,n}$ are cospectral, and hence isomorphic (see Theorem 1.3.5). $\square$

It turns out that if $n = 4$ there is exactly one exception, namely the Shrikhande graph (graph no. 69 in Table A4); see Example 1.1.14.

Consider now the general situation when m and n are not necessarily equal. The following theorem was proved by J.W. Moon [Moo], and independently by A.J. Hoffman [Hof4].

Theorem 2.6.3. *Unless* $m = n = 4$, $G = L(K_{m,n})$ *if and only if*

(i) *G has mn vertices;*

(ii) *G is regular of degree* $m + n - 2$;

(iii) *every pair of non-adjacent vertices has two common neighbours;*

(iv) *every pair of adjacent vertices has either* $m - 2$ *or* $n - 2$ *common neighbours.*

Note that Theorem 2.6.2 follows from Theorem 2.6.3 as a special case, since the spectrum of a strongly regular graph is determined by its parameters.

The following generalization of Theorem 2.6.2 appears independently in papers of M. Doob [Doo4] and D. Cvetković [Cve2].

Theorem 2.6.4. *If* $m + n > 18$ *and* $\{m, n\} \neq \{2t^2 + t, 2t^2 - t\}$ *then* $L(K_{m,n})$ *is characterized by its spectrum.*

Proof. We assume that $m > 1$ and $n > 1$ because otherwise $L(K_{m,n})$ is complete and hence characterized by its spectrum. Let G be a graph whose eigenvalues are given by (2.13). From Theorem 1.2.1 we know that G is a regular connected graph. Since its degree is greater than 18, and its least eigenvalue is -2, G is a line graph by Theorem 2.5.5. Let $G = L(H)$, where H has no isolated vertices. Since G is regular, H is either a regular graph or a semi-regular bipartite graph.

We suppose that H is a non-bipartite regular graph, for otherwise the result follows from Proposition 2.5.8. If H is regular of degree r then $2(r - 1) = m + n - 2$, whence $m + n$ is even and $r = \frac{1}{2}(m + n)$. The number of edges in H is the number of vertices in G, namely mn, and the number of vertices of H is $4\frac{mn}{m+n}$. By considering the multiplicity of -2 as an eigenvalue of $L(H)$, as given in Theorem 1.2.16, we find that

$$ mn - 4\frac{mn}{m + n} = (m - 1)(n - 1), \text{ equivalently, } \frac{(m - n)^2}{m + n} = 1. $$

We deduce that $m = n$, for otherwise $\{m, n\} = \{2t^2 + t, 2t^2 - t\}$, contrary to assumption. Now the result follows from Theorem 2.6.2.   □

We shall see this last result in a more general setting in Section 4.3. In particular, Theorem 4.3.3 specifies the cospectral mates of $L(K_{m,n})$ which arise when $m + n \leq 18$ and $\{m, n\} \neq \{2t^2 + t, 2t^2 - t\}$. In addition, M. Doob [Doo4] has shown that in the case $\{m, n\} = \{2t^2 + t, 2t^2 - t\}$ the graph $L(K_{m,n})$ has a cospectral mate if there exists a symmetric Hadamard matrix with constant diagonal of order $4t^2$. Theorem 2.6.4 shows that only finitely many graphs $L(K_{2,n})$ can have a cospectral mate, and Doob showed that in fact that no such mates arise; in other words, we have the following theorem.

Theorem 2.6.5 [Doo4]. *Each graph $L(K_{2,n})$ is characterized by its spectrum.*

3: The graphs $L(K_{n,n,n})$

In contrast to line graphs of complete bipartite graphs $K_{n,n}$, all line graphs of complete tripartite graphs of the form $K_{n,n,n}$ are characterized by their spectra. This was proved by A. J. Hoffman and B. A. Jamil [HoJa] using the forbidden subgraph technique.

Theorem 2.6.6 [HoJa]. *For each $n \in I\!N$, the graph $L(K_{n,n,n})$ is the unique graph with spectrum*

$$4n - 2, \; (2n - 2)^{3n-3}, \; (n - 2)^2, \; (-2)^{3n^2-3n}.$$

4: Line graphs of block designs

Let $\mathcal{D}$ be a balanced incomplete block design (BIBD) with parameters (v, b, k, r, λ). Thus $\mathcal{D}$ consists of b *blocks*, each of which is a k-subset of a v-set S, such that each element (or *point*) of S is contained in exactly r blocks and each pair of elements in S is contained in exactly λ blocks. We can define a semi-regular bipartite graph $H \; (= H(\mathcal{D}))$ as follows:

 (i) the vertices of H are the points and the blocks of $\mathcal{D}$;
 (ii) the adjacent pairs of vertices are precisely the incident point-block pairs.

Note that H, as a semi-regular bipartite graph, has parameters (v, b, r, k) inherited from the parameters of the design. The *line graph of a design* $\mathcal{D}$ is the graph $L(\mathcal{D}) = L(H(\mathcal{D}))$.

A BIBD is said to be *symmetric* if $b = v > k$; for example, a *projective plane* of order n is a symmetric BIBD with parameters $b = v = n^2 + n + 1$, $r = k = n + 1, \lambda = 1$.

In this section we review some spectral characterizations of line graphs of block designs. The first results in this area concerned projective and affine planes (see [HuPi, Section 3.2]). Previously, T. A. Dowling and R. Laskar [DoLa] had given a geometric characterization of projective planes. The characterization of projective planes was extended by M. Aigner and T. A. Dowling [AiDo] to all symmetric BIBDs: they noted only one exception, which occurs when $(v, k, \lambda) = (7, 4, 2)$. S. B. Rao and A. R. Rao [RaRa] extended the characterization of affine planes to asymmetric BIBDs under the restrictions $r - 2k + 1 < 0$ and $\lambda = 1$.

We shall first consider spectral characterizations of line graphs of symmetric designs.

Theorem 2.6.7 (cf. [Hof5]). *Let G be the line graph of a projective plane of order n. If G′ is a graph cospectral with G, then G′ is itself the line graph of a projective plane of order n.*

We shall see that in fact the line graphs of projective planes of order n are characterized, among the connected regular graphs of fixed order, by their distinct eigenvalues. To illustrate the proof techniques originally used by A. J. Hoffman *et al* we first establish an equivalent form of Theorem 2.6.7 (see Theorem 2.6.11 below).

Lemma 2.6.8. *A regular connected graph H on $2(n^2 + n + 1)$ vertices has as its distinct eigenvalues the numbers*

(2.16) $$n + 1, \quad \sqrt{n}, \quad -\sqrt{n}, \quad -(n + 1)$$

if and only if $H = H(\Pi)$, where Π is a projective plane of order n.

Proof. If $H = H(\Pi)$ then H has adjacency matrix

(2.17) $$A(H) = \begin{pmatrix} O & B \\ B^T & O \end{pmatrix}$$

where B is a point-line incidence matrix of Π. The eigenvalues of $A(H)$ are the eigenvalues of B and their negatives. Since the eigenvalues of B are $n + 1$ and $\sqrt{n}$ (see, for example, [Rys]) the proof in one direction follows at once.

For the converse, suppose that the distinct eigenvalues of H are given by (2.16). Since the spectrum of H is symmetric about 0, H is bipartite (see Theorem 1.2.2). Moreover, $A(H)$ has the form (2.17), where now B is a $(0, 1)$-matrix with row and column sums equal to $n + 1$, and BB^T has all but one eigenvalue equal to n. Hence $BB^T - nI$ is a non-negative integral and symmetric matrix of rank one, with every diagonal element equal to one. This implies that $BB^T - nI$ has all entries equal to one. In other words, B is the incidence matrix of a projective plane Π of order n. □

Remark 2.6.9. More generally, the graph $H(\mathcal{D})$ of a BIBD $\mathcal{D}$ with parameters (v, b, k, r, λ) has eigenvalues $\pm\sqrt{rk}$, $\pm\sqrt{r - \lambda}$ and 0 with multiplicities 1, $v - 1$ and $b - v$ respectively (see [CvDSa, p.166]). If $\mathcal{D}$ is symmetric, any graph cospectral with $H(\mathcal{D})$ is also the graph of a symmetric BIBD with the same parameters (see [CvDSa, Theorem 6.9]). □

We next consider the line graph $L(\Pi)$, where again Π is a projective plane of order n. Clearly, $L(\Pi)$ is regular of degree $2n$, with $(n + 1)(n^2 + n + 1)$ vertices and $n(n + 1)(n^2 + n + 1)$ edges. From Theorem 1.2.16, we have the following.

Lemma 2.6.10. *If Π is a projective plane of order n, then the distinct eigenvalues of the graph $L(\Pi)$ are*

$$(2.18) \qquad 2n, \quad n-1+\sqrt{n}, \quad n-1-\sqrt{n}, \quad -2.$$

We can now restate Theorem 2.6.7 as follows.

Theorem 2.6.11. *Let G be a connected regular graph, with $(n+1)(n^2+n+1)$ vertices, whose distinct eigenvalues are given by (2.18). Then $G = L(\Pi)$, for some projective plane Π of order n.*

The proof will follow from two lemmas. Throughout, G is a graph that satisfies the hypothesis of Theorem 2.6.11, A ($= (a_{ij})$) is its adjacency matrix, and $a_{ij}^{(k)}$ denotes the (i, j)-entry of A^k.

Lemma 2.6.12. *Any two adjacent vertices of G have exactly $n-1$ common neighbours. Any two non-adjacent vertices of G have at most one common neighbour.*

Proof. By Theorem 1.2.7, since G is a regular connected graph we have $P(A) = J$, where

$$(2.19) \qquad P(x) = \frac{1}{2}(x^3 - (2n-4)x^2 + (n^2 - 7n + 5)x + 2(n^2 - 3n + 1)).$$

Let i be any vertex of G. Then we have:

$$(2.20) \qquad\qquad\qquad a_{ii} = 0;$$

$$(2.21) \qquad\qquad\qquad a_{ii}^{(2)} = 2n;$$

$$(2.22) \qquad\qquad\qquad a_{ii}^{(3)} = 2n(n-1);$$

$$(2.23) \qquad a_{ii}^{(4)} = 4n - 2n(n^2 - 7n + 5) + 2n(n-1)(2n-4).$$

To prove (2.20)-(2.23), recall first that $a_{ii}^{(k)}$ is the number of $i - i$ walks of length k. Note that (2.22) follows from (2.19), and (2.23) follows from the relation $AP(A) = 2nJ$.

Since $\sum_{j \sim i} a_{ij}^{(2)} = a_{ii}^{(3)}$, we have

$$(2.24) \qquad\qquad\qquad \sum_{j \sim i} a_{ij}^{(2)} = 2n(n-1).$$

Here the left hand side is the sum of $2n$ non-negative terms each at most $n-1$, and so $a_{ij}^{(2)} = n-1$ whenever $i \sim j$. This proves the first statement. For the second, consider the matrix $B = (b_{ij})$ defined as follows:

$$(2.25) \qquad\qquad\qquad B = A^2 - 2nI - (n-1)A.$$

Clearly, B is a symmetric matrix whose entries are non-negative. We prove that B is a $(0, 1)$ matrix by showing that

$$(2.26) \qquad \sum_j b_{ij} = \sum_j b_{ij}^2.$$

Consideration of $B\mathbf{j}$ shows that

$$(2.27) \qquad \sum_j b_{ij} = 2n^2.$$

On the other hand, $\sum_j b_{ij}^2 = b_{ii}^{(2)}$, and we can use (2.27) to express B^2 as a quartic in A. From (2.20)-(2.23) we find that $b_{ii}^{(2)} = 2n^2$ also, and so (2.26) is proved. It follows from (2.27) that $a_{ij} \in \{0, 1\}$ whenever i and j are non-adjacent, and this proves the second part of the lemma. $\qquad\qquad\qquad\square$

Lemma 2.6.13. *Let* $m = 2(n^2 + n + 1)$. *Then* G *contains* m *cliques with the following properties:*

   (i) *each clique contains exactly* $n + 1$ *vertices;*
 (ii) *each vertex of* G *is contained in exactly two cliques;*
(iii) *each edge of* G *is contained in exactly one clique.*

Proof. We shall see that the collection of all maximal cliques in G satisfies conditions (i)–(iii). Consider any pair of adjacent vertices u and v. By Lemma 2.6.12, u and v have exactly $n - 1$ common neighbours, and these neighbours are pairwise adjacent. Thus uv lies in a clique of size $n + 1$, and this clique is maximal. Moreover, all maximal cliques arise in this way, and by Lemma 2.6.12, uv cannot lie in two distinct maximal cliques. It remains to establish condition (ii) and to determine the number of maximal cliques.

Consider any vertex u of G, and let v be any of its neighbours. Then u and v lie in a maximal clique C. Next, let w be a neighbour of u not in C. Then u and w lie in a maximal clique C' such that $C \cap C' = \{u\}$. Suppose that u lies in a third clique C''. Then C must contain a vertex x which lies in C' but not in C'', and a vertex y which lies in C'' but not in C'. Since x and y are adjacent, x is a common neighbour of u and y outside C''. This contradicts Lemma 2.6.12, and statement (ii) follows.

Finally, let m' be the number of maximal cliques. Counting in two ways the pairs (u, C) such that the vertex u lies in the maximal clique C, we have

$$2(n + 1)(n^2 + n + 1) = m'(n + 1),$$

whence $m = m'$ as required. $\qquad\qquad\qquad\square$

Now Theorem 2.6.11 follows immediately from Lemma 2.6.13. A more general result, due to A. J. Hoffman and D. K. Chaudhuri, reads as follows.

Theorem 2.6.14 (cf. [HoRa2]). *Let G be the line graph of a symmetric BIBD with parameters $(v, k, \lambda) \neq (4, 3, 2)$. If G' is a graph cospectral with G, then G' is the line graph of a block design with the same parameters.*

When $(v, k, \lambda) = (4, 3, 2)$, there is just one exception, namely graph no. 9 from Table A3 (cf. Theorem 4.3.4).

Finally we mention two results concerning line graphs of asymmetric designs. The following result, due to A. J. Hoffman and D. K. Ray-Chaudhuri is an analogue of Theorem 2.6.7 for affine planes.

Theorem 2.6.15 (cf. [HoRa1]). *Let G be the line graph of an affine plane of order n. If the graph G' is cospectral with G then G' is the line graph of an affine plane of order n.*

A more general result, due to M. Doob, concerns block designs with $\lambda = 1$.

Theorem 2.6.16 [Doo9]. *If $k + r > 18$, then the line graph of a BIBD with parameters $(v, b, r, k, 1)$ is characterized by its spectrum.*

3

Root Systems

In this chapter we study graphs with least eigenvalue -2 using systems of vectors which are called root systems in accordance with the theory of semi-simple Lie algebras over $\mathbb{C}$ (see [Car, Chapter 3]).

The problem of determining the graphs with least eigenvalue -2 was one of the first challenges in the theory of graph spectra. It was essentially settled by P. J. Cameron, J. M. Goethals, J. J. Seidel and E. E. Shult [CaGSS], who established a link between this problem and the theory of root systems. Their approach to the problem is outlined in Section 3.5.

Most of the results in this chapter, including the general characterization theorems in Section 3.6, stem from [CaGSS]. In contrast to the presentations in [BrCN, Chapter 3] and [CaLi, Chapter 3], we follow the derivation in [CvDo2], first outlined in [Cve7].

3.1 Gram matrices and systems of lines

Let G be a graph with n vertices, adjacency matrix A, and least eigenvalue greater than or equal to -2. Eigenvalues greater than -2 are called *principal eigenvalues*. Let m be the multiplicity of -2 as an eigenvalue, and let $r = n - m$, so that r is the number of principal eigenvalues.

The symmetric matrix $I + \frac{1}{2}A$ is positive semi-definite of rank r. Since $I + \frac{1}{2}A$ is orthogonally diagonalizable, it follows that $I + \frac{1}{2}A = B^T B$, where B is an $r \times n$ matrix of rank r, with real entries. Thus $I + \frac{1}{2}A$ is the Gram matrix (the matrix of inner products) of n unit vectors which span the Euclidean space $\mathbb{R}^r$. The angle between any two of these vectors is $60°$ or $90°$ according as the corresponding vertices of G are adjacent or non-adjacent. In this situation, n is bounded in terms of r as follows.

Proposition 3.1.1. *If* $\mathbb{R}^r$ *contains a set of n vectors at* $60°$ *or* $90°$ *then* $n \leq \frac{1}{2}r(r+3)$.

Proof. Let S be a set of unit vectors in $\mathbb{R}^r$ at $60°$ or $90°$. For each $\mathbf{v} \in S$, let $f_\mathbf{v}$ be the function $S \to \mathbb{R}$ defined by

$$f_\mathbf{v}(\mathbf{x}) = \mathbf{v}.\mathbf{x}(2\mathbf{v}.\mathbf{x} - 1).$$

The functions $f_\mathbf{v}$ ($\mathbf{v} \in S$) are linearly independent because, for $\mathbf{u} \in S$, $f_\mathbf{v}(\mathbf{u}) = 1$ if $\mathbf{u} = \mathbf{v}$ and $f_\mathbf{v}(\mathbf{u}) = 0$ if $\mathbf{u} \neq \mathbf{v}$. Now each $f_\mathbf{v}$ is the sum of homogeneous quadratic and linear functions in r variables, and the space of such functions has dimension $\frac{1}{2}r(r+3)$. The result follows. □

The one-dimensional subspaces spanned by vectors at $60°$ or $90°$ form a set of lines at $60°$ or $90°$ passing through the origin in $\mathbb{R}^r$. We refer to such sets of lines as *line systems*. The *dimension of a line system* is the dimension of the space spanned by the lines. Arguing as in Proposition 3.1.1, we can see that the number of lines in an r-dimensional line system is bounded above by a cubic function of r. (We choose a unit vector $\mathbf{v}$ along each line and consider $\frac{1}{3}\mathbf{v}.\mathbf{x}(2\mathbf{v}.\mathbf{x} - 1)(2\mathbf{v}.\mathbf{x} + 1)$ instead of $\mathbf{v}.\mathbf{x}(2\mathbf{v}.\mathbf{x} - 1)$.) In Section 3.4 we obtain improved bounds that are sharp; at this stage, we require only the finiteness property assured by an upper bound.

It is clear that line systems in Euclidean spaces are important for the study of graphs with least eigenvalue -2. There is another connection between graphs and such systems of lines. Suppose instead that G is a graph with largest eigenvalue ≤ 2. Then the matrix $I - \frac{1}{2}A$ is positive semi-definite, hence the Gram matrix of a set of unit vectors at $120°$ and $90°$ in $\mathbb{R}^r$, where now r is the number of eigenvalues of G which are smaller than 2. Again the vectors determine a line system.

For a graph G with least eigenvalue ≥ -2 and adjacency matrix A, any set of vectors with Gram matrix $\alpha(I + \frac{1}{2}A)$ ($\alpha > 0$) is called an *a-representation* of G.

In a similar way, for a graph G with largest eigenvalue ≤ 2 any set of vectors with the Gram matrix $\alpha(I - \frac{1}{2}A)$ ($\alpha > 0$) is called an *o-representation* of G. In a-representations the angle between two non-orthogonal vectors is acute, while in o-representations it is obtuse. Graphs with o-representations also have a-representations, but the converse statement is not true. (The reason is that if the largest eigenvalue is 2 at most, then the smallest is -2 at least, but the converse is false.) Accordingly, we say that the graph G can be *represented* in the line system Σ if we can arrange vectors along the lines of Σ in such a way that they form an a-representation of G.

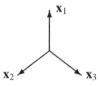

Figure 3.1: Three vectors which determine a star.

We now make our first observations about line systems in an r-dimensional Euclidean space. The case $r = 1$ is trivial: the only graph which has a representation is K_1, and this case will be excluded in later considerations.

When $r = 2$ we can construct three lines with the property that the angle between any two of them is 60°. Such lines are determined, for example, by the vectors $(1, -1, 0)^T$, $(1, 0, -1)^T$ and $(0, 1, -1)^T$ in the 2-dimensional subspace of $I\!R^3$ orthogonal to the vector $\mathbf{j}_3 = (1, 1, 1)^T$. (This means of presentation will also be used later in the context of higher-dimensional spaces.) Of course, isometries have no influence on our considerations and we may represent the same system of lines in a coordinate-free way by three vectors $\mathbf{x}_1$, $\mathbf{x}_2$, $\mathbf{x}_3$ as shown in Fig. 3.1. Here the vectors have equal length, and the angle between any two is 120°. This situation has the following algebraic description:

$$(3.1) \qquad \|\mathbf{x}_1\| = \|\mathbf{x}_2\| = \|\mathbf{x}_3\|, \quad \mathbf{x}_1 + \mathbf{x}_2 + \mathbf{x}_3 = \mathbf{0}.$$

The three lines determined by three vectors satisfying (3.1) are said to form a *star*. Given two lines from a star, the operation of adding the third line required to complete the star is called *star-closing*. A line system is said to be *star-closed* if, along with any two lines at 60°, it contains the third line of the corresponding star.

We are interested in finding (to within an isometry) all maximal sets of lines at 60° and 90° in the Euclidean spaces $I\!R^r$ ($r = 2, 3, \ldots$). The following proposition shows that such a set of lines is star-closed. (The converse statement is not always true, as will be seen in Section 3.3.)

Proposition 3.1.2. *Any r-dimensional line system is contained in an r-dimensional line system that is star-closed.*

Proof. Let Σ be an r-dimensional line system, and consider any pair of lines in Σ at 60°, say $\langle\mathbf{x}\rangle$, $\langle\mathbf{y}\rangle$ where $\mathbf{x}.\mathbf{x} = \mathbf{y}.\mathbf{y} = 2$ and $\mathbf{x}.\mathbf{y} = -1$. We show that if $\langle\mathbf{x} + \mathbf{y}\rangle \notin \Sigma$ then we may add $\langle\mathbf{x} + \mathbf{y}\rangle$ to obtain a larger line system (necessarily of the same dimension as Σ). If $\langle\mathbf{u}\rangle$ is any line of Σ other than $\langle\mathbf{x}\rangle$, $\langle\mathbf{y}\rangle$ then we may choose $\mathbf{u}$ so that $\mathbf{u}.\mathbf{u} = 2$ and $\mathbf{u}.(\mathbf{x} + \mathbf{y}) \in \{0, 1, 2\}$. If however $\mathbf{u}.(\mathbf{x} + \mathbf{y}) = 2$ then $(\mathbf{u} - \mathbf{x} - \mathbf{y}).(\mathbf{u} - \mathbf{x} - \mathbf{y}) = 0$ and we have the contradiction $\mathbf{u} = \mathbf{x} + \mathbf{y}$. Thus $\langle\mathbf{u}\rangle$ makes an angle of 60° or 90° with every line of Σ and may be added as required. If the resulting line system is not star-closed then the procedure may be repeated. In view of the finiteness property noted above, we obtain a star-closed line system of dimension r after finitely many steps. □

The star-closed line system obtained from a line system Σ by repeated star-closing as in Proposition 3.1.2 is called the *star-closure* of Σ. By the star-closure of an a-representation or o-representation R we mean the star-closure of the line system $\{\langle \mathbf{v} \rangle : \mathbf{v} \in R\}$. In discussing star-closed systems, it is convenient to consider all of the vectors of length $\sqrt{2}$ which lie along the lines of the system. The following lemma lists some obvious properties of these vectors.

Lemma 3.1.3. *Let S be the set of all vectors of length $\sqrt{2}$ that lie on the lines of a star-closed system. Then these vectors have the following properties:*

(P1) $\mathbf{x}.\mathbf{y} = 0$ *or* ± 1 *for all* $\mathbf{x}, \mathbf{y} \in S$ *such that* $\mathbf{x} \neq \pm\mathbf{y}$,
(P2) $\mathbf{x}.\mathbf{x} = 2$ *for all* $\mathbf{x} \in S$,
(P3) *if* $\mathbf{x}, \mathbf{y} \in S$ *and* $\mathbf{x}.\mathbf{y} = -1$ *then* $\mathbf{x} + \mathbf{y} \in S$, *and*
(P4) *if* $\mathbf{x} \in S$ *then* $-\mathbf{x} \in S$.

A line system is said to be *decomposable* if the lines can be partitioned into two non-empty sets Σ_1, Σ_2 with the property that every line in Σ_1 is orthogonal to every line in Σ_2. Note that a connected graph with least eigenvalue greater than or equal to -2 can always be represented in an indecomposable line system. In constructing examples of such systems below, we take $\{\mathbf{e}_1, \ldots, \mathbf{e}_n\}$ to be an orthonormal basis of $I\!R^n$, and we let $\mathbf{j}_n = \mathbf{e}_1 + \cdots + \mathbf{e}_n$.

Definition 3.1.4. *For $n > 1$, let D_n be the set of vectors of the form $\pm\mathbf{e}_i \pm \mathbf{e}_j$ ($i < j$). These $2n(n-1)$ vectors determine $n(n-1)$ lines through the origin, and these lines form an indecomposable star-closed line system $\overline{D}_n$ of dimension n.*

It is clear from the definition of star closure that if Σ is a star-closed line system in $I\!R^n$ and S is a set of vectors in $I\!R^n$ then the lines in Σ orthogonal to S also form a star-closed line system. In the case that $\Sigma = \overline{D}_n$ and $S = \{\mathbf{j}_n\}$, we obtain an $(n-1)$-dimensional indecomposable line system $\overline{A}_{n-1}$, where A_{n-1} consists of the $n(n-1)$ vectors $\mathbf{e}_i - \mathbf{e}_j$ ($i \neq j$).

Definition 3.1.5. *Let E_8 be the set of vectors in $I\!R^8$ consisting of the 112 vectors in D_8 together with the 128 vectors of the form $\pm\frac{1}{2}\mathbf{e}_1 \pm \frac{1}{2}\mathbf{e}_2 \pm \cdots \pm \frac{1}{2}\mathbf{e}_8$, where the number of positive coefficients is even. These 240 vectors determine 120 lines through the origin, and these lines form an indecomposable star-closed line system $\overline{E}_8$ of dimension 8.*

We shall see in Section 3.4 that the examples of line systems constructed in Definitions 3.1.4 and 3.1.5 embrace all possibilities in the sense that any indecomposable star-closed line system can be embedded isometrically in $\overline{E}_8$ or $\overline{D}_n$ for some n. Thus E_8 has a unique importance in the study of graphs with least eigenvalue -2, and in the next section we gather together the properties of E_8 required in this book. These properties include the following: (i) for any $\mathbf{u}_1, \mathbf{u}_2 \in E_8$, the set of vectors in E_8 orthogonal to $\mathbf{u}_1$ is isometric to the the set of vectors in E_8 orthogonal to $\mathbf{u}_2$, (ii) for any two stars S_1, S_2 in E_8, the set of vectors in E_8 orthogonal to S_1 is isometric to the the set of vectors in E_8 orthogonal to S_2. Accordingly, we may define

$E_7 = \{\mathbf{x} \in E_8 : \mathbf{x}.\mathbf{y} = 0\}$ for a given vector $\mathbf{y} \in E_8$,

$E_6 = \{\mathbf{x} \in E_8 : \mathbf{x}.\mathbf{y} = \mathbf{x}.\mathbf{z} = 0\}$ for given vectors $\mathbf{y}, \mathbf{z} \in E_8$ such that $\mathbf{y}.\mathbf{z} = -1$.

There are 126 vectors in E_7 and 72 vectors in E_6. These sets of vectors determine indecomposable star-closed line systems $\overline{E}_7$ and $\overline{E}_6$, of dimension 7 and 6 respectively, with 63 and 36 lines respectively. We refer to $A_{n-1}, D_n, E_6, E_7, E_8$ as *root systems*.

We conclude this section by mentioning some isometries which provide alternative coordinatizations of some of these root systems and show how one line system can be embedded in another; these linear maps will be defined by their action on an orthonormal basis $\{\mathbf{e}_1, \ldots, \mathbf{e}_n\}$.

As an example, define the linear transformation ϕ of $\mathbb{R}^8$ by: $\phi(\mathbf{e}_i) = \mathbf{e}_i$ ($i = 1, \ldots, 7$), $\phi(\mathbf{e}_8) = -\mathbf{e}_8$. Thus ϕ is the reflection in a hyperplane, and when applied to E_8, it yields a second description of E_8 as $D_8 \cup \{\pm\frac{1}{2}\mathbf{e}_1 \pm \frac{1}{2}\mathbf{e}_2 \pm \cdots \pm \frac{1}{2}\mathbf{e}_8\}$, where now the number of positive coefficients is odd.

For a less trivial example, define the linear map $\phi : \langle \mathbf{e}_1, \mathbf{e}_2, \ldots, \mathbf{e}_8 \rangle \to \langle \mathbf{e}_1, \mathbf{e}_2, \ldots, \mathbf{e}_9 \rangle$ by $\phi(\mathbf{e}_i) = \mathbf{e}_i - \frac{1}{6}\mathbf{j}_9 + \frac{1}{2}\mathbf{e}_9$ ($i = 1, \ldots, 8$). One verifies easily that ϕ is an isometry whose range is the hyperplane of $\langle \mathbf{e}_1, \mathbf{e}_2, \ldots, \mathbf{e}_9 \rangle$ orthogonal to $\mathbf{j}_9$, and that the image of E_8, in the second form above, consists of A_8 together with the vectors $\pm(-\frac{1}{3}\mathbf{j}_9 + \mathbf{e}_k + \mathbf{e}_l + \mathbf{e}_m)$, $1 \leq k < l < m \leq 9$. This is a useful alternative definition of E_8, which is also given in [CaGSS]. It shows that A_8 can be embedded in E_8, and we shall see that this is essentially the only way to embed A_8 in E_8.

As a final example, let $S = A_5 \cup \{\pm\frac{1}{2}\mathbf{e}_1 \pm \frac{1}{2}\mathbf{e}_2 \pm \cdots \pm \frac{1}{2}\mathbf{e}_6 \pm \frac{\sqrt{2}}{2}\mathbf{e}_7\} \cup \{\pm\sqrt{2}\mathbf{e}_7\}$, where in the second subset three of the first six coordinates are positive. Let ϕ be the linear map $\langle \mathbf{e}_1, \mathbf{e}_2, \ldots, \mathbf{e}_7 \rangle \to \langle \mathbf{e}_1, \mathbf{e}_2, \ldots, \mathbf{e}_8 \rangle$ defined by $\phi(\mathbf{e}_i) = \mathbf{e}_i$ ($i = 1, \ldots, 6$), $\phi(\mathbf{e}_7) = \frac{\sqrt{2}}{2}(\mathbf{e}_7 - \mathbf{e}_8)$. Then ϕ is an isometry which maps S to the set of vectors in E_8 orthogonal to the star with lines $\langle \mathbf{e}_7 + \mathbf{e}_8 \rangle$, $\langle \frac{1}{2}\mathbf{j}_8 - \mathbf{e}_7 - \mathbf{e}_8 \rangle$, $\langle -\frac{1}{2}\mathbf{j}_8 \rangle$. Thus our initial set S is an alternative representation of E_6, and this will be useful later.

3.2 Some properties of E_8

In this section we shall define graphs $G(E_8)$, $G(\overline{E}_8)$, a group $W(E_8)$, and a lattice $L(E_8)$, together with analogous structures for E_7 and E_6.

First, the graph $G(E_8)$ has as vertices the vectors of E_8, with two vectors adjacent if and only if the angle between them is $60°$. Note that this graph is connected.

Now the reflection in the hyperplane of $I\!R^8$ orthogonal to a vector $\mathbf{u}$ of length $\sqrt{2}$ is given by

$$\psi_{\mathbf{u}}(\mathbf{x}) = \mathbf{x} - (\mathbf{u}.\mathbf{x})\mathbf{u} \quad (\mathbf{x} \in I\!R^8).$$

We define the *Weyl group* $W(E_8)$ as the group of isometries generated by the reflections $\psi_{\mathbf{u}}$ ($\mathbf{u} \in E_8$). Each such reflection is a symmetry of E_8 and hence an automorphism of $G(E_8)$. Our first result justifies the definition of E_7 and E_6 in the previous section.

Proposition 3.2.1. *The group $W(E_8)$ is transitive on the vertices, on the edges and on the stars in $G(E_8)$.*

Proof. Since $G(E_8)$ is connected, for transitivity on vertices it suffices to show that if $\mathbf{u}$ and $\mathbf{v}$ are adjacent vectors in $G(E_8)$ then there exists $\psi \in W(E_8)$ such that $\psi(\mathbf{u}) = \mathbf{v}$. Since $\mathbf{u}.\mathbf{v} = 1$ we have $\mathbf{u} - \mathbf{v} \in E_8$ (by property (P3) of Lemma 3.1.3), and we may take $\psi = \psi_{\mathbf{u}-\mathbf{v}}$.

If $\mathbf{x} \in E_8$ then the subgraph of $G(E_8)$ induced by the 56 neighbours of $\mathbf{x}$ is connected. (By vertex-transitivity, it suffices to establish this for just one vertex, say $\mathbf{x} = \frac{1}{2}\mathbf{j}$.) It follows that for transitivity on edges, it is sufficient to show that if $\mathbf{x}, \mathbf{u}, \mathbf{v}$ are vectors in E_8 such that $\mathbf{x}.\mathbf{u} = \mathbf{x}.\mathbf{v} = \mathbf{u}.\mathbf{v} = 1$ then there exists $\psi \in W(E_8)$ such that $\psi(\mathbf{u}) = \mathbf{v}$ and $\psi(\mathbf{x}) = \mathbf{x}$. Here, $\mathbf{x}.(\mathbf{u} - \mathbf{v}) = 0$, and so again we may take $\psi = \psi_{\mathbf{u}-\mathbf{v}}$.

Finally, if $\mathbf{u}$, $\mathbf{v}$, $\mathbf{w}$ and $\mathbf{u}'$, $\mathbf{v}'$, $\mathbf{w}'$ are vectors in E_8 such that $\mathbf{u} + \mathbf{v} + \mathbf{w} = \mathbf{0} = \mathbf{u}' + \mathbf{v}' + \mathbf{w}'$ then $\mathbf{u}.(-\mathbf{v}) = \mathbf{u}'.(-\mathbf{v}') = 1$ and by edge-transitivity there exists $\psi \in W(E_8)$ such that $\psi(\mathbf{u}) = \mathbf{u}'$ and $\psi(-\mathbf{v}) = -\mathbf{v}'$. Then ψ maps $\mathbf{u}$, $\mathbf{v}$, $\mathbf{w}$ to $\mathbf{u}'$, $\mathbf{v}'$, $\mathbf{w}'$ by linearity. Thus $W(E_8)$ is transitive on stars. $\qquad\square$

Remark 3.2.2. If $\mathbf{u}$, $\mathbf{v}$ are adjacent vertices in $G(E_8)$ then the subgraph induced by their 27 common neighbours is connected. This graph, on which the stabilizer of $\mathbf{u}$ and $\mathbf{v}$ acts transitively, is a strongly regular graph with parameters $(27, 16, 10, 8)$, in fact the Schläfli graph (see Example 1.1.12). In the Schläfli graph, the subgraph induced by the 16 neighbours of a vertex is a strongly regular graph with parameters $(16, 10, 6, 6)$, and this graph is the Clebsch graph (see Example 1.1.14). $\qquad\square$

Remark 3.2.3. For $i = 6, 7$, we may define $W(E_i)$ as the subgroup of $W(E_8)$ generated by the reflections $\psi_\mathbf{u}$ ($\mathbf{u} \in E_i$), and $G(E_i)$ as the subgraph of $G(E_8)$ induced by the vectors in E_i. The graphs $G(E_i)$ are connected and we can show, exactly as in the proof of Proposition 3.2.1, that $W(E_i)$ is transitive on $G(E_i)$.   □

For $i = 6, 7, 8$, we define the graph $G(\overline{E}_i)$ as the graph whose vertices are the lines of $\overline{E}_i$, with two lines adjacent if and only if the angle between them is $60°$. Note that these graphs are connected. Moreover, $W(E_i)$ acts as a group Γ_i of automorphisms of $G(\overline{E}_i)$. We show that Γ_i is a rank 3 permutation group – that is, Γ_i is transitive and the stabilizer of a vertex u has just three orbits. In our context, the orbits are necessarily $\{u\}$, the set $\Delta(u)$ of neighbours of u and the set $\Delta^*(u)$ of the remaining vertices (which are therefore all at distance 2 from u).

Proposition 3.2.4. *For $i = 6, 7, 8$, $W(E_i)$ acts as a rank 3 group on $G(\overline{E}_i)$.*

Proof. Transitivity follows from the transitivity of $W(E_i)$ on $G(E_i)$ ($i = 6, 7, 8$). If u is a vertex of $G(\overline{E}_8)$ then the subgraphs induced by $\Delta(u)$ and $\Delta^*(u)$ are connected. The first has 56 vertices; the second is the graph $G(\overline{E}_7)$, with 63 vertices. The transitivity of the stabilizer of u on $\Delta(u)$ follows from the corresponding property for $G(E_8)$. For transitivity on $\Delta^*(u)$, it suffices to consider adjacent vertices $\langle \mathbf{x} \rangle$ and $\langle \mathbf{y} \rangle$ ($\mathbf{x}, \mathbf{y} \in E_8$). If $u = \langle \mathbf{v} \rangle$ ($\mathbf{v} \in E_8$) then $\mathbf{v}.\mathbf{x} = \mathbf{v}.\mathbf{y} = 0$ and we may choose the signs of $\mathbf{x}$ and $\mathbf{y}$ so that $\mathbf{x}.\mathbf{y} = 1$. Then $\mathbf{x} - \mathbf{y} \in E_8$ and $\psi_{\mathbf{x}-\mathbf{y}}$ maps $\langle \mathbf{x} \rangle$ to $\langle \mathbf{y} \rangle$; moreover, $\psi_{\mathbf{x}-\mathbf{y}}$ fixes $\langle \mathbf{v} \rangle$ because $\mathbf{v}.(\mathbf{x} - \mathbf{y}) = 0$.

If $v \in \Delta^*(u)$ then the subgraphs induced by $\Delta^*(u) \cap \Delta(v)$ and $\Delta^*(u) \cap \Delta^*(v)$ are connected, with 32 and 30 vertices respectively. Now we may argue as before to show that the stabilizer of u and v acts transitively on these subgraphs; in other words, $W(E_7)$ acts as a rank 3 group on $G(\overline{E}_7)$.

If $v \in \Delta(u)$ then u and v determine a star in $\overline{E}_8$ and the subgraph of $G(\overline{E}_8)$ induced by $\Delta^*(u) \cap \Delta^*(v)$ is the graph $G(\overline{E}_6)$. This is a connected graph with 36 vertices. If $w \in \Delta^*(u) \cap \Delta^*(v)$ then the subgraphs induced by $\Delta^*(u) \cap \Delta^*(v) \cap \Delta(w)$ and $\Delta^*(u) \cap \Delta^*(v) \cap \Delta^*(w)$ are connected, with 20 and 15 vertices respectively. Here again, connectedness ensures that $W(E_6)$ acts as a rank 3 group on $G(\overline{E}_6)$.   □

Any graph whose automorphism group is a rank 3 group is strongly regular. In the cases above, it is straightforward to determine the number of common neighbours of a pair of adjacent vertices, and of a pair of non-adjacent vertices. It is well known that the parameters of a strongly regular graph determine the spectrum (see Theorem 1.2.20).

Corollary 3.2.5. *The graphs* $G(\overline{E}_i)$ ($i = 6, 7, 8$) *are strongly regular, with parameters and spectra as follows:*

$G(\overline{E}_6)$ *has parameters* $(36, 20, 10, 12)$ *and spectrum* 20, 2^{20}, $(-4)^{15}$,
$G(\overline{E}_7)$ *has parameters* $(63, 32, 16, 16)$ *and spectrum* 32, 4^{27}, $(-4)^{35}$,
$G(\overline{E}_8)$ *has parameters* $(120, 56, 28, 24)$ *and spectrum* 56, 8^{35}, $(-4)^{84}$.

Let us turn now to lattices. A *lattice* in a Euclidean space V is a discrete set of vectors closed under addition and subtraction. If S is a finite subset of V, we write $L(S)$ for the lattice consisting of all $\mathbb{Z}$-linear combinations of vectors in S. The *dimension* of a lattice L is just $\dim\langle L\rangle$. In an *integral* lattice, the inner product of any two vectors is an integer. An *even* lattice is an integral lattice L such that $\mathbf{v}.\mathbf{v}$ is even for each $\mathbf{v} \in L$. Our remarks on lattices are based on [BrCN, Sections 3.9 and 3.10].

Suppose that L is an integral lattice. The *dual* lattice L^* consists of all vectors $\mathbf{v}$ in $\langle L\rangle$ such that $\mathbf{v}.\mathbf{u} \in \mathbb{Z}$ for all $\mathbf{u} \in L$. Thus $L \subseteq L^*$. If L has dimension n then an *integral basis* for L is a set of n vectors $\mathbf{v}_1, \ldots, \mathbf{v}_n$ in L such that every vector in L is a $\mathbb{Z}$-linear combination of $\mathbf{v}_1, \ldots, \mathbf{v}_n$. In this situation, if $\mathbf{v} = x_1\mathbf{v}_1 + \cdots + x_n\mathbf{v}_n \in L^*$ then $G\mathbf{x} \in \mathbb{Z}^n$, where G is the Gram matrix $(\mathbf{v}_i.\mathbf{v}_j)$ and $\mathbf{x} = (x_1, \ldots, x_n)^T$. Since G is invertible, it follows that $\mathbf{x} \in d^{-1}\mathbb{Z}^n$, where $d = \det(G)$.

Proposition 3.2.6. *For each* $i = 6, 7, 8$, $L(E_i)$ *is an even lattice of dimension* i. *Moreover,*

    (i) $L(E_8)^* = L(E_8)$;
    (ii) *if* $\mathbf{v} \in L(E_7)^* \setminus L(E_7)$ *then* $\mathbf{v}.\mathbf{v} = 2k + \frac{3}{2}$ *for some* $k \in \mathbb{Z}$;
    (iii) *if* $\mathbf{v} \in L(E_6)^* \setminus L(E_6)$ *then* $\mathbf{v}.\mathbf{v} = 2k + \frac{4}{3}$ *for some* $k \in \mathbb{Z}$.

Proof. Each lattice $L(E_i)$ is even because $\mathbf{v}.\mathbf{v} = 2$ for all $\mathbf{v} \in E_8$. For the remaining statements, let $\mathbf{v}_1, \ldots, \mathbf{v}_8$ be the following vectors in E_8:

$\mathbf{v}_1 = \frac{1}{2}(\mathbf{e}_1 + \mathbf{e}_2 + \mathbf{e}_3 - \mathbf{e}_4 - \mathbf{e}_5 - \mathbf{e}_6 - \mathbf{e}_7 + \mathbf{e}_8)$,
$\mathbf{v}_j = \mathbf{e}_j - \mathbf{e}_{j-1}$ ($j = 2, 3, 4, 5, 6, 7$),
$\mathbf{v}_8 = -\mathbf{e}_7 - \mathbf{e}_8$.

Then $\{\mathbf{v}_1, \ldots, \mathbf{v}_8\}$ is an integral basis for $L(E_8)$. Since $d = 1$ for this basis, we have $L(E_8)^* = L(E_8)$, and statement (i) follows.

Now $\{\mathbf{v}_1, \ldots, \mathbf{v}_7\}$ is an integral basis for $L(E_7)$, and $d = 2$ for this basis. Hence if $\mathbf{v} \in L(E_7)^* \setminus L(E_7)$ then there exists $\mathbf{u} \in L(E_7)$ such that $\mathbf{v} - \mathbf{u} = x_1\mathbf{v}_1 + \cdots + x_7\mathbf{v}_7$, where each x_i is 0 or $\frac{1}{2}$ and not all x_i are 0. Let $\mathbf{a} = \mathbf{v} - \mathbf{u}$. Of the 127 candidates for $\mathbf{a}$ only one is such that $\mathbf{a}.\mathbf{v}_i \in \mathbb{Z}$ for each $i = 1, \ldots, 7$, namely,

$$\mathbf{a} = \frac{1}{2}(\mathbf{v}_1 + \mathbf{v}_5 + \mathbf{v}_7) = \frac{1}{4}(\mathbf{e}_1 + \mathbf{e}_2 + \mathbf{e}_3 - 3\mathbf{e}_4 + \mathbf{e}_5 - 3\mathbf{e}_6 + \mathbf{e}_7 + \mathbf{e}_8).$$

Now $\mathbf{v}.\mathbf{v} = \mathbf{u}.\mathbf{u} + 2(\mathbf{u}.\mathbf{a}) + \mathbf{a}.\mathbf{a}$. Statement (ii) follows because $\mathbf{u}.\mathbf{u}$ is even, $\mathbf{u}.\mathbf{a} \in \mathbb{Z}$ and $\mathbf{a}.\mathbf{a} = \frac{3}{2}$.

Also, $\{\mathbf{v}_1, \ldots, \mathbf{v}_6\}$ is an integral basis for $L(E_6)$, and $d = 3$ for this basis. Proceeding as in the previous case, if $\mathbf{v} \in L(E_6)^* \setminus L(E_6)$ then there exists $\mathbf{u} \in L(E_6)$ such that $\mathbf{v} - \mathbf{u} = \mathbf{a} = x_1\mathbf{v}_1 + \cdots + x_6\mathbf{v}_6$ where each x_i is 0, $\frac{1}{3}$ or $\frac{2}{3}$ and not all x_i are 0. In this case we find that there are just two possibilities for $\mathbf{a}$, namely

$$\frac{1}{3}(2\mathbf{v}_2 + \mathbf{v}_3 + 2\mathbf{v}_5 + \mathbf{v}_6) = \frac{1}{3}(-2\mathbf{e}_1 + \mathbf{e}_2 + \mathbf{e}_3 - 2\mathbf{e}_4 + \mathbf{e}_5 + \mathbf{e}_6)$$

and

$$\frac{1}{3}(\mathbf{v}_2 + 2\mathbf{v}_3 + \mathbf{v}_5 + 2\mathbf{v}_6) = \frac{1}{3}(-\mathbf{e}_1 - \mathbf{e}_2 + 2\mathbf{e}_3 - \mathbf{e}_4 - \mathbf{e}_5 + 2\mathbf{e}_6).$$

In either case, $\mathbf{a}.\mathbf{a} = \frac{4}{3}$, and so we have statement (iii). $\qquad\square$

Remark 3.2.7. In Chapter 5, we shall make use of orthogonal vectors $\mathbf{w}$, $\mathbf{w}'$ defined as follows:

$$\mathbf{w} = \frac{1}{4}(\mathbf{e}_1 + \mathbf{e}_2 + \mathbf{e}_3 + \mathbf{e}_4 + \mathbf{e}_5 + \mathbf{e}_6 - 3\mathbf{e}_7 - 3\mathbf{e}_8),$$

$$\mathbf{w}' = \frac{1}{3}(\mathbf{e}_1 + \mathbf{e}_2 + \mathbf{e}_3 + \mathbf{e}_4 - 2\mathbf{e}_5 - 2\mathbf{e}_6).$$

Note that $\mathbf{w}.\mathbf{w} = \frac{3}{2}$ and $\mathbf{w} = \frac{1}{2}(-3\mathbf{v}_1 - 2\mathbf{v}_2 - 4\mathbf{v}_3 - 6\mathbf{v}_4 - 5\mathbf{v}_5 - 4\mathbf{v}_6 - 3\mathbf{v}_7)$ $\in L(E_7)^*$; also $\mathbf{w}'.\mathbf{w}' = \frac{4}{3}$ and $\mathbf{w}' = \frac{1}{3}(-\mathbf{v}_2 - 2\mathbf{v}_3 - 3\mathbf{v}_4 - 4\mathbf{v}_5 - 2\mathbf{v}_6) \in L(E_6)^*$. $\qquad\square$

3.3 Extensions of line systems

The basic object of study in this section is the *one-line extension* of a star-closed line system, defined as follows. Given a star-closed line system Σ, we add a new line which is at $60°$ or $90°$ to each line of Σ, and then star-close the system – that is, add the third line of a star whenever the other two lines of the star already lie in the system. If the new line is at $90°$ to all lines of Σ then a decomposable system results. Since we want to construct indecomposable systems, we shall always assume that we add a line making an angle of $60°$ with some line in Σ.

Not only shall we show that $\overline{A}_n$, $\overline{D}_n$, $\overline{E}_8$, $\overline{E}_7$ and $\overline{E}_6$ are essentially the only examples of indecomposable line systems, but we shall also see how each can be embedded in a system of higher dimension. In discussing a line system Σ we shall work with the set S of all vectors of length $\sqrt{2}$ along the lines in Σ.

In particular, if Σ is star-closed then S has the properties (P1)–(P4) listed in Lemma 3.1.3.

Let us start by considering $\overline{A}_n$: we add a new line, star-close, and then consider the vectors along the new lines with length $\sqrt{2}$. We have defined $\overline{A}_n$ in terms of an orthonormal basis $\{e_1, \ldots, e_{n+1}\}$. Adding a further line and star-closing will increase the dimension of the system by at most one, and therefore we do not need to increase the size of the basis to describe the vectors along the new lines. In view of our assumption of indecomposability, we know that the coordinates of some new vector $\mathbf{x}$ with respect to $\{e_1, \ldots, e_{n+1}\}$ are not all equal. By property (P1), any two unequal coordinates differ by 1, since $\mathbf{x}.e_i - \mathbf{x}.e_j \in \{-1, 0, 1\}$. Hence there is a real number u such that each coordinate is equal to u or $u + 1$. Now if the i-th coordinate is equal to u and the j-th is equal to $u + 1$, then $\mathbf{x}.(e_i - e_j) = -1$ and hence by (P3), $\mathbf{x} + e_i - e_j$ is also in the system. This new vector differs from $\mathbf{x}$ only in that the i-th and j-th coordinates have been interchanged. Thus once one vector with t coordinates equal to $u + 1$ and the remaining coordinates equal to u is in the system, then all such vectors with exactly t coordinates equal to $u + 1$ are in the system.

Lemma 3.3.1. *Suppose that V is the set of vectors of length $\sqrt{2}$ along the lines in a one-line extension of $\overline{A}_n$. Then*

(i) *there exists a real number u and a positive integer t such that V contains all vectors with t coordinates equal to $u + 1$ and the remaining $n + 1 - t$ coordinates equal to u,*

(ii) $(n + 1)u^2 + 2tu + t - 2 = 0$,

(iii) $t > 2$ *implies* $n \leq \frac{t^2 - t + 2}{t - 2}$, *and*

(iv) $n \geq 2t - 1$.

Proof. The conclusion (i) has already been explained, while (ii) follows directly from property (P2). Since the equation (ii) has real roots, we have $t^2 - (n + 1)(t - 2) \geq 0$, and (iii) follows. From (P4) we may assume that $t \leq \frac{n+1}{2}$, and then (iv) follows. $\square$

It follows from parts (iii) and (iv) of Lemma 3.3.1 that $t \in \{1, 2, 3, 4\}$. We consider the possibilities in turn and show that each value of t arises (for appropriate values of n). We refer to the $\binom{n+1}{t}$ vectors described in Lemma 3.3.1(i) as vectors of type I.

Case $t = 4$. Conclusions (iii) and (iv) of Lemma 3.3.1 imply that $n = 7$, and (ii) implies that $u = -\frac{1}{2}$. Thus the $\binom{8}{4}$ vectors of type I are precisely the vectors with four coordinates equal to $\frac{1}{2}$ and four equal to $-\frac{1}{2}$; and hence we have all vectors in E_8 orthogonal to $\frac{1}{2}\mathbf{j}_8$. Thus we have extended A_7 to E_7. Note that this extension of A_7 to E_7 is unique.

Case $t = 3$. We now have $5 \leq n \leq 8$. For $n = 8$ we have $u = -\frac{1}{3}$, and the $\binom{9}{3}$ vectors of type I determine 84 lines that, together with the 36 lines of $\overline{A}_8$, give us the alternative representation of E_8 mentioned previously. For $n = 5, 6, 7$, observe that if $\mathbf{v}_1, \mathbf{v}_2$ are vectors of type I which never have the value $u + 1$ in the same coordinate, then $\mathbf{v}_1.\mathbf{v}_2 = -1$ by Lemma 3.3.1(ii). Hence $\mathbf{v}_1 + \mathbf{v}_2$ is present by Lemma 3.1.3, and we have a further $\binom{n+1}{6}$ vectors, which we call vectors of type II. When $n = 7$, there are also 8 vectors with one coordinate equal to $3u + 2$ and the rest equal to $3u + 1$: each is obtained by star-closing because it has the form $\mathbf{v} + \mathbf{w}$, where $\mathbf{v}$ is of type I, $\mathbf{w}$ is of type II, and $\mathbf{v}.\mathbf{w} = -1$ by Lemma 3.3.1(ii). At this stage we have a total of 36 lines for $n = 5$, 63 lines for $n = 6$, and 120 lines for $n = 7$.

In fact, we have $\overline{E}_6$, $\overline{E}_7$ and $\overline{E}_8$. This can be verified by considering the images of our vectors under the linear maps defined as follows:

$$(n = 5) \qquad \phi(\mathbf{e}_i) = \mathbf{e}_i - \tfrac{1}{6}\mathbf{j}_6 + \tfrac{1}{6(2u+1)}(\mathbf{e}_7 - \mathbf{e}_8), \quad i = 1, \dots, 6;$$

$$(n = 6) \qquad \phi(\mathbf{e}_i) = \mathbf{e}_i - \tfrac{2u+1}{2(7u+3)}\mathbf{j}_7 + \tfrac{1}{2(7u+3)}\mathbf{e}_8, \quad i = 1, \dots, 7;$$

$$(n = 7) \qquad \phi(\mathbf{e}_i) = \mathbf{e}_i - \tfrac{2u+1}{2(8u+3)}\mathbf{j}_8, \quad i = 1, \dots, 8.$$

Note that in each case, ϕ embeds A_n in E_{n+1}. It is necessary to verify that ϕ is an isometry: this can be done by writing down the matrix M of ϕ and using Lemma 3.3.1(ii) to verify that $M^T M = I$.

Case $t = 2$. In this case we define the linear map ϕ by $\phi(\mathbf{e}_i) = \mathbf{e}_i + \tfrac{u}{2}\mathbf{j}_{n+1}$ $(i = 1, 2, \dots, n + 1)$. Then ϕ is an isometry that fixes each vector in A_n and maps the set of vectors of type I to $D_{n+1} \setminus A_n$. Thus our one-line extension of $\overline{A}_n$ is $\overline{D}_{n+1}$.

Case $t = 1$. Here we define the linear map ϕ by $\phi(\mathbf{e}_i) = \mathbf{e}_i + \tfrac{u^2}{u-1}\mathbf{j}_{n+2} + \tfrac{u}{u-1}\mathbf{e}_{n+2}$ $(i = 1, 2, \dots, n + 1)$. Then ϕ is an isometry that fixes each vector in A_n and maps the set of vectors of type I to $A_{n+1} \setminus A_n$. Thus our one-line extension of $\overline{A}_n$ is $\overline{A}_{n+1}$.

Gathering the various cases together, we have the following result:

Theorem 3.3.2. *The one-line extensions of $\overline{A}_n$ $(n \geq 2)$ are precisely those given by the following inclusions:*

(i) $\overline{A}_n \subset \overline{A}_{n+1}$, $\quad n = 2, 3, \dots$

(ii) $\overline{A}_n \subset \overline{D}_{n+1}$, $\quad n = 2, 3, \dots$

(iii) $\overline{A}_5 \subset \overline{E}_6$, $\quad \overline{A}_6 \subset \overline{E}_7$, $\quad \overline{A}_7 \subset \overline{E}_8$, $\quad \overline{A}_8 \subset \overline{E}_8$, *and*

(iv) $\overline{A}_7 \subset \overline{E}_7$.

Now, in an analogous manner, let us extend $\overline{D}_n$ by a single line $\langle \mathbf{x} \rangle$, where $\|\mathbf{x}\| = \sqrt{2}$. Since A_3 and D_3 are isomorphic, we assume that $n \geq 4$. We start by extending $\{e_1, e_2, \ldots, e_n\}$ to an orthonormal basis $\{e_1, e_2, \ldots, e_{n+1}\}$. Since D_n contains A_{n-1}, we may argue as we did for Lemma 3.3.1 that the first n coordinates of $\mathbf{x}$ are $u + 1$ or u for some real number u. We may assume that $(e_1 - e_2).\mathbf{x} = 1$. Since $e_1 + e_2 \in D_n$, property (P1) of Lemma 3.1.3 implies that $u = 0, -\frac{1}{2}$ or -1.

Suppose that $u = 0$. Then, in order to obtain a vector that is not already in the system, the final coordinate must be non-zero. This implies by (P2) that exactly one of the first n coordinates is equal to 1, and hence that the final coordinate is either 1 or -1. Using the vectors $e_i + e_j$ $(1 \le i < j \le n)$, together with properties (P3) and (P4), we obtain all the vectors in D_{n+1}.

When $u = -1$, we use property (P4) to revert to the case $u = 0$.

Finally, suppose that $u = -\frac{1}{2}$. Then the first n coordinates are $\pm\frac{1}{2}$ while the last coordinate is $\pm\frac{1}{2}\sqrt{8 - n}$. Thus $n \le 8$. If $n = 8$ then we may regard the extension as lying in $\langle e_1, e_2, \ldots, e_8 \rangle$; now the additional vectors are either all those with an even number of positive coefficients, or all those with an odd number of positive coefficients. In either case we have E_8. Now suppose that $n = 4, 5, 6$ or 7 and that each of the first n coordinates of $\mathbf{x}$ is $\pm\frac{1}{2}$. By property (P4) we may assume that the last coordinate is negative. For any two coordinates i and j with $1 \le i < j \le n$, the inner product of $\mathbf{x}$ with one of the four vectors $\pm e_i \pm e_j$ is -1; hence by (P3) there is a vector in the system identical to $\mathbf{x}$ except that the i-th and j-th coordinates have changed sign. Thus if $\mathbf{x}$ has t of the first n coordinates positive, then the system includes any other vector of the same form in which the number of positive coordinates has the same parity as t. (Note however that, in view of (P1), the system cannot contain two vectors for which the parities are different.) For $n = 7$, this gives all the vectors in E_8. For $n = 6$, let ϕ be the linear map defined by $\phi(e_i) = e_i$ $(i = 1, \ldots, 6)$ and $\phi(e_7) = \frac{\sqrt{2}}{2}(e_7 + e_8)$. For $n = 5$, let $\phi(e_i) = e_i$ $(i = 1, \ldots, 5)$ and $\phi(e_6) = \frac{1}{\sqrt{3}}(e_6 + e_7 + e_8)$. In each case, ϕ maps our set of vectors to E_{n+1}. For $n = 4$, there are 8 new vectors and the complete set of vectors is easily identified with D_5. Collecting these results together, we have the following theorem.

Theorem 3.3.3. *The one-line extensions of $\overline{D}_n$ $(n \ge 4)$ are precisely those given by the following inclusions:*

(i) $\overline{D}_n \subset \overline{D}_{n+1}$, $n = 4, 5, \ldots$, *and*
(ii) $\overline{D}_5 \subset \overline{E}_6$, $\overline{D}_6 \subset \overline{E}_7$, $\overline{D}_7 \subset \overline{E}_8$, $\overline{D}_8 \subset \overline{E}_8$.

Corollary 3.3.4. *For $n = 6, 7, 8$, the line systems $\overline{E}_n$ are maximal star-closed systems of dimension n. For other values of n, the line systems $\overline{A}_n$ and $\overline{D}_n$ are maximal star-closed systems of dimension n.*

Proof. Consider an 8-dimensional one-line extension Σ of $\overline{E}_8$. Since $D_8 \subset E_8$, Σ contains an 8-dimensional one-line extension of $\overline{D}_8$. The proof of

Theorem 3.3.3 shows that there are precisely two such extensions, and it follows that Σ contains both of them. But any new line from one meets any new line from the other at an angle not equal to $60°$ or $90°$ (since the inner product of corresponding vectors of length $\sqrt{2}$ is not an integer). This contradiction shows that $\overline{E}_8$ is maximal. For $\overline{E}_7$, the result is immediate since $\overline{E}_7$ is the unique 7-dimensional extension of $\overline{A}_7$. Finally, the system $\overline{E}_6$ contains $\overline{D}_5$, which has precisely two 6-dimensional one-line extensions, isomorphic to $\overline{D}_6$ and $\overline{E}_6$. As with $\overline{E}_8$, these extensions are mutually exclusive, and so $\overline{E}_6$ is maximal. The remaining assertions of the corollary follow immediately from Theorems 3.3.3 and 3.3.2. $\square$

Theorem 3.3.5. *The only one-line extension of $\overline{E}_6$ is $\overline{E}_7$, and the only one-line extension of $\overline{E}_7$ is $\overline{E}_8$. The system $\overline{E}_8$ has no one-line extension.*

Proof. By Corollary 3.3.4, a one-line extension Σ of $\overline{E}_8$ would be nine-dimensional. Now $\overline{D}_8 \subset \overline{E}_8$, and by Theorem 3.3.3 the only 9-dimensional one-line extension of $\overline{D}_8$ is $\overline{D}_9$. By Corollary 3.3.4 again, $\Sigma = \overline{D}_9$; but $\overline{D}_9$ has fewer lines than $\overline{E}_8$, and hence fewer than Σ, a contradiction.

A one-line extension Σ of $\overline{E}_7$ is an 8-dimensional extension of $\overline{A}_7$, and hence contains $\overline{A}_8$, $\overline{D}_8$ or $\overline{E}_8$. Since $\overline{A}_8$ and $\overline{D}_8$ both have fewer elements than $\overline{E}_7$, and $\overline{E}_8$ is a maximal 8-dimensional system, the only candidate for Σ is $\overline{E}_8$.

Finally we consider a one-line extension of $\overline{E}_6$. We saw in Section 3.1 that we may represent E_6 by the vectors in $A_5 \cup \{\pm\frac{1}{2}\mathbf{e}_1 \pm \frac{1}{2}\mathbf{e}_2 \pm \ldots \pm \frac{1}{2}\mathbf{e}_6 \pm \frac{\sqrt{2}}{2}\mathbf{e}_7\}$ $\cup \{\pm\sqrt{2}\mathbf{e}_7\}$, where in the second set three of the first six coefficients are positive. We add a new line to $\overline{E}_6$ and form a star-closed system. We use the notation $x_1\mathbf{e}_1 + \cdots + x_7\mathbf{e}_7$ for a vector $\mathbf{x}$.

We first claim that there is a line contained in the hyperplane $x_7 = 0$ not orthogonal to E_6. Suppose that $\mathbf{z}$ is a vector of length $\sqrt{2}$ on a line not orthogonal to E_6. If $z_7 \neq 0$, then $z_7 = \pm\frac{\sqrt{2}}{2}$ because $\mathbf{z}.(\pm\sqrt{2}\mathbf{e}_7) = \pm 1$; moreover, since $(\mathbf{e}_i - \mathbf{e}_j).\mathbf{z} = 0$ or ± 1 for $1 \leq i < j \leq 6$, there exist a non-negative integer t and a real number u such that t of the first six coefficients are equal to $u + 1$ and the remainder are equal to u. Replacing $\mathbf{z}$ by $-\mathbf{z}$ if necessary, we may assume that $t \leq 3$. Now let $\mathbf{x}$ be a vector given in our definition of E_6 such that $x_7 = -z_7$ and elsewhere $\mathbf{x}$ has a coefficient of $-\frac{1}{2}$ whenever $\mathbf{z}$ has a coefficient of $u + 1$. Then $\mathbf{z}.\mathbf{x} = -\frac{1}{2}(t + 1) = 0$ or ± 1, and hence $t = 1$. Further, $\mathbf{z}.\mathbf{x} = -1$, and so $\mathbf{x} + \mathbf{z}$ is in the system. Since $x_7 + z_7 = 0$ and $\mathbf{x} + \mathbf{z}$ is not orthogonal to E_6, our claim is established.

Now let $\mathbf{w}$ be a vector of length $\sqrt{2}$ such that $\langle\mathbf{w}\rangle$ lies in the system, $\mathbf{w}$ is not orthogonal to E_6, and $w_7 = 0$. Arguing as before, we find that there exist an integer t $(0 \leq t \leq 3)$ and a real number u such that t of the first six coefficients are equal to $u + 1$ and the remainder are equal to u. Now $\mathbf{w}.\mathbf{x} = -\frac{1}{2}t = 0$

or ± 1, whence $t = 0$ or 2. Since **w** is not orthogonal to E_6, we have $t = 2$. As in the proof of Lemma 3.3.1, our system contains all $\binom{6}{2}$ vectors with the same pattern of coefficients as **w**. Now star-closure ensures that our system contains $\overline{D}_6$. As a proper extension of $\overline{D}_6$, the system contains $\overline{D}_7$ or $\overline{E}_7$. The system already contains 51 lines, and so if $\overline{D}_7$ is present then the extension properly contains $\overline{D}_7$. This is not possible in a 7-dimensional space, and so the extension contains $\overline{E}_7$. Since $\overline{E}_7$ is maximal in 7-dimensional space, our system must be $\overline{E}_7$. □

3.4 Smith graphs and line systems

We have seen in Section 3.1 that graphs with largest eigenvalue ≤ 2 have o-representations in line systems. In this section we determine all such graphs and show explicitly how they are related to the line systems we encountered in Section 3.2. This enables us to find the maximal line systems of each dimension.

Some interesting classes of graphs can be obtained by prescribing an upper bound for the largest eigenvalue (or *index*). The graphs for which $\lambda_1 \leq 2$ can be determined essentially because their vertices have mean degree ≤ 2 and maximum degree ≤ 4: the maximal graphs that arise were determined by J. H. Smith and are illustrated in Fig. 3.2.

Theorem 3.4.1 [Smi]. *The connected graphs whose largest eigenvalue does not exceed 2 are precisely the induced subgraphs of the graphs shown in Fig. 3.2,*

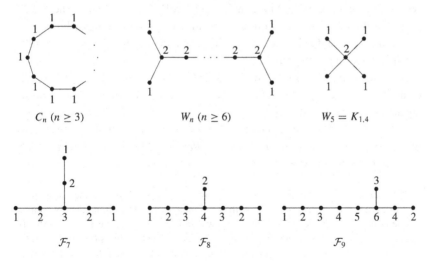

Figure 3.2: The Smith graphs.

where the graphs are ascribed names in which the subscript denotes the number of vertices.

Proof. In Fig. 3.2, the vertices of each graph are labelled with the coordinates of an eigenvector corresponding to the eigenvalue 2. Since all the coordinates are positive each graph in Fig. 3.2 has largest eigenvalue equal to 2.

Let G be a connected graph with $\lambda_1(G) \leq 2$. Since λ_1 increases strictly with the addition of vertices, provided the graph remains connected, G is either a cycle C_n or a tree; moreover $K_{1,4}$ is the only possible tree with a vertex of degree greater than 3. If the maximal degree is 3, then either G is W_n or G has a unique vertex of degree 3 with three paths attached. In the second case, either G is $\mathcal{F}_7$ or one of the three paths has length 1. If one path has length 1 then either G is $\mathcal{F}_8$ or a second path has length less than 3. In the latter case, G is an induced subgraph of $\mathcal{F}_9$ or $\mathcal{F}_8$. Finally, if the maximal degree of a vertex in G is 2 then G is a path and hence an induced subgraph of some C_n.

This completes the proof.   □

Theorem 3.4.1 and its proof are due to J. H. Smith [Smi], and accordingly the graphs of Fig. 3.2 are often called *Smith graphs* in the literature (see, for example, [CvGu], [CvDSa, p. 383]). J. J. Seidel [Sei7] proposed the name *Coxeter graphs* because the graphs in question appear implicitly in H. S. M. Coxeter's work on discrete groups generated by reflections in hyperplanes [Cox]. Since the topic of this book is a part of graph theory rather than group theory, we prefer to use the elegant graph-theoretic proof by J. H. Smith. For other proofs of Theorem 3.4.1, see [LeSe] and [Neu1]. For the role of these graphs in algebra see, for example, [GoHJ], where they appear as Dynkin diagrams. The Smith graphs will have an important role in our investigation of graphs whose least eigenvalue is bounded below by -2. First we mention in passing some other results related to Smith graphs.

The eigenvalues of Smith graphs are determined in [CvGu] by direct calculation; they are given in Table A1. It is noted in [BrCN, p. 84] that these eigenvalues can be used to determine the exponents in Coxeter systems of generators and relations; see also [Hum]. It is of interest to quote here the review of [CvGu] by J. H. Smith in *Mathematical Reviews* (MR 54#5049, reproduced by permission of the American Mathematical Society):

> The authors show that if the largest eigenvalue of a graph is ≤ 2 then all eigenvalues are of the form $2\cos \alpha\pi$, with α rational. This follows immediately from a theorem of L. Kronecker [J. Reine Angew. Math. 53 (1857), 173–175] but the authors use the known listing of such graphs and direct calculations (not included) to get the spectra explicitly. The listing of those graphs for which the adjacency matrix, A, satisfies the equivalent condition that $I - \frac{1}{2}A$ is non-negative definite is part

of a classical result in the theory of Coxeter groups [e.g., N. Bourbaki, Éléments de mathématique. XXXIV. Groupes et algèbres de Lie. Chapitre VI; Systèmes de racines, especially p. 199, Actualités Sci. Indust., No. 1337, Hermann, Paris, 1968; MR 39#1590; Russian translation, Izdat. "Mir", Moscow, 1972; MR 50#7404; C. T. Benson and L. C. Grove, Finite reflection groups, especially p. 60, Bogden and Quigley, Tarrytown-on-Hudson, N.Y., 1971; MR 52#4099]; the authors refer to a more graph theoretical approach of the reviewer. The authors then choose a subset of this set of graphs whose spectra provide a basis for all formal linear combinations of spectra of the graphs, and use this basis to derive criteria for such a formal linear combination to be the spectrum of an actual graph.   *(John H. Smith)*

It is proved in [Cve11] that Smith graphs are characterized by their eigen-values and angles. Eigenvalues alone do not suffice since there are many cospectral graphs among those whose largest eigenvalue does not exceed 2 (see [CvGu]).

Let $p_k(x)$ be the characteristic polynomial of a path P_k on k vertices. A graph G with the adjacency matrix A is called *path-positive* if the matrix $p_k(A)$ is non-negative for all k. It was proved in [BaLa1] that a graph G is path-positive if and only if $\lambda_1(G) \geq 2$. A more general theorem in terms of a semi-simple algebra appears in [HaWe]. A generalization in another direction is given in [BaLa2]. A graph is called a *path-zero graph* if $p_k(A) = 0$ for some k. The path-zero graphs are determined in [Sei8] and they are all induced subgraphs of Smith graphs. Smith graphs appear also in analysis of certain games of numbers called Mozes' games [Err].

The *spectral spread* of a graph is the difference between the largest and smallest eigenvalues. Smith graphs have spectral spread at most 4, and all the graphs with spectral spread at most 4 are determined in [Pet1].

We return to the main theme of this section. The connected proper induced subgraphs of the Smith graphs are called *reduced Smith graphs*; they are shown in Fig. 3.3, with names in which the subscript always denotes the number of vertices.

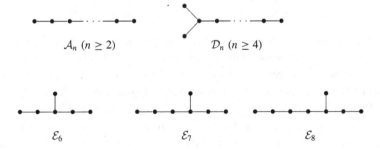

Figure 3.3: The reduced Smith graphs.

If G is a reduced Smith graph with n vertices and adjacency matrix A, then both $A + 2I$ and $A - 2I$ have rank n. Thus n is the least dimension of a line system in which G has an a-representation or in which G has an o-representation. We describe next some sets of vectors that determine star-closed line systems related to o-representations\ of the reduced Smith graphs.

Theorem 3.4.2. *The star closure of any o-representation of the reduced Smith graph $\mathcal{A}_n$ ($n \geq 2$), $\mathcal{D}_n$ ($n \geq 4$), $\mathcal{E}_6$, $\mathcal{E}_7$, $\mathcal{E}_8$ is the line system $\overline{A}_n$ ($n \geq 2$), $\overline{D}_n$ ($n \geq 4$), $\overline{E}_6$, $\overline{E}_7$, $\overline{E}_8$, respectively.*

Proof. The graph $\mathcal{A}_n$ has an $(n + 1)$-dimensional o-representation consisting of the vectors $\mathbf{e}_i - \mathbf{e}_{i+1}$ ($i = 1, 2, \ldots, n$), and the star-closure of this representation is $\overline{A}_n$.

The graph $\mathcal{D}_n$ has an n-dimensional o-representation consisting of the vectors $\mathbf{e}_i - \mathbf{e}_{i+1}$ ($i = 1, 2, \ldots, n - 1$) and $\mathbf{e}_{n-1} + \mathbf{e}_n$, and star-closure yields the line system $\overline{D}_n$.

The graphs $\mathcal{E}_6$, $\mathcal{E}_7$ and $\mathcal{E}_8$ have the following 8-dimensional representations: in each case, the vector $\frac{1}{2}\mathbf{j}_8 - (\mathbf{e}_1 + \mathbf{e}_2 + \mathbf{e}_3 + \mathbf{e}_8)$ is present, and we add the vectors:

$\mathbf{e}_i - \mathbf{e}_{i+1}$ ($i = 1, 2, \ldots, 5$) for $\mathcal{E}_6$;
$\mathbf{e}_i - \mathbf{e}_{i+1}$ ($i = 1, 2, \ldots, 6$) for $\mathcal{E}_7$;
$\mathbf{e}_i - \mathbf{e}_{i+1}$ ($i = 1, 2, \ldots, 6$) and $\mathbf{e}_7 + \mathbf{e}_8$ for $\mathcal{E}_8$.

The star closures of these o-representations are the exceptional line systems $\overline{E}_6$, $\overline{E}_7$, $\overline{E}_8$ respectively. $\qquad\square$

We say that an indecomposable system of lines Σ is *maximal* if it is not properly contained in another system of the same dimension. We use the abbreviation IMSL for an indecomposable maximal system of lines at 60° and 90°. We are now in a position to determine all IMSLs.

Theorem 3.4.3. *If Σ is an IMSL of dimension $n > 1$, then there exists a reduced Smith graph with n vertices which has an o-representation in Σ.*

Proof. Note that K_2, the smallest reduced Smith graph, has an o-representation in Σ. Let Q be a reduced Smith graph with the largest number of vertices which has such a representation. Suppose, by way of contradiction that $|V(Q)| = k < n$. Let R be the star-closure of an o-representation of Q in Σ. Then $\dim R = k$ and (by Proposition 3.1.2), R is a line system. We know that $R \neq \overline{E}_8$ because $\overline{E}_8$ cannot be extended to Σ. Let $\alpha \in \Sigma \backslash R$, where the line α is not orthogonal to at least one line from R. The star-closure of $R \cup \{\alpha\}$ is a line system R' of dimension $k + 1$ or k. The results of Section 3.3 show that if $\dim R' = k + 1$

then there is a reduced Smith graph with $k + 1$ vertices which has an o-representation in R' (and hence in Σ), a contradiction. If $\dim R' = k$ then we know from Section 3.3 that either $k = 7$ and $R = \overline{A}_7$ or $k = 8$ and R' is one of $\overline{A}_8$, $\overline{D}_8$. If $R = \overline{A}_8$ or $\overline{D}_8$ then $R' = \overline{E}_8$, a contradiction because then R' cannot be extended to Σ. If $R = \overline{A}_7$ then $R' = \overline{E}_7$ and $\Sigma = \overline{E}_8$. This is a final contradiction because the reduced Smith graph $\mathcal{E}_8$ has an o-representation in $\overline{E}_8$. □

Theorem 3.4.4. *Any IMSL is determined by a root system.*

Proof. Let Σ be an IMSL and let $\dim \Sigma = n$. By Theorem 3.4.3 there is a reduced Smith graph with n vertices which has an o-representation in Σ. The star-closure of this representation is an n-dimensional line system $R \subseteq \Sigma$. Now either $\Sigma = R$ or R is one of $\overline{A}_7$, $\overline{A}_8$, $\overline{D}_8$. In the first of these last three cases we have $\Sigma = \overline{E}_7$ while in the remaining two cases we have $\Sigma = \overline{E}_8$. This completes the proof. □

Theorem 3.4.5. *The only IMSLs in $\mathbb{R}^n$ $(n > 1)$ are those determined by the root systems:*

  (i) A_n *for* $n = 2, 3$;

 (ii) A_n, D_n *for* $n = 4, 5$;

(iii) A_6, D_6 *and* E_6 *for* $n = 6$;

 (iv) D_7 *and* E_7 *for* $n = 7$;

  (v) E_8 *for* $n = 8$;

 (vi) A_n, D_n *and* E_8 *for* $n > 8$.

Proof. By Theorem 3.4.4 any IMSL is determined by a root system, and so the result follows from the list of inclusions given in Theorems 3.3.2, 3.3.3 and 3.3.5. □

3.5 An alternative approach

In this section we outline an alternative proof of Theorem 3.4.5 due to P. J. Cameron, J. M. Goethals, J. J. Seidel and E. E. Shult. This original means of classifying indecomposable star-closed line systems is perhaps the most elegant, but it too requires some detailed calculations related to $\overline{E}_6$, $\overline{E}_7$ and $\overline{E}_8$.

Let Σ be an indecomposable star-closed system of lines at $60°$ or $90°$, and let S be a star in Σ. In what follows, we choose a vector $\mathbf{w}$ of length $\sqrt{2}$ along each line of Σ, and write $\Sigma = \{\langle \mathbf{w} \rangle : \mathbf{w} \in W\}$. Let $S = \{\langle \mathbf{x} \rangle, \langle \mathbf{y} \rangle, \langle \mathbf{z} \rangle\}$, so that

$\mathbf{x} + \mathbf{y} + \mathbf{z} = \mathbf{0}$. Note that each line of Σ outside S is orthogonal to exactly one or three lines in S, and so $W = S \;\dot\cup\; A_{\mathbf{x}} \;\dot\cup\; A_{\mathbf{y}} \;\dot\cup\; A_{\mathbf{z}} \;\dot\cup\; B$, where

$$A_{\mathbf{x}} = \{\mathbf{w} \in W : \mathbf{w}.\mathbf{x} = 0, \; \mathbf{w}.\mathbf{y} = 1, \mathbf{w}.\mathbf{z} = -1\},$$
$$A_{\mathbf{y}} = \{\mathbf{w} \in W : \mathbf{w}.\mathbf{x} = -1, \; \mathbf{w}.\mathbf{y} = 0, \mathbf{w}.\mathbf{z} = 1\},$$
$$A_{\mathbf{z}} = \{\mathbf{w} \in W : \mathbf{w}.\mathbf{x} = 1, \; \mathbf{w}.\mathbf{y} = -1, \mathbf{w}.\mathbf{z} = 0\},$$
$$B = \{\mathbf{w} \in W : \mathbf{w}.\mathbf{x} = 0, \; \mathbf{w}.\mathbf{y} = 0, \mathbf{w}.\mathbf{z} = 0\}.$$

Lemma 3.5.1. Σ *is determined by* $S \;\dot\cup\; A_{\mathbf{x}}$.

Proof. By calculating scalar products, we find that $A_{\mathbf{y}} = \{\mathbf{w} + \mathbf{z} : \mathbf{w} \in A_{\mathbf{x}}\}$, $A_{\mathbf{z}} = \{\mathbf{w} - \mathbf{y} : \mathbf{w} \in A_{\mathbf{x}}\}$ and $B \supseteq \{\mathbf{u} - \mathbf{v} : \mathbf{u}, \; \mathbf{v} \in A_{\mathbf{x}}\}$. It suffices to establish the reverse of this last inclusion. Let $\mathbf{t} \in B$, and note that if $\mathbf{t}.\mathbf{w} = 0$ for all $\mathbf{w} \in A_{\mathbf{x}}$ then $\mathbf{t}.\mathbf{w} = 0$ for all $\mathbf{w} \in W$, contrary to the assumption of indecomposability. Accordingly, $\mathbf{t}.\mathbf{a} \in \{-1, 1\}$ for some $\mathbf{a} \in A_{\mathbf{x}}$. Now we find that $\mathbf{a} + \mathbf{t} \in A_{\mathbf{x}}$ if $\mathbf{t}.\mathbf{a} = -1$ and $\mathbf{a} - \mathbf{t} \in A_{\mathbf{x}}$ if $\mathbf{t}.\mathbf{a} = 1$. It follows that $\mathbf{t} \in B$ as required. $\qquad\square$

In view of Lemma 3.5.1, it suffices to investigate $A_{\mathbf{x}}$. To this end we define Γ as the graph with vertex-set $A_{\mathbf{x}}$ in which $\mathbf{w}_1 \sim \mathbf{w}_2$ if and only if $\mathbf{w}_1.\mathbf{w}_2 = 0$.

Lemma 3.5.2. *The Gram matrix of* $A_{\mathbf{x}}$ *is determined by* Γ.

Proof. It suffices to rule out the possibility that $\mathbf{w}_1.\mathbf{w}_2 = -1$ for non-adjacent vertices $\mathbf{w}_1, \mathbf{w}_2$ of Γ. In this situation we find that $(\mathbf{w}_1 + \mathbf{w}_2 - \mathbf{y}).(\mathbf{w}_1 + \mathbf{w}_2 - \mathbf{y}) = 0$; but then $\mathbf{w}_1 + \mathbf{w}_2 = \mathbf{y}$ and $\mathbf{x}.\mathbf{y} = 0$, a contradiction. $\qquad\square$

From the definition of $A_{\mathbf{x}}$, we already know the scalar products of vectors in S with vectors in $A_{\mathbf{x}}$, and so by Lemma 3.5.2, the Gram matrix of $S \;\dot\cup\; A_{\mathbf{x}}$ is determined by Γ. Thus the embedding of $S \;\dot\cup\; A_{\mathbf{x}}$ in a Euclidean space is determined to within an isometry; by Lemma 3.5.1, the same is true of our original line system Σ.

Lemma 3.5.3. *The graph* Γ *has the property*

($\star$) *any edge* $\mathbf{uv}$ *lies in a unique triangle* $\{\mathbf{u}, \mathbf{v}, \mathbf{w}\}$ *and any further vertex is adjacent to exactly one of* $\mathbf{u}, \mathbf{v}, \mathbf{w}$.

Proof. Let $\mathbf{w} = \mathbf{y} - \mathbf{z} - \mathbf{u} - \mathbf{v}$. We find that $\mathbf{w}.\mathbf{w} = 2$, $\mathbf{w}.\mathbf{x} = 0$, $\mathbf{w}.\mathbf{y} = 1$ and $\mathbf{w}.\mathbf{z} = -1$, so that $\mathbf{w} \in A_{\mathbf{x}}$; moreover, $\mathbf{w}.\mathbf{u} = \mathbf{w}.\mathbf{v} = 0$ and so $\{\mathbf{u}, \mathbf{v}, \mathbf{w}\}$ is a triangle. If $\mathbf{t}$ is a further vector in $A_{\mathbf{x}}$ then $\mathbf{t}.\mathbf{u} + \mathbf{t}.\mathbf{v} + \mathbf{t}.\mathbf{w} = 2$ and so $\mathbf{t}.\mathbf{u}, \mathbf{t}.\mathbf{v}, \mathbf{t}.\mathbf{w} = 1, 1, 0$ in some order. Thus $\mathbf{t}$ is adjacent to exactly one of $\mathbf{u}, \mathbf{v}, \mathbf{w}$. $\qquad\square$

It turns out that a graph Γ satisfies condition $(\star)$ of Lemma 3.5.3 if and only if it is one of the following:

(a) the empty graph,
(b) a graph of the form $K_1 \nabla r K_2$ (called a *windmill*),   •
(c) the graph $K_3 \times K_3$,
(d) the graph $\overline{L(K_6)}$,
(e) Sch_{10}, the complement of the Schläfli graph.

Details appear in [CaGSS], and we do not reproduce them here. Note that if $\mathbf{u} \in V(\Gamma)$ then it follows immediately from $(\star)$ that the subgraph induced by $\mathbf{u}$ and its neighbours is a windmill $K_1 \nabla r K_2$. If there is no vertex at distance 2 from $\mathbf{u}$ then we have case (b). Otherwise it can be shown that $r \in \{2, 3, 5\}$ and Γ is strongly regular with parameters $(6r - 3, 2r, 1, r)$; moreover the graphs (c),(d),(e) are the only strongly regular graphs with these parameters. Now we can retrace the arguments which showed that Σ is determined by Γ to establish that in cases (a),(b),(c),(d),(e), the line system Σ is $\overline{A}_n$, $\overline{D}_n$, $\overline{E}_6$, $\overline{E}_7$, $\overline{E}_8$ respectively. We summarize the result as follows.

Theorem 3.5.4 [CaGSS]. *The indecomposable star-closed systems of lines at $60°$ or $90°$ are $\overline{A}_n$ $(n \geq 1)$, $\overline{D}_n$ $(n \geq 4)$, $\overline{E}_6$, $\overline{E}_7$ and $\overline{E}_8$.*

By inspecting the dimensions of the line systems here, we can recover Theorem 3.4.5.

3.6 General characterization theorems

The theory of maximal line systems makes it possible to formulate a general characterization theorem as follows.

Theorem 3.6.1. *A connected graph G has least eigenvalue ≥ -2 if and only if it can be represented in the root system D_n, for some n, or in the exceptional root system E_8.*

Proof. It follows from our considerations in Section 3.1 that G has least eigenvalue ≥ -2 if and only if it has an a-representation in an indecomposable maximal system of lines at $60°$ or $90°$. Such line systems are determined in Theorem 3.4.5. In view of Theorems 3.3.2, 3.3.3 and 3.3.5 we have $A_2 \subset A_3 \subset D_4$, $A_n \subset D_{n+1}$ $(n = 4, 5, \ldots)$ and $E_6 \subset E_7 \subset E_8$; accordingly such a representation is contained in D_n (for some n) or E_8. $\square$

Theorem 3.6.2. *A graph can be represented in A_n if and only if it is the line graph of a bipartite graph.*

Proof. Suppose first that H is a bipartite graph on $U \, \dot\cup \, V$. Then the line graph $L(H)$ is represented by the vectors $\mathbf{e}_i - \mathbf{e}_j$ $(i \in U, \; j \in V)$. Conversely, if the graph G has an a-representation R in A_n, then whenever a basis vector $\mathbf{e}_i$ appears in a vector from R, it appears with the same sign. Let U be the set of vectors appearing with a positive sign, and let V be the set of vectors appearing with a negative sign. Let H be the bipartite graph on $U \, \dot\cup \, V$, with $\mathbf{e}_i \sim \mathbf{e}_j$ if and only if $\mathbf{e}_i - \mathbf{e}_j \in R$. Then $G = L(H)$.   □

Theorem 3.6.3. *A graph can be represented in D_n if and only if it is a generalized line graph.*

Proof. The generalized line graph $L(H; a_1, \ldots, a_n)$ has an a-representation in terms of an orthonormal basis $\{\mathbf{e}_{i,j} : 1 \le i \le n, \; 0 \le j \le a_i\}$, by means of the vectors

$$\{\mathbf{e}_{i,0} + \mathbf{e}_{j,0} : \{i, j\} \text{ is an edge of } H\} \cup \{\mathbf{e}_{i,0} \pm \mathbf{e}_{i,k} : 1 \le k \le a_i, \; 1 \le i \le n\}.$$

Conversely, if the graph G has an a-representation R in D_n, consider the set S of vectors $\mathbf{e}_i$ which always appear in R with the same sign. We may suppose that the sign is always positive, for if $\mathbf{e}_i, \mathbf{e}_j \in S$ and $\mathbf{e}_i + \mathbf{e}_h, -\mathbf{e}_j - \mathbf{e}_k \in R$ then $h \ne k$ and we may change the signs of $\mathbf{e}_j$ and $\mathbf{e}_k$.

If $\mathbf{e}_i \notin S$, then (since all inner products are non-negative) R contains a pair of vectors $\{\mathbf{e}_i + \mathbf{e}_j, -\mathbf{e}_i + \mathbf{e}_j\}$ or $\{\mathbf{e}_i - \mathbf{e}_j, -\mathbf{e}_i - \mathbf{e}_j\}$. If this vector $\mathbf{e}_j$ does not lie in S then R contains a pair of vectors $\{\mathbf{e}_j + \mathbf{e}_k, -\mathbf{e}_j + \mathbf{e}_k\}$ or $\{\mathbf{e}_j - \mathbf{e}_k, -\mathbf{e}_j - \mathbf{e}_k\}$. Neither of the first two pairs is compatible with either of the second two pairs, and so $\mathbf{e}_j \in S$.

Now let H be the graph on S, and let $\hat{H}$ be the multigraph obtained from H by adding, for each $\mathbf{e}_j \in S$, petals joining $\mathbf{e}_j$ and $\mathbf{e}_i$ whenever R contains the pair $\{\mathbf{e}_i + \mathbf{e}_j, -\mathbf{e}_i + \mathbf{e}_j\}$. Then G is the generalized line graph $L(\hat{H})$.   □

Corollary 3.6.4. *A graph is exceptional if and only if it is not a generalized line graph but can be represented in the root system E_8.*

Example 3.6.5. The line graph $L(K_8)$ is represented by the vectors $\mathbf{e_i} + \mathbf{e_j}$ $(1 \le i < j \le 8)$. Let S be a set of edges in K_8. If G is the graph obtained from $L(K_8)$ by switching with respect to S then G is represented by the vectors $\mathbf{e}_i + \mathbf{e_j}$ $(ij \notin S)$ and the vectors $\frac{1}{2}\mathbf{j}_8 - \mathbf{e}_i - \mathbf{e_j}$ $(ij \in S)$. We may add the vector $\frac{1}{2}\mathbf{j}_8$ to obtain a representation of the cone over G; such a cone is an exceptional graph because it has a vertex of degree 28 and hence cannot be represented in D_8. (A graph representable in D_8 has maximal degree ≤ 24.)   □

Remark 3.6.6. We saw in Theorem 3.6.3 that a generalized line graph $L(H;$ $a_1, \ldots, a_n)$ can be represented by vectors in D_m, where $m = \sum_{i=1}^n (1 + a_i)$. The least eigenvalue of such a graph is therefore bounded below by -2. This bound is attained if the number of vertices is greater than m because the corresponding Gram matrix is singular. □

Since an exceptional graph G can be represented in E_8, we deduce not only that G is an induced subgraph of $G(E_8)$ but also that it satisfies the following conditions.

Theorem 3.6.7. *If an exceptional graph has n vertices and mean degree d, then $n \leq \min\{36, 2d + 8\}$; moreover, every vertex has degree at most 28.*

Proof. Note first that an exceptional graph G is an induced subgraph of $G(\overline{E}_8)$, a strongly regular graph with spectrum $56, 8^{35}, (-4)^{84}$ (see Corollary 3.2.5). If G has more than 36 vertices then, by interlacing, its least eigenvalue is -4, a contradiction.

Now let us write the adjacency matrix of $G(\overline{E}_8)$ in the form

$$\begin{pmatrix} A_{11} & A_{12} \\ A_{21} & A_{22} \end{pmatrix},$$

where A_{11} is the adjacency matrix of G. Let $B = (b_{ij})$, where b_{ij} is the average row sum for the matrix A_{ij}. Thus if G has n vertices with average degree d, then

$$B = \begin{pmatrix} d & 56 - d \\ \dfrac{(56 - d)n}{120 - n} & d - \dfrac{(56 - d)n}{120 - n} \end{pmatrix},$$

with eigenvalues d and $d - \dfrac{(56 - d)n}{120 - n}$. By Theorem 1.2.22, these eigenvalues lie between -4 and 56. In particular,

$$d - \frac{(56 - d)n}{120 - n} \geq -4,$$

equivalently, $n \leq 2d + 8$.

Finally, in view of Proposition 3.2.1, we may represent a given vertex v by the vector $\frac{1}{2}\mathbf{j}_8$ in E_8. The vectors $\mathbf{v} \in E_8$ such that $\mathbf{v}.\frac{1}{2}\mathbf{j}_8 = 1$ are the 28 vectors $\mathbf{v}_{ij} = \mathbf{e}_i + \mathbf{e}_j$ and the 28 vectors $\mathbf{j}_8 - \mathbf{v}_{ij}$ ($1 \leq i < j \leq 8$). For given i, j, not both $\mathbf{v}_{ij}$ and $\mathbf{j}_8 - \mathbf{v}_{ij}$ can be present, and so v has at most 28 neighbours. □

An alternative proof of Theorem 3.6.7 is given in [BrCN, p. 107]. A proof without recourse to root systems is given in Section 5.3. Both bounds from the theorem are attained in the graph obtained from $L(K_9)$ by switching with respect to the vertices of an 8-clique [CaGSS]. This graph is one of the 473 maximal exceptional graphs described in Chapter 6.

3.7 Comments on some results from Chapter 2

In this section some of the results presented in Chapter 2 will be discussed from the point of view of the root system technique.

First, let us consider the 573 exceptional graphs with least eigenvalue greater than -2 which appear in Theorem 2.3.20. These graphs have only principal eigenvalues and since they are exceptional they are represented in E_8. Moreover, the star-closure of such a representation is E_6, E_7 or E_8, and so we say that the graph *generates* E_6, E_7 or E_8. Thus the number of principal eigenvalues is 6, 7 or 8, and accordingly the graphs have 6, 7 or 8 vertices.

Let G be one of these 573 graphs, with adjacency matrix A. If G is represented in the corresponding root system by a set S of vectors of norm $\sqrt{2}$, then the corresponding Gram matrix is $A + 2I$. The vectors in S are linearly independent and form a basis of the corresponding integral lattice. The matrix $A + 2I$ is non-singular and the determinant $\det(A + 2I)$ is equal to the volume of the parallelepiped generated by S, the basic cell of the lattice. This determinant is equal to 3, 2 and 1 for E_6, E_7 and E_8, respectively. For D_n, which represents a generalized line graph, the determinant is equal to 4 [MiHu] (cf. Proposition 1.1.8); see Lemma 7.5.2 for further details.

The minimal non-generalized line graphs are described by Corollary 2.3.17. We analyse how they fit into the theory developed in this chapter. By Theorems 3.6.1 and 3.6.3, a minimal non-generalized line graph G with $\lambda(G) > -2$ is determined by vectors from the root system E_8. In fact, since $A + 2I$ is non-singular, G must have six, seven, or eight vertices which correspond to graphs that generate E_6, E_7, or E_8. Now by Theorem 2.3.20 any such graph must contain a graph on six vertices that generates E_6. There are twenty such graphs, and they are displayed in Table A2. For each of these graphs, a subgraph induced by any five vertices must generate D_5, or A_5 and hence must be a generalized line graph. Thus the twenty graphs must be minimal non-generalized line graphs. In fact these twenty minimal graphs were originally derived from a computer search of independent sets of vectors in the root system E_6.

Now we can give a more sophisticated proof (from [CvDS2]) of Lemma 2.3.7, which shows that there are no minimal non-generalized line graphs G with $\lambda(G) = -2$. Suppose by way of contradiction that G is such a graph.

Then we know that G generates E_6, E_7, or E_8. If we can show that some proper subgraph also generates E_6, E_7, or E_8, then we will know that G is not minimal. We use the following lemma.

Lemma 3.7.1. *Suppose that the span of the vectors* $x_1, x_2, \ldots, x_n$ *has dimension* d, *and that the root system generated by any* d *linearly independent vectors* $x_{i_1}, x_{i_2}, \ldots, x_{i_d}$ *is* D_d. *Then the root system generated by* $\{x_1, x_2, \ldots, x_n\}$ *is also* D_d.

Proof. Assume that the vectors are in $\mathbb{R}^d$. The determinant of the matrix with $x_{i_1}, x_{i_2}, \ldots, x_{i_d}$ as columns is ± 2 (see [MiHu]). We may renumber the columns if necessary and assume that $x_1, x_2, \ldots, x_d$ are linearly independent. Now $x_k = \sum_{i=1}^{d} r_i x_i$ for any $k = 1, 2, \ldots, n$. Solving for r_i by Cramer's rule, we find that all non-zero determinants are ± 2. Hence if r_i is nonzero, it must be that $r_i = \pm 1$. Thus each x_k is in the integral span of $\{x_1, x_2, \ldots, x_d\}$ and hence $\{x_1, x_2, \ldots, x_n\}$ generates D_d. $\square$

Corollary 3.7.2. *If G is a non-generalized line graph with $\lambda(G) = -2$, then G is not a minimal non-generalized line graph.*

Proof. Let $A(G) + 2I$ be the Gram matrix of the vectors $x_1, x_2, \ldots, x_n$. Since $\lambda(G) = -2$, these vectors linearly dependent. If G were indeed minimal then the hypotheses of Lemma 3.7.1 would be satisfied, and the conclusion would imply that G is a generalized line graph, a contradiction. $\square$

The root system technique, developed in this chapter, affords much shorter proofs of some results originally obtained by the forbidden subgraph technique and described in Chapter 2. In particular, Theorem 2.5.1 can now be proved as follows. Consider a graph G which satisfies hypotheses (i)–(iii) of this theorem. By Theorem 3.6.7, G cannot be an exceptional graph since its vertex degrees exceed 28. Hence G is a generalized line graph by Theorem 3.6.3. Suppose that $G = L(H; a_1, \ldots, a_n)$ where we have $a_i > 0$ for at least one i. Then G contains an induced subgraph isomorphic to a generalized cocktail party graph $GCP(p, q)$ with $p, q > 0$. Let x be a vertex of type a and y a vertex of type b in the generalized cocktail party graph. Now $\deg(x) \geq p - 1$, $\deg(y) = p - 2$, and x, y have exactly $p - 3$ common neighbours. This contradicts condition (iii) of the theorem and so G is a line graph.

Finally, we see that the conclusion of Theorem 2.5.1 still holds when condition (ii) is relaxed to $\delta(G) > 28$.

4

Regular Graphs

Some general properties of regular exceptional graphs, including a relation between the order and degree of such graphs, are given in Section 4.1. In Section 4.2 we establish a spectral characterization of regular line graphs: we use the methods of [CvDo2], and a reformulation of results from [BuCS1] and [BuCS2], to provide a computer-free proof. Some characterizations of special classes of line graphs are given in Section 4.3. The regular exceptional graphs are determined in Section 4.4: they were first found in [BuCS1], but are derived here in a more economical way, partly using arguments from [BrCN].

4.1 Regular exceptional graphs

In this section we determine some general properties of regular exceptional graphs, beginning with the following observation.

By Proposition 1.1.9, a regular connected generalized line graph is either a line graph or a cocktail party graph. In view of Theorems 3.6.1 and 3.6.3, the following theorem is now straightforward.

Theorem 4.1.1. *If G is a regular connected graph with least eigenvalue* -2, *then one of the following holds:*

(i) *G is a line graph,*
(ii) *G is a cocktail party graph, or*
(iii) *G is an exceptional graph (with a representation in* E_8*).*

The next result imposes strong restrictions on the possibilities which can arise in Theorem 4.1.1(iii). Recall that a *principal* eigenvalue is an eigenvalue greater than -2.

Proposition 4.1.2. *The number of principal eigenvalues of an exceptional graph is 6, 7 or 8.*

Proof. Let G be an exceptional graph with r principal eigenvalues. We know from Section 3.6 that G has an a-representation R in E_8, but no such representation in D_n for any n. Now the star-closure of R is an r-dimensional line system contained in an IMSL, and from Section 3.4, this star-closure can only be $\overline{E}_6$, $\overline{E}_7$ or $\overline{E}_8$, of dimension 6, 7 or 8, respectively. $\qquad\square$

Proposition 4.1.3 [DoCv]. *There are no regular exceptional graphs with least eigenvalue greater than* -2.

Proof. By Proposition 4.1.2, an exceptional graph G has 6, 7 or 8 principal eigenvalues. There are no non-principal eigenvalues, and so G has 6, 7 or 8 vertices. The degree of G or $\overline{G}$ is at most 3, and so the spectra of the relevant graphs are easily found, using Corollary 1.2.14 where necessary. (The spectra of the cubic graphs on 8 vertices are given in [CvDSa, Appendix, Table 3].) Alternatively, one can use a table of regular graphs up to 10 vertices from [CvRa2]. We find that no example arises. $\qquad\square$

In view of Proposition 4.1.3 we introduce the following definition.

Definition 4.1.4. *We define $\mathcal{G}$ as the set of all connected regular graphs which have least eigenvalue* -2, *and which are neither line graphs nor cocktail party graphs.*

A. J. Hoffman [Hof6] posed the problem of determining $\mathcal{G}$ in 1969; he and D. K. Ray-Chaudhuri [HoRa3] showed that graphs in $\mathcal{G}$ cannot have degree ≥ 17 (cf. Theorem 2.5.5). This result, which also appears as Theorem 4.4 of [CaGSS], follows from Theorem 4.1.5 below.

All of the graphs in $\mathcal{G}$ were found by F. C. Bussemaker, D. Cvetković and J. J. Seidel [BuCS1, BuCS2] in 1975, using a mixture of mathematical reasoning and a computer search (described in Section 4.4). The report [BuCS1] contains a table of all 187 graphs from $\mathcal{G}$, while the paper [BuCS2] is an announcement of results. In this book, the table from [BuCS1] is reproduced in a quite different form as Table A3, and in the sequel we refer to graphs in $\mathcal{G}$ by the identification numbers there.

The results in this section, proved without recourse to a computer, may now be taken as the starting point for a computer search; the original search was conducted under weaker conditions.

The next theorem was first established in [BuCS1, BuCS2]; the computer-free proof given here is taken from [BrCN].

Theorem 4.1.5. *If G is a graph in $\mathcal{G}$ with n vertices and degree r then one of the following holds:*

(a) $n = 2(r + 2) \leq 28$,

(b) $n = \frac{3}{2}(r + 2) \leq 27$ *and G is an induced subgraph of the Schläfli graph,*

(c) $n = \frac{4}{3}(r + 2) \leq 16$ *and G is an induced subgraph of the Clebsch graph.*

Proof. Let S be a representation of G in E_8, and let $L = L(S)$. Then L is one of $L(E_6)$, $L(E_7)$, $L(E_8)$. Let $\mathbf{u}$ be the vector $\frac{1}{r+2}\Sigma_{\mathbf{v}\in S}\mathbf{v}$. Since $\mathbf{u}.\mathbf{v} = 1$ for each $\mathbf{v} \in S$, we know that $\mathbf{u}$ lies in the dual lattice L^*. Now $\mathbf{u}.\mathbf{u} = n/(r + 2)$ and it follows that $\mathbf{u}.\mathbf{u} < 3$, for otherwise $3(r + 2) \leq n \leq 2r + 8$ by Theorem 3.6.7; then $r \leq 2$ and G would be a line graph. We now exploit the properties of L^* established in Proposition 3.2.6.

If $L = L(E_8)$ then $L^* = L$. In this case the squared length of a vector in L^* is an even integer, and so $\mathbf{u}.\mathbf{u} = 2$, $n = 2(r + 2)$. By transitivity of $G(E_8)$ (Proposition 3.2.1) we may assume that $\mathbf{u} = \frac{1}{2}\mathbf{j}$. Since $\mathbf{u}.\mathbf{v} = 1$ for each $\mathbf{v} \in S$, the vectors in S are among those which fall into 28 pairs $\{\mathbf{e}_i + \mathbf{e}_j, \mathbf{u} - \mathbf{e}_i - \mathbf{e}_j\}$. Since at most one vector from each pair can be present in S, we have $n \leq 28$, and (a) follows.

If $L = L(E_7)$ then the squared length of a vector in L^* is of the form $2k$ or $2k + \frac{3}{2}$ ($k \in \mathbb{Z}$), and so $\mathbf{u}.\mathbf{u} = 2$ or $\frac{3}{2}$ in this case. If $\mathbf{u}.\mathbf{u} = 2$ then we may proceed as in (a). If $\mathbf{u}.\mathbf{u} = \frac{3}{2}$ then we may extend S to a representation of $K_2 \bigtriangledown G$ in E_8 by representing the two additional vertices by $\mathbf{u} \pm \mathbf{w}$, where $\mathbf{w}$ is a vector orthogonal to E_7 such that $\mathbf{w}.\mathbf{w} = \frac{1}{2}$. From Section 3.2 we know that in $G(E_8)$, the common neighbours of two adjacent vertices induce a Schläfli graph, and (b) follows.

If $L = L(E_6)$ then the squared length of a vector in L^* is of the form $2k$ or $2k + \frac{4}{3}$ ($k \in \mathbb{Z}$), and so $\mathbf{u}.\mathbf{u} = 2$ or $\frac{4}{3}$ in this case. Again, if $\mathbf{u}.\mathbf{u} = 2$ then we may proceed as in (b). If $\mathbf{u}.\mathbf{u} = \frac{4}{3}$ then we may extend S to a representation of $K_3 \bigtriangledown G$ in E_8 by representing the three additional vertices by $\mathbf{u} + \mathbf{w}_1, \mathbf{u} + \mathbf{w}_2, \mathbf{u} + \mathbf{w}_3$, where the vectors $\mathbf{w}_1, \mathbf{w}_2, \mathbf{w}_3$ are orthogonal to E_6, $\mathbf{w}_i.\mathbf{w}_i = \frac{2}{3}$ ($i = 1, 2, 3$) and $\mathbf{w}_i.\mathbf{w}_j = -\frac{1}{3}$ ($i \neq j$). (Thus $\mathbf{w}_1, \mathbf{w}_2, \mathbf{w}_3$ determine a star.) From Section 3.2 we know that in $G(E_8)$, the common neighbours of the vertices in a 3-cycle induce a Clebsch graph, and (c) follows. $\square$

By inspecting conditions (a), (b), (c) of Theorem 4.1.5, we can now verify Theorem 2.5.5:

Corollary 4.1.6. *A regular exceptional graph has degree at most 16, and this bound is attained in the Schläfli graph. In particular, a regular connected graph with least eigenvalue -2 and degree ≥ 17 is a line graph or a cocktail party graph.*

In accordance with [BuCS1], a graph is said to be in the *first, second* or *third layer* of $\mathcal{G}$ according as it satisfies condition (a), (b) or (c) of Theorem 4.1.5, respectively. From Table A3, we see that there are 163 graphs (nos. 1–163) in the first layer, 21 graphs (nos. 164–184) in the second, and 3 graphs (nos. 185–187) in the third.

We note a useful consequence of the proof of Theorem 4.1.5. Here we say that a graph is *even (odd)* if each vertex has even (odd) degree. An *even-odd bipartite* graph is a graph in which every edge joins a vertex of even degree to a vertex of odd degree.

Proposition 4.1.7. *If $G \in \mathcal{G}$ then G is switching-equivalent to the line graph of a graph H on 8 vertices. If G is of even order then each component of H is an even graph, an odd graph or an even-odd bipartite graph.*

Proof. From the proof of Theorem 4.1.5, we know that G has a representation in E_8 among the neighbours of $\frac{1}{2}\mathbf{j}$ in $G(E_8)$, since in each case we may take $\frac{1}{2}\mathbf{j}$ to represent one of the additional vertices considered. If we switch with respect to the set of vertices represented by vectors of the form $\frac{1}{2}\mathbf{j} - \mathbf{e}_i - \mathbf{e}_j$ then we obtain a graph represented entirely by vectors of the form $\mathbf{e}_i + \mathbf{e}_j$: such a graph is evidently the line graph of a graph H on 8 vertices (cf. Example 3.6.5).

Suppose that G is r-regular of order n, obtained from $L(H)$ by switching with respect to X, where $|X| = k$. Denote by d_{ij} the degree of an edge ij of H. After switching, the degree of ij is congruent modulo 2 to $d_{ij} + (n - k)$ if $ij \in X$, and $d_{ij} + k$ if $ij \notin X$. Hence $d_{ij} \equiv r + k \pmod 2$ for all $ij \in E(H)$. The second assertion of the Proposition follows because $d_{ij} = \deg(i) + \deg(j) - 2$. $\qquad\square$

4.2 Characterizing regular line graphs by their spectra

Our goal in this section is to prove Theorem 4.2.9, which shows that with just 17 exceptions, the property that a graph is a regular, connected line graph can be recognized from its spectrum. We first present a number of results on which the proof is based.

Proposition 4.2.1. *A regular connected line graph can have an exceptional cospectral mate only if it is the line graph of one of the following 30 graphs:*

(i) $K_{m,n}$, $\{m, n\} = \{2, 5\}, \{2, 6\}, \{2, 7\}, \{3, 4\}, \{3, 5\}, \{3, 6\}, \{4, 4\}$ or $\{4, 5\}$;

(ii) K_n, $n = 6, 7, 8$;

(iii) $CP(n)$, $n = 3, 4$;

(iv) $\overline{C}_n$, $n = 6, 7, 8$;

(v) $\overline{C_m \dot{\cup} C_n}$, $\{m, n\} = \{3, 4\}, \{3, 5\}$ or $\{4, 4\}$;

(vi) G or $\overline{G}$ where G is one of the five connected cubic graph with 8 vertices;

(vii) the semi-regular bipartite graph with parameters

$(m, n, r_1, r_2) = (6, 3, 2, 4)$.

Proof. Suppose that the line graph $L(G)$ is regular and connected. We may suppose that G is connected, and then G is either regular or semi-regular bipartite. We know from Theorem 2.2.4 (see also Theorems 1.2.16 and 1.2.17) that if G has m edges and n vertices, then the multiplicity of -2 as an eigenvalue of $L(G)$ is $m - n + 1$ if G is bipartite, and $m - n$ otherwise. Hence the number of principal eigenvalues of $L(G)$ is $n - 1$ or n respectively. From Proposition 4.1.2 we see that the number of vertices of G is at least 6 and at most 9. The only such graphs are those listed.   □

We now wish to use Proposition 1.1.11 to produce new graphs cospectral with a given regular line graph $L(H)$. In other words, we wish to switch with respect to a set X of edges of H which induce a regular graph in $L(H)$; thus X induces a regular or semi-regular bipartite subgraph of H. In particular, if we use the value $k = 0$ from Proposition 1.1.10, then the edges in X are disjoint. If we use the value $k = 1$, then the edges lie in a subgraph of the root graph consisting of copies of $K_{1,2}$.

Proposition 4.2.2. *Suppose that G is an r-regular graph with n vertices, where $n > 2r$. If G contains a 1-factor and a 4-cycle that intersect in a single edge, then there exists a graph H such that H and $L(G)$ are cospectral but not isomorphic. Thus $L(G)$ is not characterized by its spectrum.*

Proof. Let X be a subset of the 1-factor with cardinality $n - 2r$ that contains the edge of the 4-cycle, and apply Proposition 1.1.10 with $k = 0$. Then switching $L(G)$ with respect to X produces a cospectral graph H that contains an induced subgraph $K_{1,3}$. Hence H is not even a line graph, and certainly not isomorphic to $L(G)$.   □

Proposition 4.2.3. *The line graphs of the following 17 graphs are cospectral with an exceptional graph:*

(i) $K_{4,4}$, $K_{3,6}$,

(ii) $CP(4)$,

(iii) K_8,

(iv) $\overline{C}_8$,

(v) $\overline{C_m \dot{\cup} C_n}$, $\{m, n\} = \{3, 4\}, \{4, 4\}$, (vi-a) *G where G is one of four connected cubic graphs on* 8 *vertices (see Table A3),*

(vi-b) $\overline{G}$ *where G is one of the five connected cubic graphs on* 8 *vertices (see Fig. 1 of Table A4), and*

(vii) *the semi-regular bipartite graph with parameters* $(m, n, r_1, r_2) =$ $(6, 3, 2, 4)$.

Proof. In each of cases (i)–(vi), there is a 1-factor and a 4-cycle satisfying the hypothesis of Proposition 4.2.2. In the remaining case, we apply Proposition 1.1.10 with $k = 1$. In all cases, we obtain a cospectral graph H with $K_{1,3}$ as an induced subgraph, and then H is neither a line graph nor a cocktail party graph. □

We have seen that 17 of the 30 possible regular line graphs possess cospectral mates. Now we show that the remaining 13 graphs from Proposition 4.2.1 are characterized by their spectra. This will complete the picture in relation to spectral characterizations of regular line graphs.

Proposition 4.2.4. *The graphs* (a) $L(K_{m,n})$ $((m, n) = (2, 5), (2, 7), (3, 4),$ $(3, 5), (4, 5)),$ $L(K_7), L(\overline{C_7}), L(C_3 \dot\cup C_4),$ (b) $L(K_{2,6}),$ (c) $L(K_6),$ *and* (d) $L(CP(3))$ *are characterized by their spectra.*

Proof. Theorem 4.1.5 shows that the graphs in part (a) cannot have exceptional cospectral mates. It can easily be checked that these graphs cannot have cospectral mates among line graphs or cocktail party graphs. The proof for (b), (c) and (d) follows from Theorems 2.6.5, 2.6.1 and 2.6.6, respectively. □

To show that the remaining two graphs are characterized by their spectra, we use the forbidden subgraph technique in conjunction with the following straightforward result.

Proposition 4.2.5. *Let* $\lambda_1 \geq \lambda_2 \geq \cdots \geq \lambda_n$ *be the eigenvalues of a graph G. Let* $\overline{d} = \frac{1}{n} \sum_{i=1}^{n} \lambda_i^2$, *and* $\overline{t} = \frac{1}{2n} \sum_{i=1}^{n} \lambda_i^3$. *Then* (i) $\overline{d}$ *is the average degree, and* (ii) $\overline{t}$ *is the average number of triangles containing a given vertex.*

Recall that $\overline{d} = \lambda_1$ if and only if G is regular (see Theorem 1.2.1), and so we can recognize regularity from the spectrum. In particular, a graph cospectral with an r-regular graph is itself r-regular. Note also that $\overline{t}$ is the average number of edges in G_v, where for $v \in V(G)$, G_v denotes the subgraph of G induced by the vertices adjacent to v. We can now outline the arguments which avoid a brute force approach by computer.

Proposition 4.2.6. *The graph* $L(\overline{C_6})$ *is characterized by its spectrum.*

Sketch proof. Suppose that G is a graph with the spectrum of $L(\overline{C}_6)$, namely $4, 2, 1^2, (-1)^2, (-2)^3$. By Proposition 4.2.5, G is 4-regular with $\overline{t} < 3$. Thus $|E(G_v)| \le 2$ for some vertex v of G. A forbidden subgraph $G^{(1)}, G^{(2)}, G^{(3)}$ or $G^{(7)}$ from Fig. 2.4 arises unless $E(G_v)$ consists of two independent edges. Now none of the four remaining vertices vertices can be adjacent to three of the five original vertices of G_v, again because of forbidden subgraphs. Thus these four vertices form a cycle of length four, and the remaining edges can be added in only two ways to avoid forbidden subgraphs. One yields $L(\overline{C}_6)$ and the other yields $L(K_{3,3})$, with spectrum $4, 1^4, (-2)^4$. This completes the proof.   □

Let S_5 be the cubic graph formed from two copies of the graph with 4 vertices and 5 edges by adding two edges to produce a connected regular graph (see Table A4, Fig.1).

Proposition 4.2.7. *The graph $L(S_5)$ is characterized by its spectrum.*

Sketch proof. Suppose that G is a graph with the spectrum of $L(S_5)$, namely $4, 1 + \sqrt{5}, 2, 0^4, 1 - \sqrt{5}, (-2)^4$. Then by Proposition 4.2.5, the average number of edges in G_v is 3. Let us suppose first that every subgraph G_v has three edges. Then each G_v is $K_{1,3}$, $K_3 \,\dot\cup\, K_1$ or P_4. Suppose that v is a vertex for which G_v is $K_{1,3}$, and let u be one of the vertices of degree 1 in G_v. Then G_u has fewer than three edges, contrary to assumption. If G_v is $K_3 \,\dot\cup\, K_1$ for some vertex v, then G_v is $K_3 \,\dot\cup\, K_1$ for every vertex v. Then each vertex lies in exactly one complete graph with four vertices, and G is covered by three copies of K_4. The only graph with this property is the line graph of a semi-regular bipartite graph and it has $4, (1 + \sqrt{2})^2, 0^3, (1 - \sqrt{2})^2, (-2)^4$ as its spectrum.

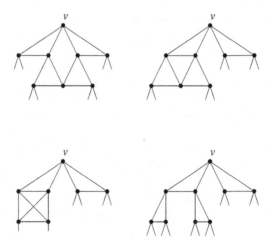

Figure 4.1: The graphs from Proposition 4.2.7.

If G_v is P_4 for every v, then there is only one way to complete the graph avoiding forbidden subgraphs, and the spectrum is 4, $(1 + \sqrt{3})^2$, 0^3, $(1 - \sqrt{3})^2$, $(-2)^4$.

These contradictions show that there is a vertex v such that G_v has fewer than three edges. As in Proposition 4.2.6, a forbidden subgraph arises unless $E(G_v)$ consists of two independent edges. Now Fig. 4.1 illustrates all the possible ways of adding further vertices adjacent to neighbours of v. In each case, it is straightforward to complete the graph. Among the graphs obtained in this way, only $L(S_5)$ has the given spectrum. Details are left to the reader; to show that a graph does not have the given spectrum, it suffices to count the numbers of triangles, quadrilaterals and pentagons (see [CvDSa, p. 97]). □

Propositions 4.2.4, 4.2.6 and 4.2.7 serve to prove the following.

Proposition 4.2.8. *The* 13 *graphs from Proposition 4.2.1 not listed in Proposition 4.2.3 are characterized by their spectra.*

We now have the tools to prove our main theorem.

Theorem 4.2.9. *The spectrum of a graph G determines whether or not it is a regular connected line graph except for* 17 *cases. The exceptional cases are those in which G has the spectrum of L(H) where H is one of the 3-connected regular graphs on 8 vertices or H is a connected semi-regular bipartite graph on* $6 + 3$ *vertices.*

Proof. If H is a 3-connected regular graph on 8 vertices or a connected semi-regular bipartite graph on $6 + 3$ vertices, then $L(H)$ is one of the 17 graphs from Proposition 4.2.3. We saw there that, in these 17 cases, it is not possible to tell whether the graph is a line graph or a cospectral mate obtained by Seidel switching. But in all other cases we see from Proposition 4.2.8 and Theorem 1.2.1 that we can recognize a graph as being a regular connected line graph. □

This theorem was first announced in [Cve7]. Its proof in this section follows [CvDo2], and does not require a computer search.

It is still possible, of course, for two non-isomorphic cospectral regular line graphs to arise from non-isomorphic root graphs. Theorem 4.3.1 from the next section specifies the possibilities.

It turns out that there are exactly 68 regular exceptional graphs which are cospectral with the 17 line graphs from Theorem 4.2.9. They are easily identified from the list of all 187 regular exceptional graphs obtained as described in Section 4.1. As we note in Section 4.4, the techniques used in the search have been adapted to construct the 68 graphs without recourse to a computer.

4.3 Special characterization theorems

Here we shall present a few theorems giving spectral characterizations of special classes of regular line graphs. We start with the following theorem, due to F. C. Bussemaker, D. Cvetković and J. J. Seidel.

Theorem 4.3.1 [BuCS1, BuCS2]. *Let $L(G_1)$, $L(G_2)$ denote cospectral, connected, regular line graphs of the connected graphs G_1, G_2. Then one of the following holds:*

(i) *G_1 and G_2 are cospectral regular graphs with the same degree,*
(ii) *G_1 and G_2 are cospectral semi-regular bipartite graphs with the same parameters,*
(iii) *$\{G_1, G_2\} = \{H_1, H_2\}$, where H_1 is regular and H_2 is semi-regular bipartite; in addition there exist positive integers $s > 1$ and $t \leq \frac{1}{2}s$, and real numbers λ_i, $0 \leq \lambda_i < t\sqrt{s^2 - 1}$, $i = 2, 3, \ldots, \frac{1}{2}s(s-1)$, such that H_1 has $s^2 - 1$ vertices, degree st, and the eigenvalues*

$$st, \ \pm\sqrt{\lambda_i^2 + t^2}, \ -t \quad (\text{of multiplicity } s);$$

H_2 has s^2 vertices, parameters $n_1 = \frac{1}{2}s(s+1)$, $n_2 = \frac{1}{2}s(s-1)$, $r_1 = t(s-1)$, $r_2 = t(s+1)$, and the eigenvalues $\pm t\sqrt{s^2 - 1}$, $\pm\lambda_i$, 0 (of multiplicity s).

Proof. We know that if the graph G is connected and the line graph $L(G)$ is regular then either G is regular or G is a semi-regular bipartite graph. If G_1 and G_2 from the theorem are both regular or both semi-regular bipartite we have cases (i) and (ii) from the theorem; this follows readily from Theorem 1.2.16, and from Theorem 1.2.17 and Proposition 1.2.18.

Suppose therefore that $\{G_1, G_2\} = \{H_1, H_2\}$ where H_2 is semi-regular bipartite with parameters (n_1, n_2, r_1, r_2) $(n_1 > n_2)$, and H_1 is regular non-bipartite of degree r with n vertices. Since $L(H_1)$ and $L(H_2)$ are cospectral they must have the same degree, the same number of vertices and the same multiplicity of the eigenvalue -2. This yields the following relations

$$r_1 + r_2 - 2 = 2r - 2, \quad n_1 r_1 = \frac{nr}{2}(= n_2 r_2),$$

$$n_1 r_1 - n_1 - n_2 + 1 = \frac{nr}{2} - n,$$

which may be rewritten as follows:

(4.1) $$r = \frac{r_1 + r_2}{2},$$

$$(4.2) \qquad\qquad nr = 2n_1 r_1 = 2n_2 r_2,$$

$$(4.3) \qquad\qquad n = n_1 + n_2 - 1.$$

If we use (4.1) and (4.3) to substitute for r and n in (4.2), we obtain

$$(4.4) \qquad\qquad n_1 - n_2 = \frac{r_1 + r_2}{r_2 - r_1}.$$

Let $\lambda_1, \lambda_2, \ldots, \lambda_{n_2}$ be the first n_2 largest eigenvalues of H_2. From the proof of Proposition 1.2.18, we know that H_2 has also the eigenvalues $-\lambda_1, -\lambda_2, \ldots, -\lambda_{n_2}$ and $n_1 - n_2$ eigenvalues equal to 0, where $\lambda_1 = \sqrt{r_1 r_2}$. Since the sum of squares of eigenvalues is twice the number of edges we have

$$2r_1 r_2 + 2 \sum_{i=2}^{n_2} \lambda_i^2 = 2n_1 r_1,$$

equivalently,

$$(4.5) \qquad\qquad \sum_{i=2}^{n_2} \lambda_i^2 = n_1 r_1 - r_1 r_2.$$

Now, by Theorem 1.2.16 and Proposition 1.2.18, the eigenvalues of H_1 are $\frac{1}{2}(r_1 + r_2)$ with multiplicity 1 (the largest eigenvalue), $\frac{1}{2}(r_1 - r_2)$ with multiplicity $n_1 - n_2$ and $\pm\sqrt{\lambda_i^2 + \frac{1}{4}(r_1 - r_2)^2}$ ($i = 2, 3, \ldots, n_2$). The sum of eigenvalues must be 0 and this yields again the relation (4.4). Considering the sum of squares we have

$$(\frac{r_1 + r_2}{2})^2 + (n_1 - n_2)(\frac{r_1 - r_2}{2})^2 + 2 \sum_{i=2}^{n_2} (\lambda_i^2 + \frac{(r_1 - r_2)^2}{4}) = 2n_1 r_1.$$

Using (4.5) we obtain

$$(4.6) \qquad\qquad n_1 + n_2 = \left(\frac{r_1 + r_2}{r_2 - r_2}\right)^2.$$

Let $s = \dfrac{r_1 + r_2}{r_2 - r_1}$. Then s is an integer greater than 1, and relations (4.4) and (4.6) yield

$$n_1 = \frac{s^2 + s}{2} \quad \text{and} \quad n_2 = \frac{s^2 - s}{2}.$$

By equation (4.1), r_1 and r_2 are of the same parity, and since $r_2 > r_1$ we can take $r_2 = r_1 + 2t$, where t is a positive integer. Then

$$r_1 = t(s - 1) \quad \text{and} \quad r_2 = t(s + 1).$$

Since $r_1 \leq n_2$ and $r_2 \leq n_1$ we see that $t \leq \frac{s}{2}$. If we now express the spectra of H_1 and H_2 in terms of s, t and the λ_i, the proof of the theorem is complete.   □

Remark 4.3.2. When $s = 2$ we have $H_1 = K_3$ and $H_2 = K_{1,3}$, but then $L(H_1)$ and $L(H_2)$ are not only cospectral, but also isomorphic. (By Theorem 2.1.6, $\{K_3, K_{1,3}\}$ is the only pair of non-isomorphic connected graphs having isomorphic line graphs.) When $s = 3$, H_2 is the first graph in Fig. 2 of Table A4; but then H_1 does not exist. For $s = 4$ and $t = 2$ we have $H_2 = K_{10,6}$ and $H_1 = L(K_6)$; and, of course, $L(K_{10,6})$ and $L(L(K_6))$ are cospectral but not isomorphic. In the case $s = 4, t = 1$, H_2 belongs to the design with the parameters $v = 6, b = 10, r = 5, k = 3, \lambda = 2$, and H_1 is the Petersen graph. For higher values of s, in the known examples H_2 is the graph of a 2-design. We do not know whether there exists a pair of graphs H_1, H_2 such that (i) H_2 is not the graph of a 2-design, and (ii) H_1, H_2 satisfy the conditions of Theorem 4.3.1(iii) with $s > 4$.   □

Theorems 4.3.1 and 4.2.9 have some interesting consequences when applied to special classes of regular line graphs. In particular, we can now strengthen results of Section 2.6 which were proved when only the forbidden subgraph technique was available. In the following results the exceptional graphs are referred to by their numbers in Table A3. Theorems 4.3.3 and 4.3.4 are also due to F. C. Bussemaker, D. Cvetković and J. J. Seidel.

Theorem 4.3.3 [BuCS1, BuCS2]. *The graph* $L(K_{m,n})$ *is characterized by its spectrum unless*

   (i) $m = n = 4$, *where graph no. 69 provides the only exception,*
  (ii) $m = 6, n = 3$, *where graph no. 70 provides the only exception,*
 (iii) $m = 2t^2 + t, n = 2t^2 - t$, *and there exists a symmetric Hadamard matrix with constant diagonal of order* $4t^2$.

Proof. The possible cospectral mates of $L(K_{m,n})$ may or may not be line graphs. If they are not line graphs, then they can be identified immediately from Table A4, and we have cases (i) and (ii) of the theorem. The exceptions which are line graphs are described by Theorem 4.3.1, and we recover Theorem 2.6.14: from $n_1 = r_1 = m$ and $n_2 = r_2 = n$ we have $t = \frac{s}{2}$ and $n_1 = 2t^2 + t, n_2 = 2t^2 - t$. Since the eigenvalues of $K_{m,n}$ are $\pm\sqrt{mn}$ and 0, the spectrum of the graph H_1 in Theorem 4.3.1 consists of eigenvalues $2t^2$, $\pm t$, and its adjacency matrix A satisfies $A^2 = t^2(I + J)$. Replacing the zeros of A by (-1)s, and bordering the matrix with (-1)s, we obtain a symmetric Hadamard matrix with diagonal $-I$. This completes the proof.   □

We may deal similarly with the line graphs of 2-designs.

Theorem 4.3.4 [BuCS1, BuCS2]. *Let G_1 be the line graph of a 2-design with parameters (v, k, b, r, λ). Let G_2 be a graph cospectral with G_1. Then one of the following holds:*

(i) *G_2 is the line graph of a 2-design having the same parameters;*

(ii) *$(v, k, b, r, \lambda) = (3, 2, 6, 4, 2)$ and G_2 is graph no. 6;*

(iii) *$(v, k, b, r, \lambda) = (4, 3, 4, 3, 2)$ and G_2 is graph no. 9;*

(iv) *$(v, k, b, r, \lambda) = (4, 4, 4, 4, 4)$ and G_2 is graph no. 69;*

(v) *$(v, k, b, r, \lambda) = (3, 3, 6, 6, 6)$ and G_2 is graph no. 70;*

(vi) *$v = \frac{1}{2}s(s-1)$, $k = t(s-1)$, $b = \frac{1}{2}s(s+1)$, $r = t(s+1)$, $\lambda = \frac{2t(st-t-1)}{s-2}$, where s and t are integers with st even, $t \leq \frac{1}{2}s$, $(s-2)|2t(t-1)$, and $G_2 = L(H)$ where H is a regular graph on $s^2 - 1$ vertices with the eigenvalues st, $\pm\sqrt{ts(s-1-t)(s-2)^{-1}}$, $-t$ of multiplicities 1, $\frac{1}{2}(s-2)(s+1)$, $\frac{1}{2}(s-2)(s+1)$, s, respectively.*

Proof. Assume first that G_2 is a line graph. Then cases (ii) and (iii) of Theorem 4.3.1 apply, with case (ii) there corresponding to case (i) here. In case (iii) of Theorem 4.3.1 we have $n_1 = b$, $n_2 = v$, $r_1 = k$, $r_2 = r$, and the well-known relation $\lambda(v-1) = r(k-1)$ yields $(s-2)\lambda = 2t(st-t-1)$. Of course, λ must be an integer and we obtain the condition $(s-2)|2t(t-1)$. We have $G_2 = L(H)$ where H is a regular graph on $s^2 - 1$ vertices, and the eigenvalues of H can be found in the form stated by using the relation $\lambda_i = \sqrt{r - \lambda}$ $(i = 2, 3, \ldots, \frac{1}{2}s(s-1))$ (see Theorem 4.3.1 and Remark 2.6.9).

It remains to show that st is even, and we prove this (as in [CaGSS]) by considering a diagonal entry of A^3, where A is the adjacency matrix of H. The minimal polynomial of A is $(x - st)(x + t)(x^2 - q)$, where $q = st(s-t-1)/(s-2)$. Since H is regular, we have by Theorem 1.2.7:

$$(A + tI)(A^2 - qI) = \frac{(st+t)(s^2t^2 - q)}{(s^2 - 1)}J.$$

From this equation we find that each diagonal entry of A^3 is st^3. Since this number is twice the number of triangles with a given vertex, we deduce that st is even.

Finally, if G_2 is not a line graph then we find from Table A4 that cases (ii)–(v) are the only possibilities that arise. This completes the proof. $\square$

Next we present a generalization of Theorem 4.3.3.

Let K_ℓ^n denote the complete ℓ-partite graph having n vertices in each part. The following theorem appears in [RaSV2] as a reformulation of some results from [BuCS1].

Theorem 4.3.5. *The graph $L(K_\ell^n)$ is characterized by its spectrum except for the following cases:*

(i) $\ell = 2, n = 4$, *where graph no. 69 provides the only exception;*

(ii) $\ell = 4, n = 2$, *in which case there are exactly eight exceptions (graph nos. 153–160).*

We conclude this section with a theorem of J. J. Seidel which enumerates the strongly regular graphs with least eigenvalue -2. The theorem was originally proved using the Seidel adjacency matrix.

Theorem 4.3.6 [Sei1]. *Let G be a strongly regular graph with least eigenvalue -2. Then G is one of the following graphs:*

(a) $L(K_n)$, $n = 4, 5, \ldots;$

(b) $L(K_{n,n})$, $n = 2, 3, \ldots;$

(c) $CP(n)$, $n = 2, 3, \ldots;$

(d) *one of the three Chang graphs (graphs nos. 161–163);*

(e) *the Schläfli graph (graph no. 184);*

(f) *the Shrikhande graph (graph no. 69);*

(g) *the Clebsch graph (graph no. 187);*

(h) *the Petersen graph (graph no. 5).*

Proof. If G is a generalized line graph, then by Proposition 1.1.9, it is either a line graph $L(H)$ or a cocktail party graph. Since cocktail party graphs are strongly regular the second case yields (c), while in the first case H is either regular or semi-regular bipartite (cf. Proposition 1.1.5). Now by Theorem 1.2.20, G has three distinct eigenvalues. It follows that if H is regular then (by Theorem 1.2.16) either H has two distinct eigenvalues and is not bipartite, or H has three distinct eigenvalues and is bipartite. In the first case, H is a complete graph and we obtain (a), while in the second case we obtain (b). If H is semi-regular bipartite then we apply Theorem 1.2.17 and no new cases arise.

If G is an exceptional graph then it appears in Table A3. On identifying the graphs in this table with three distinct eigenvalues, we easily obtain cases (d)–(h). □

4.4 Regular exceptional graphs: computer investigations

In this section we derive the results underlying the computer search for the graphs in $\mathcal{G}$. We start by exploring in more detail the situation described in Proposition 4.1.7. This proposition tells us that if $G \in \mathcal{G}$ then there exists a graph H of order 8 and a subset X of $E(H)$ such that G is obtained from

$L(H)$ by switching with respect to X. Let H_1 be the spanning subgraph of H with $E(H_1) = X$, and let H_2 be the spanning subgraph of H with $E(H_2) = E(H) \setminus X$. Thus H is factorized into two factors H_1 and H_2.

Definition 4.4.1. *The factorization of a graph H on 8 vertices into spanning subgraphs H_1 and H_2 is said to be* line-regular *if the graph obtained from $L(H)$ by switching with respect to $L(H_1)$ (equivalently, $L(H_2)$) is regular and representable in E_8.*

We shall now describe line-regular factorizations in terms of parameters $s_1, \ldots, s_8$, where s_i is the sum of the i-th coordinates of vectors in a representation S in E_8 of a graph from $\mathcal{G}$.

Proposition 4.4.2. *Let r, n be positive integers such that $n \leq 2(r+2)$, and let H be a graph with n edges on 8 vertices. Suppose that H has a factorization into spanning subgraphs H_1 and H_2 having p and q edges, respectively, where $p \geq q$ and $p + q = n$. Let the vertex degrees in H_1 and H_2 be $x_1, \ldots, x_8$ and $y_1, \ldots, y_8$, respectively, and let $s_i = q + 2(x_i - y_i)$ ($i = 1, \ldots, 8$).*

The factorization of H is line-regular, and yields a regular graph of degree r, if and only if the integers s_i satisfy the following conditions:

(i) $\sum_{i=1}^{8} s_i = 4n$, $\quad \sum_{i=1}^{8} s_i^2 = 4n(r+2)$;
(ii) $s_i + s_j = 2(r+2)$ *for each edge ij of H_1;*
(iii) $s_i + s_j = 2(n - r - 2)$ *for each edge ij of H_2.*

Proof. Suppose first that the factorization is line-regular. Let $\{\mathbf{e}_1, \ldots, \mathbf{e}_8\}$ be the standard orthonormal basis of $\mathbb{R}^8$. We have seen that we may represent each edge ij of H_1 by the vector $2\mathbf{e}_i + 2\mathbf{e}_j$ and each edge ij of H_2 by $\mathbf{j} - 2\mathbf{e}_i - 2\mathbf{e}_j$. The vectors in such a representation S are the columns of an $8 \times n$ matrix (α_{ij}) whose row sums are $s_1, \ldots, s_8$. (To see this, note that in the i-th row, 2 appears x_i times, -1 appears y_i times, and 1 appears $q - y_i$ times.) It is clear that $\sum_{i=1}^{8} s_i = 4n$. For the second equality in (i), note that

$$\sum_{i=1}^{8} s_i^2 = \sum_{j=1}^{n}(\sum_{i=1}^{8} \alpha_{ij}^2) + \sum_{j \neq k}(\sum_{i=1}^{8} \alpha_{ij}\alpha_{ik}) = 8n + (4n)r,$$

since $\sum_{i=1}^{8} \alpha_{ij}\alpha_{ik}$ is 4 or 0 according as the j-th and k-th edges of H are or are not adjacent. The second condition follows by taking the inner product of $2\mathbf{e}_i + 2\mathbf{e}_j$ with the sum of all other vectors in S: the non-zero summands are equal to 4 and arise from $x_i - 1$ vectors with i-th entry 2, $x_j - 1$ vectors with j-th entry 2, and $q - y_i - y_k$ vectors with i-th and j-th entries equal to 1. Condition (iii) is obtained similarly.

Suppose now that the s_i satisfy (i), (ii) and (iii), and let $d_1, \ldots, d_8$ be the vertex degrees in H. From $x_i + y_i = d_i$ and $x_i - y_i = \frac{1}{2}(s_i - q)$ we obtain the formulas

$$(4.7) \qquad x_i = \frac{1}{2}d_i + \frac{1}{4}(s_i - q), \qquad y_i = \frac{1}{2}d_i - \frac{1}{4}(s_i - q).$$

Consider an edge ij of H_1. The corresponding vertex of $L(H)$ will have the following degree after switching $L(H)$ with respect to $L(H_2)$:

$$x_i + x_j - 2 + q - y_i - y_j = (x_i - y_i) + (x_j - y_j) + q - 2$$

$$= \frac{1}{2}(s_i - q) + \frac{1}{2}(s_j - q) + q - 2 = \frac{1}{2}(s_i + s_j) - 2 = r.$$

For an edge ij in H_2 we have similarly:

$$y_i + y_j - 2 + p - x_i - x_j = p - 2 - (x_i - y_i) - (x_j - y_j)$$

$$= p - 2 - \frac{1}{2}(s_i - q) - \frac{1}{2}(s_j - q) = n - \frac{1}{2}(s_i + s_j) = r.$$

Hence the factorization is line-regular and yields a regular graph of degree r. This completes the proof. $\qquad\qquad\qquad\qquad\qquad\qquad\qquad\qquad\square$

We now turn our attention to graphs in the first layer. In this situation, we can sharpen some of our previous results.

Proposition 4.4.3. *If G lies in the first layer of $\mathcal{G}$ then*

(i) $s_1 = s_2 = \cdots = s_8 = r + 2,$
(ii) *there exists an odd or even graph H on 8 vertices such that G is switching equivalent to $L(H)$.*

Proof. Since $n = 2(r + 2)$, Proposition 4.4.2(i) implies that

$$\left(\sum_{i=1}^{8} s_i\right)^2 = 8 \sum_{i=1}^{8} s_i^2 = 64(r + 2)^2,$$

and (i) follows. Moreover, the relation $x_i - y_i = \frac{1}{2}(s_i - q)$ implies that all the differences $x_i - y_i$ are equal. Since $x_i - y_i$ and $x_i + y_i$ are of the same parity and since $d_i = x_i + y_i$, all vertex degrees in H are of the same parity. This completes the proof. $\qquad\qquad\qquad\qquad\qquad\qquad\qquad\qquad\square$

Proposition 4.4.4. *Let H be a graph on 8 vertices with $2(r + 2)$ edges. Suppose that H has a line-regular factorization into graphs H_1, H_2, with p, q edges respectively, which yields a regular graph of degree r. Then*

(i) $p \equiv q \equiv r + 2 \pmod 4$ *if H is even;*
(ii) $p \equiv q \equiv r \pmod 4$ *if H is odd.*

Proof. By (4.7) and Proposition 4.4.3(i), the vertex degrees of H_1 and H_2 are given by

$$(4.8) \qquad x_i = \frac{1}{2}d_i + \frac{1}{4}(r + 2 - q), \qquad y_i = \frac{1}{2}d_i - \frac{1}{4}(r + 2 - q)$$

and the Proposition follows since $p + q = 2r + 4$. $\square$

From Proposition 4.4.2 and the relation (4.8) we have the following observation.

Proposition 4.4.5. *Let H be an odd or even graph on 8 vertices with $2(r + 2)$ edges. A factorization of H into spanning subgraphs H_1 and H_2 is line-regular, and yields a regular graph of degree r, if and only if the vertex degrees $x_1, \ldots, x_8$ of H_1 and the vertex degrees $y_1, \ldots, y_8$ of H_2 satisfy the relations*

$$(4.9) \qquad x_i = \frac{1}{2}d_i + \alpha, \qquad y_i = \frac{1}{2}d_i - \alpha \qquad (i = 1, \ldots, 8),$$

for some α.

In view of Proposition 4.4.3(ii), in order to find all graphs in the first layer of $\mathcal{G}$, we should start with all odd and even graphs on 8 vertices with an even number of edges, and look for all their line-regular factorizations. From Section 2.4 we know all cubic graphs in the first layer (they have $r = 3$, $n = 10$), and so it is sufficient to consider odd and even graphs with 12 or more edges. Clearly there are no odd or even graphs with 26 edges (since such a graph would be obtained by removing two edges from K_8) and so there are no graphs in the first layer with $n = 26$. When $n = 28$, the only graphs in the first layer are those switching-equivalent to $L(K_8)$, and these are the Chang graphs. A list of all odd and even graphs on 8 vertices with an even number n ($12 \leq n \leq 24$) of edges is given in Table 7.7 of [BuCS1]. The set of these graphs will be denoted by $\mathcal{P}$.

Since different graphs from $\mathcal{P}$ can give rise to the same graph in the first layer, the following definition is useful.

Definition 4.4.6. *The graphs G_1 and G_2 are said to be* line-switching equivalent *if their line graphs $L(G_1)$ and $L(G_2)$ are switching-equivalent.*

Obviously, line-switching equivalent graphs yield the same graphs in the first layer. Therefore the line-switching equivalence classes of the set $\mathcal{P}$ have been found and a representative for each class has been chosen (the first graph listed for each equivalence class in Table 7.7 of [BuCS1]). The search for line-switching equivalent graphs was undertaken by computer, using Proposition 1.1.17.

The following two useful propositions follow from inspection of the equivalence classes in $\mathcal{P}$.

Proposition 4.4.7. *For each odd (even) graph H from $\mathcal{P}$ there exists an even (odd) graph H' in $\mathcal{P}$ such that H and H' are line-switching equivalent.*

This means that we can start with only even or only odd graphs when searching for the graphs in the first layer. However the representatives of classes in Table 7.7 of [BuCS1] were chosen according to other criteria, which will be explained later.

Line graphs of line-switching equivalent graphs necessarily have the same Seidel spectrum, but in $\mathcal{P}$ the converse also holds.

Proposition 4.4.8. *Each line-switching equivalence class of graphs in $\mathcal{P}$ is uniquely determined by the Seidel spectrum of the line graph of any graph from the class.*

Remark 4.4.9. For regular graphs, a knowledge of the Seidel spectrum is equivalent to a knowledge of the spectrum of the adjacency matrix. Accordingly Proposition 4.4.8 is useful in finding the 68 exceptional graphs cospectral with line graphs (see Theorem 4.2.9 and Proposition 4.2.3). Note that such graphs lie in the first layer of $\mathcal{G}$. By Proposition 4.4.3, each such graph is switching-equivalent to the line graph of an 8-vertex graph H^*.

From Section 4.2 we know that there are 17 graphs H for which the line graph $L(H)$ has a cospectral mate; fifteen of these graphs are 3-connected regular graphs on 8 vertices, and two are connected semi-regular bipartite graphs on $6 + 3$ vertices. In the former case, we may take $H = H^*$ because both $L(H)$ and $L(H^*)$ are cospectral with G. From Proposition 4.4.5 we know that if $H_1 \cup H_2$ is a line-regular factorization of H then the factors H_1 and H_2 are themselves regular. Thus the graphs cospectral with $L(H)$ are easily determined once we know all the line-regular factorizations of H into two regular factors. Accordingly, representatives of the line-switching equivalence classes in Table A3 are chosen regular wherever possible. The same observation is the basis for a computer-free construction of the 68 regular exceptional graphs outlined in [CvRa1], the relevant factorizations having already been found in [Rad]; see Table A4. Each of these 68 graphs is cospectral with one of the 17 line graphs $L(H)$ listed in Proposition 4.2.3. It transpires that 66 of the 68 graphs arise in this way from a regular graph H.

A second criterion for choosing representatives of line switching equivalence classes in $\mathcal{P}$ is that the minimum degree of H should be as small as possible. In view of (4.8), this restricts the number of possible values of q, and hence the number of line-regular factorizations. For example, only one value of q is possible when one of the vertex degrees of H is 0 or 1.   □

Algorithm 4.4.10. (An algorithm for finding graphs in the first layer.) *Find a set of representatives of the line-switching equivalence classes of $\mathcal{P}$. For any H from that set, and for each possible α, find all factorizations of H into spanning subgraphs H_1 and H_2, whose vertex degrees are given by (4.9). Switch $L(H)$ with respect to the vertices corresponding to edges of H_2. The graph obtained is regular and lies in the first layer of $\mathcal{G}$. All graphs from the first layer can be obtained in this way.*

A computer was used to implement the algorithm, resulting in the following proposition.

Proposition 4.4.11. *There are exactly* 163 *graphs in the first layer of $\mathcal{G}$; they are listed in Table A3.*

Inspection of Table A3 reveals that only graph no. 5 (the Petersen graph) has 6 principal eigenvalues, equivalently has a representation in E_6. The graphs numbered 1, 9 and 69 are those with 7 principal eigenvalues, and they have a representation in E_7.

We turn now to the second layer.

Proposition 4.4.12. *All graphs in the second layer can be obtained if we start with the following solution of equations from condition (i) of Proposition 4.4.2:* $s_1 = s_2 = n, s_3 = s_4 = \cdots = s_8 = \frac{1}{3}n$.

Proof. Let G be a graph from the second layer. By Theorem 4.1.5, G is an induced subgraph of the Schläfli graph. From Section 3.2 we know that $G \cup K_1$ can be represented by vectors of the form $2\mathbf{e}_i + 2\mathbf{e}_j$ or $\mathbf{j}_8 - 2\mathbf{e}_i - 2\mathbf{e}_j$. Let us represent the isolated vertex by the vector $\mathbf{y} = \mathbf{j}_8 - 2\mathbf{e}_1 - 2\mathbf{e}_2$. Consider the row sums s_i for the representation S of G defined in this way. Now the sum of inner products of $\mathbf{y}$ with all vectors from S is zero, that is,

$$-s_1 - s_2 + \sum_{i=3}^{8} s_i = 0.$$

This relation, together with Proposition 4.4.2(i), gives $s_1 + s_2 = 2n$. Suppose now that $s_1 = n + \alpha$, $s_2 = n - \alpha$ for some α. We show that $\alpha = 0$. From Proposition 4.4.2(i) we have

$$\sum_{i=3}^{8} s_i = 2n \text{ and } \sum_{i=3}^{8} s_i^2 = 4n(r+2) - 2n^2 - 2\alpha^2.$$

For any solutions s_i we have

$$\left(\sum_{i=3}^{8} s_i\right)^2 \le 6 \sum_{i=3}^{8} s_i^2,$$

and therefore we obtain

$$4n^2 \le 6(4n(r+2) - 2n^2 - 2\alpha^2),$$

which implies that

$$n \le \frac{3}{2}(r+2) - \frac{3\alpha^2}{4n}.$$

Since G is in the second layer, we have $n = \frac{3}{2}(r+2)$ and $\alpha = 0$. Hence, $s_1 = s_2 = n$ and

$$\sum_{i=3}^{8} s_i = 6\left(\frac{n}{3}\right), \quad \sum_{i=3}^{n} s_i^2 = 6\left(\frac{n}{3}\right)^2,$$

which implies $s_3 = s_4 = \cdots = s_8 = \frac{n}{3}$.

This completes the proof.   □

Now we explain how to find all graphs in the second layer of $\mathcal{G}$. Here, $n = \frac{3}{2}(r+2)$ and we write $n = 3v$ ($3 \le v \le 9$). Since a graph G in the second layer is an induced subgraph of the Schläfli graph, we know from Section 3.2 that G has a representation S in $\mathbb{R}^8$ by vectors of the form (a) $2\mathbf{e}_1 + 2\mathbf{e}_i$ ($i = 3, 4, \ldots, 8$) or (b) $2\mathbf{e}_2 + 2\mathbf{e}_i$ ($i = 3, 4, \ldots, 8$), and $\mathbf{j}_8 - 2\mathbf{e}_i - 2\mathbf{e}_j$ ($i < j$; $i, j = 3, 4, \ldots, 8$).

Suppose that there are p vectors of type (a) and q vectors of type (b) in S. Since $s_1 = s_2$, p is even and S contains $\frac{1}{2}p$ vectors having 2 as the first coordinate, and $\frac{1}{2}p$ vectors having 2 as the second coordinate.

The last 6 rows of the representation S are of three types: (i) those with two entries equal to 2, (ii) those with one entry equal to 2, (iii) those with no entry equal to 2. Let s, t, u be the number of rows of type (i), (ii), (iii) respectively. Then

$$p \equiv 0 \pmod 2, \quad p + q = 3v, \quad s + t + u = 6, \quad 2s + t = p.$$

Since each row sum is v, the number of entries equal to -1 in a row of type (i), (ii), (iii) is j, k, ℓ respectively, where

$$j = v - \frac{1}{2}p + 2, \quad k = v - \frac{1}{2}p + 1, \quad \ell = v - \frac{1}{2}p.$$

Let $H_1 \cup H_2$ be the line-regular factorization of the 8-vertex graph H associated with S, and consider the subgraph F of H_2 obtained by deleting vertices

1 and 2. The vertex degrees of F are just the numbers j, k, ℓ (and they appear with multiplicities s, t, u respectively). In particular, $\ell \leq 5$ and we have $j = k + 1 = \ell + 2 \leq 7$. Now it is easy to determine all possible values of the various parameters. Given v, we can choose some p and then we also know j, k and ℓ. Then we can choose some s which yields t and u. All of the possible sets of parameters are given in Table A3, together with the corresponding vertex degree sequence of F. The value $n = 27$ is excluded since in this case the only graph in the second layer is the Schläfli graph itself.

Using vertex degree sequences it is easy to construct all possible graphs F on 6 vertices. (One can make use of the table of graphs on 6 vertices from [CvPe].) Given a graph F, the corresponding graph H can easily be constructed, and the corresponding line-regular factorization determined. Different graphs F can yield the same graph in the second layer, and so it is necessary to check for isomorphisms between the graphs that arise. The outcome of a computer search based on the above procedure is as follows.

Proposition 4.4.13. *There are exactly* 21 *graphs from* $\mathcal{G}$ *in the second layer. They are displayed in Table A3.*

Finally, let us find all of the graphs in the third layer of $\mathcal{G}$. By Theorem 4.1.1, all such graph are induced subgraphs of the Clebsch graph. The only possible values for the parameters (n, r) are $(16, 10)$, $(12, 7)$ and $(8, 4)$.

In the first case the only solution is the Clebsch graph itself. In the third case we have the complement of a cubic graph on 8 vertices. Using the table of cubic graphs in [CvDSa] we easily find that the only solution is the complement of the Möbius ladder on 8 vertices, in other words, the total graph of a 4-cycle.

For the remaining case consider the Clebsch graph as depicted in Fig. 4.2 (cf. Example 1.1.14 and Fig. 1.3). We have to remove a set X of 4 vertices to leave a graph G with 12 vertices. In this situation, there are exactly 36 edges between G and X. It follows that the subgraph induced by X contains exactly 2 edges, and hence is $K_{1,2} \,\dot\cup\, K_1$ or $2K_2$. In the first case, no such X can be found,

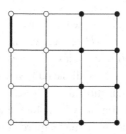

Figure 4.2: Construction of the Clebsch graph and one of its subgraphs.

and in the second case, there is essentially only one choice of two independent edges, as indicated by the bold lines in Fig. 4.2. Accordingly there is just one possibility for G. Now we have:

Proposition 4.4.14. *There are exactly* 3 *graphs in the third layer of* $\mathcal{G}$*: the Clebsch graph, the graph from Fig.* 4.2, *and the graph* $T(C_4)$.

We summarize our results as follows.

Theorem 4.4.15. *The graphs in* $\mathcal{G}$ *are the* 187 *graphs listed in Table A3.*

An immediate corollary of Theorem 4.4.15 is the following.

Theorem 4.4.16. *A regular connected graph with least eigenvalue* -2 *is a line graph, or a cocktail party graph, or one of the of the* 187 *graphs displayed in Table A3.*

We have already noted in Section 4.3 what this classification tells us about cospectral line graphs. In particular, we obtain the following result by inspecting the graphs in Table A3.

Theorem 4.4.17. *There are exactly* 68 *connected regular graphs which are not line graphs but which are cospectral with a line graph. They are displayed in Table A4.*

Our objective in the remainder of this section is to sharpen statement (a) of Theorem 4.1.5: we shall see that with five exceptions, a graph in the first layer of $\mathcal{G}$ is an induced subgraph of a Chang graph. The proof rests ultimately on an inspection of Table A3, but first we need another observation concerning factorization.

Lemma 4.4.18. *Let* H *be an even graph on* 8 *vertices, with an even number of edges, different from graphs (a), (b), (c) in Fig. 4.3. Then the edges of* H *can be coloured by two colours (say, red and blue) so that for each vertex the number of red edges incident to that vertex equals the number of blue edges incident to that vertex.*

Proof. Suppose first that H is connected. Since H is even, its edges may be partitioned into cycles; and since H has an even number of edges, the number of odd cycles is even. If two odd cycles have a vertex in common then they form a closed trail of even length. If all the odd cycles can be paired in this way, then the edges of H may be partitioned into closed trails of even length. In this case, the edges of each such trail may be coloured alternately red and blue to obtain an edge-colouring that meets our requirements. Otherwise, deletion of the edges in a maximal set of such trails leaves an even number of odd cycles with disjoint vertex sets. Since H has 8 vertices, the only possibility is that there remain just two disjoint odd cycles, say Z_1 and Z_2.

Since H is connected we can find among the deleted trails a sequence $T_1, T_2, \ldots, T_k$ of trails such that Z_1 has a common vertex x_1 with T_1, T_1 has a common vertex x_2 with T_2, and so on until T_k has a common vertex x_{k+1} with Z_2. Colour the two edges of Z_1 incident with x_1 in red, and two edges from T_1 incident with x_1 in blue. Starting from these edges and proceeding in both directions around T_1, colour the edges of T_1 alternately red and blue. Finally, two edges of T_1 incident with x_2 will be coloured by the same colour, say red. The edges from T_2 incident with x_2 will then be coloured blue. Continuing in this way we can colour the edges contained in the trails $Z_1, T_1, \ldots, T_k, Z_2$ by two colours so that at each vertex the colours are balanced. The edge-colouring of H is completed by colouring the edges of any remaining trails alternately red and blue.

Now suppose that H is disconnected. We may assume that H has at least two non-trivial components, because the foregoing argument holds for graphs with only one such component. Since H is even, any nontrivial component has at least 3 vertices, and so the number of non-trivial components is exactly two. The only even graphs with 8 vertices, an even number of edges and exactly two non-trivial components are those shown in Fig. 4.3. The graph (d) of Fig. 4.3 can be coloured in the desired manner, but the others cannot.

This completes the proof.   □

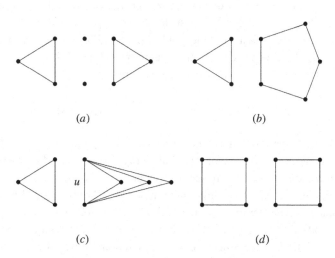

(a)   (b)

(c)   (d)

Figure 4.3: Even graphs with two non-trivial components.

Remark 4.4.19. The graph (c) of Fig. 4.3 nevertheless has a line-regular factorization, in which the edge u belongs to one factor, and all remaining edges to the other. □

Theorem 4.4.20. *With five exceptions, a graph from $\mathcal{G}$ is an induced subgraph of the Schläfli graph or of one of the three Chang graphs. The exceptions are switching-equivalent 9-regular graphs on 22 vertices, numbered 148–152 in Table A3.*

Proof. Since the Clebsch graph is an induced subgraph of the Schläfli graph, it suffices to check whether a graph in $\mathcal{G}$ satisfying $n = 2(r + 2)$ is an induced subgraph of one of the Chang graphs (see Theorem 4.1.5).

For $n = 28$, this result is immediate, and there is no graph in the first layer of $\mathcal{G}$ with $n = 26$. For $n = 10$, we may check directly that each of the five cubic graphs from the first layer is an induced subgraph of a Chang graph. By Proposition 4.4.7, each of the remaining graphs in the first layer can be obtained by applying Algorithm 4.4.10 to an odd graph H from $\mathcal{P}$. Let $H_1 \cup H_2$ be a line-regular factorization of H.

If H_2 is regular then by Theorem 1.1.13 a Chang graph is obtained when we switch $L(K_8)$ with respect to $L(H_2)$. Since $L(H)$ is an induced subgraph of $L(K_8)$, an induced subgraph of this Chang graph is obtained when we switch $L(H)$ with respect to $L(H_2)$.

In the case that H_2 is not regular, consider the even graph $\overline{H}$. By Proposition 4.4.5, H (and hence also $\overline{H}$) is non-regular. As before, let $x_1, \ldots, x_8$ and $y_1, \ldots, y_8$ be the vertex degrees in H_1 and H_2, respectively. By Proposition 4.4.5, there exists a constant α such that $x_i - y_i = \alpha$ $(i = 1, \ldots, 8)$. By Lemma 4.4.18, if $\overline{H}$ is neither of the graphs (a), (c) in Fig. 4.3 then $\overline{H}$ has a factorization $H_1' \cup H_2'$ for which the vertex degrees in H_2' are $\frac{1}{2}(7 - x_i - y_i)$ $(i = 1, \ldots, 8)$. Thus in $H_2 \cup H_2'$ the vertex degrees are

$$y_i + \frac{1}{2}(7 - x_i - y_i) = \frac{7}{2} - \alpha \quad (i = 1, \ldots, 8).$$

Now a Chang graph is obtained when we switch $L(K_8)$ with respect to $L(H_2 \cup H_2')$, and an induced subgraph of this Chang graph is obtained when we switch $L(H)$ with respect to $L(H_2)$. (To see this, picture an adjacency matrix of $L(K_8)$ with partitions of rows and columns determined by $H_1' \cup H_1 \cup H_2 \cup H_2'$.)

It remains to consider the cases in which $\overline{H}$ is the graph (a) or (c) of Fig. 4.3. In the second case $\overline{H}$ is the graph (c) of Fig. 4.3 with the degree sequence $2^6, 4^2$. By Remark 4.4.19, $\overline{H}$ has a line-regular factorization $H_1' \cup H_2'$ in which H_1' and H_2' have degree sequences $2^6, 3^2$ and $0^6, 1^2$. Now H has degree sequence $5^6, 3^2$ and in a line-regular factorization $H = H_1 \cup H_2$, the degree sequences of H_1

and H_2 would be either 3^6, 2^2 and 2^6, 1^2 or 4^6, 3^2 and 1^6, 0^2. Hence, $H_2 \cup H'_2$ is regular of degree 2 in the first case and of degree 1 in the second case, and we arrive at the same conclusion as above. In the remaining case $\overline{H}$ is the graph (a) of Fig. 4.3 and this has no line-regular factorization.

We show that the factorization $H'_1 \cup H'_2$ of $\overline{H}$ is line-regular. If $d_1, \ldots, d_8$ are the vertex degrees of H, then the vertex degrees of $\overline{H}$ are $d'_1 = 7 - d_1, \ldots, d'_8 = 7 - d_8$. The vertex degrees y_i of H_2 are given by (4.8). Suppose that $H_2 \cup H'_2$ is regular of degree f. Then the vertex degrees y'_1 of H'_2 are

$$y'_i = f - y_i = f - \left(\frac{1}{2}d_i - \frac{1}{4}(r+2-q)\right)$$

$$= \frac{1}{2}(7 - d_i) + \frac{1}{4}(r+2-q) + f - \frac{7}{2} = \frac{1}{2}d'_i + \alpha \quad (i = 1, \ldots, 8),$$

where α does not depend on i. Hence $\overline{H}$ is even and the factorization $\overline{H} = H'_1 \cup H'_2$ is line-regular.

Thus all the graphs G are induced subgraphs of some Chang graph except in the last case when H is the graph no. 207 from Table 7.7 of [BuCS1]. This graph belongs to the class no. 37 in Table A3 and we obtain 5 exceptional graphs as mentioned in the theorem.

This completes the proof.   □

5

Star complements

Let G be a graph of order n with μ as an eigenvalue of multiplicity k. In Section 5.1, we define a star complement for μ in terms of the orthogonal projection of $I\!R^n$ onto the eigenspace $\mathcal{E}(\mu)$. We show that the star complements for μ in G are just the induced subgraphs of G of order $n - k$ that do not have μ as an eigenvalue, and we derive the properties of star complements required in this book. (For a survey of star complements, see [Row5].) In Section 5.2 we introduce the notion of a foundation for the root multigraph of a generalized line graph: it is used to characterize star complements for -2 in generalized line graphs, and at the same time to describe the eigenspace of -2. In Section 5.3, we show that a graph is exceptional if and only if it has an exceptional star complement for -2. By interlacing, such a star complement has least eigenvalue greater than -2 and hence is one of 573 known graphs (see Table A2 and Theorem 2.3.20). It follows that the exceptional graphs can be constructed, as extensions of star complements, without recourse to root systems. In Section 5.4 we show how certain graphs with least eigenvalue -2 can be characterized by star complements for -2. Finally, in Section 5.5 we discuss the role of switching in the construction of exceptional graphs from star complements.

5.1 Basic properties

Let G be a graph with vertex set $V(G) = \{1, \ldots, n\}$ and adjacency matrix A. Let $\{\mathbf{e}_1, \ldots, \mathbf{e}_n\}$ be the standard orthonormal basis of $I\!R^n$ and let P be the matrix which represents the orthogonal projection of $I\!R^n$ onto the eigenspace $\mathcal{E}(\mu)$ of A with respect to $\{\mathbf{e}_1, \ldots, \mathbf{e}_n\}$. Since $\mathcal{E}(\mu)$ is spanned by the vectors $P\mathbf{e}_j$ ($j = 1, \ldots, n$) there exists $X \subseteq V(G)$ such that the vectors $P\mathbf{e}_j$ ($j \in X$) form a basis for $\mathcal{E}(\mu)$. Such a subset X of $V(G)$ is called a *star set* for μ in G.

(The terminology reflects the fact that the vectors $P\mathbf{e}_1, \ldots, P\mathbf{e}_n$ form a eutactic star in the sense of Seidel [Sei5].)

Proposition 5.1.1. *Let G be a graph with μ as an eigenvalue of multiplicity $k > 0$. The following conditions on a subset X of $V(G)$ are equivalent:*

    (i) *X is a star set for μ;*

    (ii) *$\mathbb{R}^n = \mathcal{E}(\mu) \oplus \mathcal{V}$, where $\mathcal{V} = \langle \mathbf{e}_i : i \notin X \rangle$;*

    (iii) *$|X| = k$ and μ is not an eigenvalue of $G - X$.*

Proof. ((i) $\Rightarrow$ (ii)) Since $\dim \mathcal{E}(\mu) = k$ and $\dim \mathcal{V} = n - k$, it suffices to show that $\mathcal{E}(\mu) \cap \mathcal{V} = \{\mathbf{0}\}$. Let $\mathbf{x} \in \mathcal{E}(\mu) \cap \mathcal{V}$. Then $\mathbf{x} = P\mathbf{x}$ and $\mathbf{x}^T \mathbf{e}_j = 0$ for all $j \in X$. Hence $\mathbf{x}^T (P\mathbf{e}_j) = \mathbf{x}^T (P^T \mathbf{e}_j) = (P\mathbf{x})^T \mathbf{e}_j = 0$ for all $j \in X$. Thus $\mathbf{x} \in \langle P\mathbf{e}_j : j \in X \rangle^\perp = \mathcal{E}(\mu)^\perp$ and $\mathbf{x} = \mathbf{0}$.

((ii) $\Rightarrow$ (iii)) Suppose that $\mathbb{R}^n = \mathcal{E}(\mu) \oplus \mathcal{V}$. We consider an adjacency matrix A of G in the form $\begin{pmatrix} * & * \\ * & A' \end{pmatrix}$, where A' is the adjacency matrix of $G - X$. Suppose that $A'\mathbf{x}' = \mu\mathbf{x}'$. If $\mathbf{y} = \begin{pmatrix} \mathbf{0} \\ \mathbf{x}' \end{pmatrix}$, then

$$A\mathbf{y} = \begin{pmatrix} * & * \\ * & A' \end{pmatrix} \begin{pmatrix} \mathbf{0} \\ \mathbf{x}' \end{pmatrix} = \begin{pmatrix} * \\ \mu\mathbf{x}' \end{pmatrix}.$$

Now let $\mathbf{x} \in \mathcal{V}$. Then $\mathbf{x}^T$ has the form $(\mathbf{0}^T \; \mathbf{z}^T)$, and $\mathbf{x}^T A\mathbf{y} = \mu\mathbf{z}^T\mathbf{x}' = \mu\mathbf{x}^T\mathbf{y}$. Hence $(A - \mu I)\mathbf{y} \in \mathcal{V}^\perp$. On the other hand, if $\mathbf{x} \in \mathcal{E}(\mu)$, then $\mathbf{x}^T A\mathbf{y} = \mathbf{x}^T A^T \mathbf{y} = (A\mathbf{x})^T \mathbf{y} = (\mu\mathbf{x})^T \mathbf{y} = \mu\mathbf{x}^T\mathbf{y}$ and so $(A - \mu I)\mathbf{y} \in \mathcal{E}(\mu)^\perp$. Hence $(A - \mu I)\mathbf{y} \in \mathcal{V}^\perp \cap \mathcal{E}(\mu)^\perp = (\mathcal{E}(\mu) + \mathcal{V})^\perp$, which is the zero subspace. Therefore, $\mathbf{y} \in \mathcal{E}(\mu)$. But $\mathbf{y} \in \mathcal{V}$, and since $\mathcal{E}(\mu) \cap \mathcal{V} = \{\mathbf{0}\}$ we have $\mathbf{y} = \mathbf{0}$. Hence $\mathbf{x}' = \mathbf{0}$ and μ is not an eigenvalue of $G - X$.

((iii) $\Rightarrow$ (i)) Here, it suffices to prove that $\langle P\mathbf{e}_j : j \in X \rangle = \mathcal{E}(\mu)$. Suppose, by way of contradiction, that $\langle P\mathbf{e}_j : j \in X \rangle \subset \mathcal{E}(\mu)$. Then there is a non-zero vector $\mathbf{x} \in \mathcal{E}(\mu) \cap \langle P\mathbf{e}_j : j \in X \rangle^\perp$. Thus $\mathbf{x}^T P\mathbf{e}_j = 0$ for all $j \in X$. Hence $(P\mathbf{x})^T \mathbf{e}_j = (\mathbf{x}^T P)\mathbf{e}_j = 0$ for all $j \in X$. Consequently $P\mathbf{x} \in \langle \mathbf{e}_j : j \in X \rangle^\perp = \langle \mathbf{e}_s : s \notin X \rangle = \mathcal{V}$. But $\mathbf{x} = P\mathbf{x}$ and so we have a non-zero vector $\mathbf{x} \in \mathcal{E}(\mu) \cap \mathcal{V}$. Since $\mathbf{x} = \begin{pmatrix} \mathbf{0} \\ \mathbf{x}' \end{pmatrix}$ with $\mathbf{x}' \neq \mathbf{0}$ it follows that $\mathbf{x}'$ is an eigenvector of $G - X$, a contradiction. $\square$

Here $G - X$ is the subgraph of G induced by the complement of X; it is called the *star complement* for μ corresponding to X. (Such graphs are called *μ-basic* subgraphs in [Ell].) Statement (iii) of Proposition 5.1.1 provides a characterization of star sets and star complements which is often the most

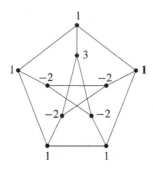

Figure 5.1: The Petersen graph (Example 5.1.2).

useful in practice. For instance, the claims in the following example are easily verified by these means.

Example 5.1.2. In Fig. 5.1, the vertices of the Petersen graph are labelled with eigenvalues in such a way that the vertices labelled μ form a star set for μ. For example, -2 is an eigenvalue of multiplicity 4, and if we delete the 4 vertices labelled -2 we obtain a subgraph H consisting of a 5-cycle with a single pendant edge attached. Since H does not have -2 as an eigenvalue, this subgraph is a star complement for -2.   □

Remark 5.1.3. It can be shown that if G is a graph with $\mu_1, \ldots, \mu_m$ as its distinct eigenvalues then $V(G)$ has a partition $X_1 \,\dot{\cup}\, \cdots \,\dot{\cup}\, X_m$ such that X_i is a star set for μ_i ($i = 1, \ldots, m$). Such a partition is called a *star partition*, and in this context the star sets X_i are called *star cells*. The existence of star partitions is established, along with related results, in [CvRS2, Chapter 7].   □

Proposition 5.1.4. *Let X be a star set for μ in G, and let $\overline{X} = V(G) \setminus X$.*
(i) If $\mu \neq 0$ then $\overline{X}$ is a dominating set for G;
(ii) If $\mu \neq -1$ or 0 then $\overline{X}$ is a location-dominating set for G — that is, the $\overline{X}$-neighbourhoods of distinct vertices in X are distinct and non-empty.

Proof. The matrix P, which represents the orthogonal projection onto $\mathcal{E}(\mu)$ is a polynomial function of the adjacency matrix A (see Section 1.1), and so $AP = PA$. For each vertex u of G we have

$$(5.1) \qquad \mu P\mathbf{e}_u = AP\mathbf{e}_u = PA\mathbf{e}_u = P\Big(\sum_{i\sim u} \mathbf{e}_i\Big) = \sum_{i\sim u} P\mathbf{e}_i.$$

For part (i), we have to show that any vertex u in X is adjacent to a vertex in $\overline{X}$. Since $\mu \neq 0$, we know from equation (5.1) that the vectors in $\{P\mathbf{e}_u\} \cup \{P\mathbf{e}_i : i \sim u\}$ are linearly dependent. Since the vectors $P\mathbf{e}_j$ ($j \in X$)

are linearly independent, it follows that there is a vertex adjacent to u which lies outside X.

For part (ii), let $\Gamma(u)$ be the set of neighbours of u in X. Suppose by way of contradiction that u and v are vertices in X with the same neighbourhoods in $\overline{X}$. From equation (5.1) and its counterpart for v we have

$$\mu P e_u - \mu P e_v - \sum_{j \in \Gamma(s)} P e_j + \sum_{j \in \Gamma(v)} P e_j = 0.$$

This is a relation on vectors in $\{P e_j : j \in X\}$. Since these vectors are linearly independent, it follows that either $\mu = -1$ and $u \sim v$, or $\mu = 0$ and $u \not\sim v$, contrary to assumption. $\square$

It follows from Proposition 5.1.4(i) that if $G - X$ is connected then so is G. It follows from Proposition 5.1.4(ii) that there are only finitely many graphs with a prescribed star complement for an eigenvalue $\mu \neq -1$ or 0, for if $|\overline{X}| = t$ then $|X| < 2^t$. This exponential bound will be improved to a quadratic bound in due course.

To prove the next result, we require the following observation.

Lemma 5.1.5. *If the column space of the symmetric matrix* $\begin{pmatrix} C & D^T \\ D & E \end{pmatrix}$ *has the columns of* $\begin{pmatrix} C \\ D \end{pmatrix}$ *as a basis, then the columns of C are linearly independent.*

Proof. Since each column of $\begin{pmatrix} D^T \\ E \end{pmatrix}$ is a linear combination of the columns of $\begin{pmatrix} C \\ D \end{pmatrix}$, there exists a matrix L such that $D^T = CL$, equivalently $D = L^T C$.

Thus if $C\mathbf{x} = \mathbf{0}$ then $\begin{pmatrix} C \\ D \end{pmatrix} \mathbf{x} = \mathbf{0}$, whence $\mathbf{x} = \mathbf{0}$ as required. $\square$

Theorem 5.1.6. *Let μ be an eigenvalue of the connected graph G, and let K be a connected induced subgraph of G not having μ as an eigenvalue. Then G has a connected star complement for μ containing K.*

Proof. Let $|V(K)| = r$. Since G is connected we may label its vertices $1, 2, \ldots, n$ so that each vertex after the first is adjacent to a predecessor. Since K is connected we may take $1, \ldots, r$ to be the vertices of K. Let A be the adjacency matrix of G, with columns $\mathbf{c}_1, \ldots, \mathbf{c}_n$, and let $\{\mathbf{c}_k : k \in Y\}$ be the basis of the column space of $\mu I - A$ obtained by deleting each column which is a linear combination of its predecessors. Note that $\{1, \ldots, r\} \subseteq Y$ because μ is not an eigenvalue of K. By Lemma 5.1.5, the principal submatrix of $\mu I - A$

determined by Y is invertible. Since $|Y| = \mathrm{codim}\,\mathcal{E}(\mu)$, $\overline{Y}$ is a star set for μ and the subgraph H induced by Y is a star complement. ·

We prove that H is connected by showing that each vertex y of Y with $y > 1$ is adjacent to a previous vertex j of Y. We take j to be least element of $\{1, 2, \ldots, n\}$ such that j is adjacent to y in G. Then $j < y$ and the y-th entry of $\mathbf{c}_j$ is -1. On the other hand, the y-th entry of each $\mathbf{c}_i$ $(i < j)$ is 0, and so $\mathbf{c}_j$ is not a linear combination of its predecessors. Accordingly $j \in Y$ as required.   □

The next result is well-known in the context of Schur complements (cf. [Gan, p. 47]); in the graph-theoretical context, it is known as the Reconstruction Theorem and its converse (see [CvRS6, Theorem 4.6] or [Ell, Theorem 1.1]).

Theorem 5.1.7. *Let X be a set of k vertices in the graph G and suppose that G has adjacency matrix $\begin{pmatrix} A_X & B^T \\ B & C \end{pmatrix}$, where A_X is the adjacency matrix of the subgraph induced by X. Then X is a star set for μ in G if and only if μ is not an eigenvalue of C and*

$$(5.2) \qquad\qquad \mu I - A_X = B^T (\mu I - C)^{-1} B.$$

In this situation, the eigenspace of μ consists of the vectors $\begin{pmatrix} \mathbf{x} \\ (\mu I - C)^{-1} B\mathbf{x} \end{pmatrix}$, where $\mathbf{x} \in \mathbb{R}^k$

Proof. Suppose first that X is a star set for μ. Then μ is not an eigenvalue of C, and we have

$$\mu I - A = \begin{pmatrix} \mu I - A_X & -B^T \\ -B & \mu I - C \end{pmatrix},$$

where $\mu I - C$ is invertible. In particular, if $|V(G)| = n$ then the matrix $(-B \mid \mu I - C)$ has rank $n - k$; but $\mu I - A$ has rank $n - k$ and so the rows of $(-B \mid \mu I - C)$ form a basis for the row space of $\mu I - A$. Hence there exists a $k \times (n - k)$ matrix L such that $(\mu I - A_X \mid -B^T) = L(-B \mid \mu I - C)$. Now $\mu I - A_X = -LB$, $-B^T = L(\mu I - C)$ and equation (5.2) follows by eliminating L.

Conversely, if μ is not an eigenvalue of C and equation (5.2) holds, then it is straightforward to verify that the vectors specified lie in $\mathcal{E}(\mu)$. They form a k-dimensional space, and, by interlacing, the multiplicity of μ is exactly k. Hence X is a star set for μ.   □

Note that if X is a star set for μ then the corresponding star complement $H(= G - X)$ has adjacency matrix C, and equation (5.2) tells us that G is determined by μ, H and the H-neighbourhoods of vertices in X (cf. [Row1, Theorem 4.8]). If $\mu \neq -1$ or 0 then by Proposition 5.1.4(ii), there is a one-one correspondence between the vertices in X and their H-neighbourhoods. Accordingly we have the following result.

Corollary 5.1.8. *Let G, G' be graphs with H, H' respectively as star complements for μ, where $\mu \neq -1$ or 0. If ψ is an isomorphism $H \to H'$ such that ψ maps the neigbourhoods $\Delta_H(v)$ ($v \notin V(H)$) onto the neighbourhoods $\Delta_{H'}(v')$ ($v' \notin V(H')$) then ψ extends to an isomorphism $G \to G'$, defined outside H by $\psi(\Delta_H(v)) = \Delta_{H'}(\psi(v))$.*

To find all the graphs with a prescribed star complement for μ, we have to find all solutions A, B of equation (5.2), given μ and C. In this situation, let $|V(H)| = t$ and define a bilinear form on $\mathbb{R}^t$ by

$$\langle\!\langle \mathbf{x}, \mathbf{y} \rangle\!\rangle \; = \mathbf{x}^T (\mu I - C)^{-1} \mathbf{y} \;\; (\mathbf{x}, \mathbf{y} \in \mathbb{R}^t).$$

If we denote the columns of B by $\mathbf{b}_u$ ($u \in X$) and equate matrix entries in equation (5.2), we obtain the following consequence of Theorem 5.1.7 (cf. [Ell, Corollary 2.1]).

Corollary 5.1.9. *Suppose that μ is not an eigenvalue of the graph H. There exists a graph G with a star set X for μ such that $G - X = H$ if and only if the vectors $\mathbf{b}_u$ ($u \in X$) satisfy*
   (i) $\langle\!\langle \mathbf{b}_u, \mathbf{b}_u \rangle\!\rangle = \mu$ *for all $u \in X$, and*
   (ii) $\langle\!\langle \mathbf{b}_u, \mathbf{b}_v \rangle\!\rangle \in \{-1, 0\}$ *for all pairs u, v of distinct vertices in X.*

We can now establish a quadratic upper bound for the order of G when $\mu \neq -1$ or 0.

Proposition 5.1.10 [BeRo]. *Let G be a graph of order n, and let μ be an eigenvalue of G, $\mu \notin \{-1, 0\}$. If the eigenspace of μ has codimension t then either (a) $n \leq \frac{1}{2}t(t + 1)$ or (b) $\mu = 1$ and $G = K_2$ or $2K_2$.*

Proof. Suppose first that G is connected. Using the notation of Theorem 5.1.7, we let $S = (B|C - \mu I)$, with columns $\mathbf{s}_u$ ($u = 1, \ldots, n$). Then

$$\mu I - A = S^T (\mu I - C)^{-1} S,$$

and we have, for all vertices u, v of G,

$$\langle\!\langle \mathbf{s}_u, \mathbf{s}_v \rangle\!\rangle = \begin{cases} \mu & \text{if } u = v \\ -1 & \text{if } u \sim v \\ 0 & \text{otherwise} \end{cases}.$$

We define quadratic functions $\varphi_1, \ldots, \varphi_n$ as follows:

$$\varphi_u(\mathbf{x}) = \langle\langle \mathbf{s}_u, \mathbf{x} \rangle\rangle^2 \quad (\mathbf{x} \in \mathbb{R}^t).$$

It is easily checked that if $k = \dim\mathcal{E}(\mu)$ and $\mathbf{x} = (x_{k+1}, \ldots, x_n)^T$ then $\varphi_u(\mathbf{x}) = x_u^2$ $(u = k+1, \ldots, n)$.

We show that $\varphi_1, \ldots, \varphi_n$ are linearly independent unless $\mu = 1$ and $G = K_2$. If μ is the index of G, then $k = 1$ and $\varphi_u(\mathbf{x}) = x_u^2$ $(u = 2, \ldots, n)$. If $\varphi_1, \ldots, \varphi_n$ are linearly dependent, then, since φ_1 is the square of a linear function, φ_1 must be a multiple of one of $\varphi_2, \ldots, \varphi_n$, say of φ_v. The continuity of the functions $\mathbf{x} \mapsto \langle\langle \mathbf{x}, \mathbf{s}_1 \rangle\rangle$ and $\mathbf{x} \mapsto \langle\langle \mathbf{x}, \mathbf{s}_v \rangle\rangle$ ensures that $\langle\langle \mathbf{x}, \mathbf{s}_1 \rangle\rangle$ is a constant multiple of x_v, and therefore $\mathbf{s}_1$ is a multiple of the v-th column of $\mu I - C$. But the entries of $\mathbf{s}_1$ and of C are all either 0 or 1; and since $\mu \neq -1, 0$, we deduce that the vertices 1 and v are adjacent to each other but to no other vertices of G. Since G is connected we have $G = K_2$ and $\mu = 1$.

Now let μ_1 be the index of G, and consider the case in which $\mu \neq \mu_1$. Let $\mathbf{w}$ be an eigenvector of G corresponding to μ_1, with all entries of $\mathbf{w}$ positive. Let $\mathbf{w} = (w_1, \ldots, w_n)^T$, and let $\mathbf{w}^* = (w_{k+1}, \ldots, w_n)^T$. Since $\mathbf{w}$ lies in $\mathcal{E}(\mu)^\perp$, it follows from Theorem 5.1.6 that

$$\langle\langle \mathbf{s}_u, \mathbf{w}^* \rangle\rangle = -w_u \quad (u = 1, \ldots, n).$$

Suppose that $\sum_u \alpha_u \varphi_u = 0$, that is, $\sum_u \alpha_u \langle\langle \mathbf{s}_u, \mathbf{x} \rangle\rangle^2 = 0$ for all $\mathbf{x} \in \mathbb{R}^t$. Taking $\mathbf{x} = \mathbf{s}_i$, we obtain $\mu^2 \alpha_i + \sum_{u \sim i} \alpha_u = 0$ $(i = 1, \ldots, n)$. Thus

$$(\mu^2 I + A)\mathbf{a} = \mathbf{0}, \quad \text{where } \mathbf{a} = (\alpha_1, \ldots, \alpha_n)^T.$$

From $\sum_u \alpha_u \langle\langle \mathbf{s}_u, \mathbf{x} + \mathbf{y} \rangle\rangle^2 = \mathbf{0}$, we obtain $\sum_u \alpha_u \langle\langle \mathbf{s}_u, \mathbf{x} \rangle\rangle \langle\langle \mathbf{s}_u, \mathbf{y} \rangle\rangle = \mathbf{0}$ for all $\mathbf{x}, \mathbf{y} \in \mathbb{R}^t$. Taking $\mathbf{x} = \mathbf{s}_i$ and $\mathbf{y} = \mathbf{w}^*$, we obtain $\mu \alpha_i w_i - \sum_{u \sim i} \alpha_u w_u = 0$ $(i = 1, \ldots, n)$. Thus

$$(\mu I - A)\mathbf{a}' = \mathbf{0}, \quad \text{where } \mathbf{a}' = (\alpha_1 w_1, \ldots, \alpha_n w_n)^T.$$

Because $\mu \neq -1, 0$, we have $\mu \neq -\mu^2$, and so $\mathbf{a}^T \mathbf{a}' = 0$, that is, $\alpha_1^2 w_1 + \cdots + \alpha_n^2 w_n = 0$. It follows that $\alpha_u = 0$ for all u, and so $\varphi_1, \ldots, \varphi_n$ are linearly independent. Now the functions φ_u lie in the space of all homogeneous quadratic functions on $\mathbb{R}^t$, and since this space has dimension $\frac{1}{2}t(t+1)$, we have $n \leq \frac{1}{2}t(t+1)$.

Finally, suppose that G is not connected. It is clear that, for any vertex u, $\varphi_u(\mathbf{x})$ involves only those entries of $\mathbf{x}$ which correspond to vertices in the same component as u. Thus, if in each component the φ_u are linearly independent, then all the φ_u are linearly independent. It follows that the bound holds except possibly when $G = rK_2$ for some r. In this case $n = 2r$, $t = r$, and the inequality holds whenever $r \geq 3$. This completes the proof. $\qquad\square$

The bound in Proposition 5.1.10 is attained in the graph obtained from $L(K_9)$ by switching with respect to K_8: here $\mu = -2$ and $t = 8$. Apart from a few trivial exceptions, the bound is not attained in any regular graph; in fact if G is r-regular, $\mu \notin \{-1, 0, r\}$ and $t > 2$, then $n \leq \frac{1}{2}t(t+1) - 1 = \frac{1}{2}(t-1)(t+2)$ [BeRo, Theorem 3.1]

In practice it is often convenient to write equation (5.2) in the form

$$(5.3) \qquad m(\mu)(\mu I - A_X) = B^T m(\mu)(\mu I - C)^{-1} B$$

where $m(x)$ is the minimal polynomial of C. This is because $m(\mu)(\mu I - C)^{-1}$ is given explicitly as follows. The proof is straightforward.

Proposition 5.1.11. *Let C be a square matrix with minimal polynomial*

$$m(x) = x^{d+1} + c_d x^d + c_{d-1}x^{d-1} + \cdots + c_1 x + c_0.$$

If μ is not an eigenvalue of C then

$$m(\mu)(\mu I - C)^{-1} = a_d C^d + a_{d-1}C^{d-1} + \cdots + a_1 C + a_0 I$$

where $a_d = 1$ and for $0 < i \leq d$,

$$a_{d-i} = \mu^i + c_d \mu^{i-1} + c_{d-1}\mu^{i-2} + \cdots + c_{d-i+1}.$$

If G has H as a star complement for μ, with a corresponding star set X of size k, then the deletion of any r vertices in X results in a graph with μ as an eigenvalue of multiplicity $k - r$. The reason is that the multiplicity of an eigenvalue changes by 1 at most when any vertex is deleted (see Theorem 1.2.21). It follows that each induced subgraph $G - Y$ ($Y \subset X$) also has H as a star complement for μ. Moreover any graph with H as a star complement for μ is an induced subgraph of such a graph G for which X is maximal, because H-neighbourhoods determine adjacencies among vertices in a star set. Accordingly, in determining all the graphs with H as a star complement for μ, it suffices to describe those for which a star set X is maximal. By Proposition 5.1.4(ii), such maximal graphs always exist when $\mu \neq -1$ or 0. In fact, the values -1 and 0 for μ afford little obstruction to arguments using star complements (see, for example, [CvRo2]), but here we shall normally take $\mu = -2$ or 1.

Example 5.1.12. As a simple example, let us find the graphs having a 5-cycle 123451 as a star complement H for -2. In the notation of Proposition 5.1.11, C is the circulant matrix with first row 01001, $\mu = -2$ and $m(x) = (x - 2)$

$(x^2 + x - 1)$. Here $m(\mu) = -4$ and the Proposition yields

$$4(2I + C)^{-1} = C^2 - 3C + 3I = \begin{pmatrix} 5 & -3 & 1 & 1 & -3 \\ -3 & 5 & -3 & 1 & 1 \\ 1 & -3 & 5 & -3 & 1 \\ 1 & 1 & -3 & 5 & -3 \\ -3 & 1 & 1 & -3 & 5 \end{pmatrix}.$$

Now we apply Corollary 5.1.9(i). From equation (5.3) we know that $\langle\langle \mathbf{b}_u, \mathbf{b}_u \rangle\rangle = -2$ if and only if $\mathbf{b}_u^T(C^2 - 3C + 3I)\mathbf{b}_u = 8$. In this situation the neighbours of u in H constitute a set S such that the i-th entry of $\mathbf{b}_u$ is 1 if $i \in S$, 0 if $i \notin S$. Accordingly we have to find the subsets S of $\{1, 2, 3, 4, 5\}$ such that the sum of entries in the principal submatrix of $C^2 - 3C + 3I$ determined by S is equal to 8. It is straightforward to verify that this occurs precisely when $|S| = 4$. All five possibilities for S occur simultaneously in $L(K_5)$, which is therefore the unique maximal graph that arises. The graphs with a 5-cycle as a star complement for -2 are therefore the induced subgraphs of $L(K_5)$ containing C_5. Since $C_5 = L(C_5)$, these graphs are just the graphs $L(G)$, where G is a Hamiltonian graph of order 5. $\qquad\square$

The arguments of Example 5.1.12 have been generalized by F. K. Bell [Bel1] to show that for any odd $t > 3$, $L(K_t)$ is the unique maximal graph with a t-cycle as a star complement for -2. Determination of the possible subsets S requires substantial effort in the general case. An inspection of $L(K_t)$ reveals easily that such sets include those consisting of two pairs of consecutive vertices on the t-cycle, and the work lies in proving that there are no other possibilities for S. (Since the graphs in question have least eigenvalue -2, the forbidden subgraph technique provides an alternative approach here; see [BeSi].) Bell [Bel2] has also investigated the graphs in which the path P_t is a star complement for -2: when $t \geq 3$ and $t \neq 7, 8$, such graphs are precisely the line graphs of bipartite graphs of order $t + 1$ (other than P_{t+1}) which have a Hamiltonian path. See also Theorems 5.4.2, 5.4.4 and 5.4.5.

In Example 5.1.12, there was no need to apply part (ii) of Corollary 5.1.9 because we had prior knowledge of a graph in which all possible vertices were added to the prescribed star complement. We cannot expect that a unique maximal graph will always exist, and in the general case, where a graph H occurs as a star complement for an eigenvalue μ, it is useful to consider a *compatibility graph* defined as follows (cf. [Ell]). The vertices are those $\mathbf{b}_u$ for which $\langle\langle \mathbf{b}_u, \mathbf{b}_u \rangle\rangle = \mu$, and $\mathbf{b}_u$ is adjacent to $\mathbf{b}_v$ if and only if $\langle\langle \mathbf{b}_u, \mathbf{b}_v \rangle\rangle \in \{-1, 0\}$. It is convenient to represent the edge $\mathbf{b}_u \mathbf{b}_v$ by a full line if $\langle\langle \mathbf{b}_u, \mathbf{b}_v \rangle\rangle = -1$, and by a broken line if $\langle\langle \mathbf{b}_u, \mathbf{b}_v \rangle\rangle = 0$. If each vertex $\mathbf{b}_u$ is labelled instead with the

H-neighbourhood of u, then this same graph is called the *extendability graph* $\Gamma(H, \mu)$ (cf. [Row3]). Note that there is a one-one correspondence between cliques in $\Gamma(H, \mu)$ and graphs with H as a star complement for μ; moreover, the full lines in a clique determine the subgraph induced by the corresponding star set. In particular, if we use a computer to find the maximal graphs with H as a star complement for μ, we can invoke an algorithm for finding the maximal cliques in a graph. The next example (taken from [Row3]) illustrates the procedure in a small case.

Example 5.1.13. Here we find the graphs having a 5-cycle 123451 as a star complement H for 1. In this case, Proposition 5.1.11 yields

$$(I - C)^{-1} = 3I - C^2 = \begin{pmatrix} 1 & 0 & -1 & -1 & 0 \\ 0 & 1 & 0 & -1 & -1 \\ -1 & 0 & 1 & 0 & -1 \\ -1 & -1 & 0 & 1 & 0 \\ -0 & -1 & -1 & 0 & 1 \end{pmatrix}.$$

First, we apply Corollary 5.1.9(i). From equation (5.2) we know that $\langle\!\langle \mathbf{b}_u, \mathbf{b}_u \rangle\!\rangle = 1$ if and only if $\mathbf{b}_u^T (3I - C^2) \mathbf{b}_u = 1$. Now we have to find the subsets S of $\{1, 2, 3, 4, 5\}$ such that the sum of entries in the principal submatrix of $3I - C^2$ determined by S is equal to 1. It is straightforward to verify that this occurs if and only if S consists of a single vertex or three consecutive vertices of the 5-cycle. Next we apply part (ii) of Corollary 5.1.9 to construct the extendability graph $\Gamma(C_5, 1)$ shown in Fig. 5.2. The automorphism group of H has three orbits of maximal cliques (with orders 2,3 and 5). These determine the three maximal

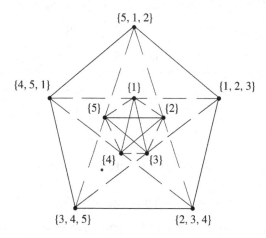

Figure 5.2: The extendability graph $\Gamma(C_5, 1)$

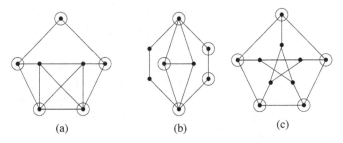

Figure 5.3: The maximal graphs with C_5 as a star complement for 1.

graphs illustrated in Fig. 5.3, where the vertices of H are circled. The Petersen graph has already featured in Example 5.1.2. Alternatively, its occurrence here could have been predicted from Example 5.1.12, where C_5 is a star complement for -2 in $L(K_5)$: since -2 is not a main eigenvalue of $L(K_5)$, we deduce from the Corollary 1.2.14 that $\overline{C_5}$ is a star complement for $-(-2)-1$ in $\overline{L(K_5)}$ — that is, C_5 is a star complement for 1 in the Petersen graph.   □

The set of procedures illustrated above is known as the *star complement technique* for constructing and characterizing graphs with a prescribed star complement for a prescribed eigenvalue. The technique has its origins in papers of M. N. Ellingham [Ell] and P. Rowlinson [Row1], published independently in 1993. We give one further illustration of the technique: the example is relevant to the discussion in Chapter 6 and it gives a hint of the complexities that can arise with the simplest of star complements.

Example 5.1.14. Suppose that K_8 is a star complement H for -2. In this situation we have $C = J - I$, $m(x) = (x+1)(x-7)$ and equation (5.3) becomes

$$9(2I + A_X) = B^T(9I - J)B.$$

Equating (u, u)-entries here, we have $18 = 9h - h^2$, where $h = |\Delta_H(u)|$. Hence $h = 3$ or 6. Equating non-diagonal entries, we find that the following conditions on H-neighbourhoods are necessary and sufficient for the simultaneous addition of two vertices u and v:

$$\text{if } |\Delta_H(u)| = |\Delta_H(v)| = 3 \text{ then } |\Delta_H(u) \cap \Delta_H(v)| \in \{1, 2\};$$

$$\text{if } |\Delta_H(u)| = 3 \text{ and } |\Delta_H(v)| = 6 \text{ then } |\Delta_H(u) \cap \Delta_H(v)| \in \{2, 3\};$$

$$\text{if } |\Delta_H(u)| = |\Delta_H(v)| = 6 \text{ then } |\Delta_H(u) \cap \Delta_H(v)| \in \{4, 5\}.$$

Note that the third condition is satisfied automatically because there $\Delta_H(u)$ and $\Delta_H(v)$ are 6-subsets of an 8-set. For the maximal graphs G with K_8 as a star complement for -2, we need to find the maximal families of 3-sets and 6-sets satisfying the other two conditions. We give just three of many examples of such a family $\mathcal{F}$.

(a) $\mathcal{F}$ consists of all 28 subsets of $V(H)$ of size 6; in this case, the maximal graph G is the graph obtained from $L(K_9)$ by switching with respect to K_8. In fact, G is the unique largest maximal exceptional graph (denoted by $G473$ in Chapter 6).

(b) $\mathcal{F}$ consists of all 21 subsets of size 3 containing a fixed vertex of H; in this case, G is the cone over $L(K_8)$ (a maximal exceptional graph, denoted by $G006$ in Chapter 6).

(c) $\mathcal{F}$ consists of all 7 subsets of size 6 containing a fixed vertex w of H, together with 7 subsets of size 3 which form the lines of a geometry $PG(3, 2)$ on $V(H) \setminus \{w\}$; in this case, G is the unique smallest maximal exceptional graph (the graph denoted by $G001$ in Chapter 6). $\qquad\square$

In order to describe the general form of a maximal family of neighbourhoods in Example 5.1.14, we give some further definitions. Suppose that $\mathcal{F}$ is a family of 3-subsets of $\{1, 2, \ldots, 8\}$, and let $\mathcal{F}^{(2)}$ be the family of 2-sets which are contained in some 3-set of $\mathcal{F}$. We say that $\mathcal{F}$ is an *intersecting* family if $U \cap V \neq \emptyset$ for all $U, V \in \mathcal{F}$; and such a family $\mathcal{F}$ is *complete* if there does not exist an intersecting family of 3-sets $\mathcal{F}_0$ such that $\mathcal{F} \subset \mathcal{F}_0$ and $\mathcal{F}^{(2)} = \mathcal{F}_0^{(2)}$. (For example, if $\mathcal{F} = \{138, 157, 568\}$ then $\mathcal{F}$ is not complete because we can take $\mathcal{F}_0 = \mathcal{F} \cup \{158\}$.) The final result of this section shows that a maximal exceptional graph with K_8 as a star complement for -2 is determined by a complete intersecting family of 3-subsets of $\{1, 2, \ldots, 8\}$, and *vice versa*. Here we take $V(H) = \{1, 2, \ldots, 8\}$ and write $\overline{ij}$ for the complement of $\{i, j\}$ in $V(H)$.

Theorem 5.1.15 [Row4]. *Let G be a graph with K_8 as a star complement for -2, say $H = G - X \cong K_8$. Then G is a maximal exceptional graph if and only if the family of H-neighbourhoods $\Delta_H(u)$ $(u \in X)$ has the form $\mathcal{F}_3 \cup \mathcal{F}_6$ where $\mathcal{F}_3$ is a complete intersecting family of 3-sets and $\mathcal{F}_6 = \{\overline{ij} : ij \notin \mathcal{F}_3^{(2)}\}$.*

M. Lepović has shown by computer that there are exactly 363 maximal exceptional graphs with K_8 as a star complement for -2. We shall see in Chapter 6 that they include all 43 maximal exceptional graphs that are not cones.

5.2 Graph foundations

Let G be a generalized line graph with root multigraph $\hat{H}$, so that $G = L(\hat{H})$. Let μ be an eigenvalue of G, and let Y be a set of edges of $\hat{H}$. We say that Y is a *line star set* for μ in $\hat{H}$ if it is a star set for μ in $L(\hat{H})$. In this situation, $\hat{H} \setminus Y$ (the spanning subgraph of $\hat{H}$ obtained by deleting the edges in Y) is the corresponding *line star complement* for μ in G. In particular, if $\mu = -2$ we call a line star complement a *foundation* for $\hat{H}$.

The results in this section stem from [CvRS4]. We discuss line graphs first.

Example 5.2.1. The graph $L(K_5)$ has spectrum 6, 1^4, $(-2)^5$, and a star complement for -2 has the form $L(F)$ where the foundation F is one of the graphs of Fig. 5.4. Here the graphs are shown in increasing order of the largest eigenvalue (or *index*).   □

Theorem 5.2.2. (i) *Let G be a connected graph. Then the least eigenvalue of $L(G)$ is greater than -2 if and only if G is a tree or an odd-unicyclic graph.*

(ii) *Let G be a connected bipartite graph such that $L(G)$ has least eigenvalue -2. Then the subgraph F of G is a foundation of G if and only if F is a spanning tree for G.*

(iii) *Let G be a connected non-bipartite graph such that $L(G)$ has least eigenvalue -2. Then the subgraph F of G is a foundation of G if and only if F is a spanning subgraph in which each component is an odd-unicyclic graph.*

Proof. By Corollary 2.2.5, the least eigenvalue of $L(G)$ is greater than -2 when G is either a tree or an odd-unicyclic graph. In these cases, the vertex-edge incidence matrix of G has full rank; see Lemma 2.2.1.

Now let G be a connected graph with n vertices and m edges, and suppose that G is neither a tree nor an odd-unicyclic graph. Let k be the minimum number of edges whose removal results in a graph which is either a tree or a graph in which each component is odd-unicyclic. Let E be a set of k edges with this property. If G is bipartite then $k = m - n + 1$ and $G \setminus E$ is a spanning

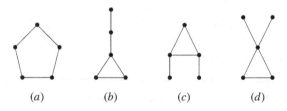

(a)   (b)   (c)   (d)

Figure 5.4: The foundations for K_5.

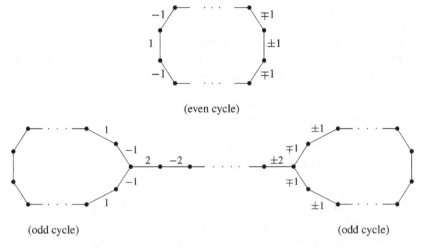

(even cycle)

(odd cycle)   (odd cycle)

Figure 5.5: A construction for eigenvectors of a line graph.

tree, while if G is non-bipartite then $k = m - n$ and each component of $G \setminus E$ is odd-unicyclic. For each edge $e \in E$ we shall define a vector $\mathbf{v}_e$ which is an eigenvector for $L(G)$, with corresponding eigenvalue -2. This will prove statement (i) of the theorem. It follows that a foundation of a connected bipartite graph is a spanning tree, and a foundation of a connected non-bipartite graph is a graph in which every component is odd-unicyclic. Conversely, a subgraph of the type specified in (ii) or (iii) is of the form $G \setminus E$ with $|E| = k$. Now the vectors $\mathbf{v}_e$ $(e \in E)$ turn out to be linearly independent, and so -2 has multiplicity at least k. By interlacing, this multiplicity is precisely k, and $G \setminus E$ is a foundation for G. This is sufficient to complete the proof of statements (ii) and (iii).

It remains to construct the vectors $\mathbf{v}_e$ $(e \in E)$. We fix e and we let x_l $(l \in E(G))$ be the coordinates of $\mathbf{v}_e$. We write F for the foundation $G \setminus E$.

If G is bipartite then $F + e$ contains a unique cycle Z, and Z is of even length. We take x_l to be 1 and -1 for alternate edges l of Z, with $x_e = 1$, and we define $x_l = 0$ for all $l \notin E(Z)$; see Fig. 5.5.

If G is not bipartite and the addition of e to F creates an even cycle Z, then Z is the only even cycle in $F + e$ and we repeat the construction above. Otherwise, the addition of e creates either an odd cycle or a link between two components of F. In either case, a component of $F + e$ has just two cycles, say Z and Z'; they have odd length and are edge-disjoint. Let P be the unique path of least length (possibly zero) between a vertex of Z and a vertex of Z'. If P has non-zero length then we take x_l to be 2 and -2 for alternate edges l of P. Then we take $x_l = \pm 1$ for $l \in E(Z) \cup E(Z')$ as shown for the dumb-bell shape in Fig. 5.5. Finally we define $x_l = 0$ for all remaining edges

l of G. Reversing all signs if necessary, we may take $x_e > 0$ to determine $\mathbf{v}_e$ uniquely.

In all cases, it is straightforward to check that $\mathbf{v}_e$ is an eigenvector of $L(G)$, with corresponding eigenvalue -2. These eigenvectors are linearly independent because the l-coordinate of $\mathbf{v}_e$ is 1 if $e = l$, and 0 if $e \neq l$. $\square$

We call the vectors $\mathbf{v}_e$ ($e \in E$) the eigenvectors of $L(G)$ *constructed from* F. As a corollary of the above proof, we have the following result.

Theorem 5.2.3. *The eigenspace corresponding to the eigenvalue -2 of a line graph $L(G)$ has as a basis the set of eigenvectors constructed from any foundation of G.*

Further details, with an interpretation of Theorem 5.2.3 in a matroid context, may be found in [Doo8]; see also Theorem 2.2.9.

We now turn to generalized line graphs that are not line graphs, and in this context the following definitions will be helpful. An *orchid* is a graph which is either odd-unicyclic or a tree with one petal; an *orchid garden* is a graph whose components are orchids.

Example 5.2.4. Let $\hat{H}$ be the root multigraph of the generalized line graph $L(K_3; 1, 1, 0)$. Thus $\hat{H}$ consists of a triangle with single petals added at two vertices. All non-isomorphic foundations of $\hat{H}$ are shown in Fig. 5.6. Note that each foundation is an orchid garden. $\square$

Figure 5.6: A multigraph and its foundations.

Theorem 5.2.5. *Let the connected multigraph $\hat{H}$ be the root graph of a generalized line graph G which is not a line graph.*

(i) *The graph G has least eigenvalue greater than -2 if and only if $\hat{H}$ is an orchid.*

(ii) *Suppose that the least eigenvalue of G is -2. Then F is a foundation of $\hat{H}$ if and only if F is an orchid garden which spans $\hat{H}$.*

Proof. The proof mirrors that of Theorem 5.2.2. First, it is straightforward to check that if $\hat{H}$ is an orchid then the least eigenvalue of $L(\hat{H})$ is greater than -2. Now suppose that $\hat{H}$ is not an orchid. Let k be the minimum number of edges whose removal from $\hat{H}$ results in an orchid garden, and let E be a set of k edges with this property. If G has the form $L(H; a_1, \ldots, a_n)$, then

$k = m - n + \sum_{i=1}^{n} a_i$, where m is the number of edges of H. Below we define k linearly independent vectors $\mathbf{v}_e$ ($e \in E$), which are eigenvectors of G with corresponding eigenvalue -2. This establishes statement (i) and shows that every foundation is a spanning orchid garden. Conversely, any spanning orchid garden has the form $\hat{H} \setminus E$ with $|E| = k$. By interlacing, $\hat{H} \setminus E$ is a foundation for $\hat{H}$, and statement (ii) follows.

It remains to construct the vectors $\mathbf{v}_e$ ($e \in E$). We fix e and we let x_l ($l \in E(\hat{H})$) be the coordinates of $\mathbf{v}_e$. We write F for the foundation $\hat{H} \setminus E$, and we use the term *supercycle* to mean either an odd cycle or a petal. There are $m - n + \sum_{i=1}^{n} a_i$ edges of $\hat{H}$ not in F, and, since F is an orchid garden, three possibilities arise when such an edge e is added to F: (1) the edge creates an even cycle, (2) the edge creates a supercycle (that is, it creates an odd cycle or doubles one pendant edge), (3) the edge joins a vertex of one orchid to a vertex of another orchid. We now ascribe weights x_l to the edges of $\hat{H}$ as follows.

In case (1) all weights are 0 except for 1 and -1 alternately on edges of the even cycle. In cases (2) and (3), $F + e$ contains a unique shortest path P between vertices of two different supercycles, and we first ascribe weights of 2 and -2 alternately to the edges of P. To within a unique choice of sign, weights are ascribed to the edges of the two supercycles as illustrated in Fig. 5.7, and all remaining weights are 0. (In all cases the construction may be seen as ascribing weights ± 1 alternately to the edges in a closed trail, with the assumption that double edges are assigned the same value; in edges traversed twice, the values are added.)

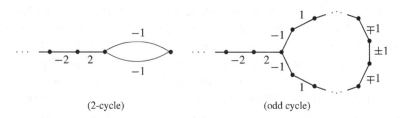

(2-cycle)   (odd cycle)

Figure 5.7: A construction for eigenvectors of a generalized line graph.

In each case, we choose signs so that $x_e > 0$. The weights x_l of edges in $\hat{H}$ are taken as coordinates of a vector $\mathbf{v}_e$ whose entries are indexed by the corresponding vertices of $L(\hat{H})$. It is straightforward to check that each such vector is an eigenvector of $L(\hat{H})$ corresponding to -2. Again, we call the vectors $\mathbf{v}_e$ the eigenvectors of $L(\hat{H})$ *constructed from* F. These $m - n + \sum_{i=1}^{n} a_i$ vectors are linearly independent because each of the aforementioned closed trails contains an edge not present in any of the others. $\qquad\square$

Again we call the vectors $\mathbf{v}_e$ ($e \in E$) the eigenvectors *constructed from* F. Accordingly we have proved the following result (formulated to subsume Theorem 5.2.3).

Theorem 5.2.6. *The eigenspace for the eigenvalue* -2 *of a generalized line graph has as a basis the set of eigenvectors constructed from any foundation of the corresponding root multigraph.*

Remark 5.2.7. In the situation of Theorem 5.2.5 we can construct a foundation F for the root multigraph of $L(H; a_1, \ldots, a_n)$ from a foundation F' of H (if any) as follows. If H is not bipartite then F' is an orchid garden which spans H and we may take F to consist of F' together with a_i (single) pendant edges attached at vertex v_i ($i = 1, \ldots, n$). If H is bipartite then F' is a tree which spans H: here we first modify F' by adding a_i pendant edges at vertex v_i ($i = 1, \ldots, n$) and then obtain F by replacing one of these pendant edges by a petal. Of course, not all foundations for the root multigraph of $L(H; a_1, \ldots, a_n)$ can be constructed in this way.   □

Corollary 5.2.8. *A connected generalized line graph has least eigenvalue equal to* -2 *if and only if the corresponding root multigraph contains either an even cycle or two supercycles connected by a path (possibly of length 0).*

Corollary 5.2.9. *Let* $L(H; a_1, \ldots, a_n)$ *be a connected generalized line graph with least eigenvalue* -2 *and* $(a_1, \ldots, a_n) \neq (0, \ldots, 0)$. *Then* -2 *is a main eigenvalue if and only if the root multigraph of* $L(H; a_1, \ldots, a_n)$ *has an odd cycle or two 2-cycles connected by a path of odd length.*

Proof. Since $\sum_{i=1}^{n} a_i > 0$, the root multigraph $\hat{H}$ of $L(H; a_1, \ldots, a_n)$ has a 2-cycle C. If an odd cycle Z is also present then $\hat{H}$ has a foundation F and an edge e not in F such that both C and Z lie in a component of $F + e$. Now as above we can construct an eigenvector corresponding to -2 not orthogonal to $\mathbf{j}$. Similar remarks apply when $\hat{H}$ contains two 2-cycles connected by a path of odd length. Conversely, if $\hat{H}$ has no odd cycles and no pair of 2-cycles connected by a path of odd length then each vector $\mathbf{v}_e$ in a basis for the eigenspace of -2 is orthogonal to $\mathbf{j}$.   □

Remark 5.2.10. The authors of the papers [CvDS2], [BrMS], [Bra], [CvRS4] established independently, and in quite different settings, some of the properties of eigenvectors corresponding to the eigenvalue -2 in generalized line graphs. While [CvDS2], [CvRS4] continue earlier mathematical research on graphs with least eigenvalue -2, the papers [BrMS], [Bra] deal with a practical problem that arises in relation to the security of statistical databases; specifically, some necessary and sufficient conditions are formulated for a class of query sets in statistical databases to be compromise-free. See [BrCv] for some further results along these lines. We explain briefly the viewpoint of [BrMS], [Bra].

Let B be the vertex-edge incidence matrix of a graph H. It is known from [CvDSa, Theorem 3.38] that a non-zero vector $\mathbf{x}$ is an eigenvector of -2 in $L(H)$ if and only if $B\mathbf{x} = \mathbf{0}$. By Proposition 2.2.2, this necessary and sufficient condition extends to generalized line graphs with a suitably chosen incidence matrix C. In this case the matrix contains also entries equal to -1 (for one of the two edges of each petal; cf. Remark 1.1.7).

Instead of petals, the authors of [BrMS], [Bra] used "semi-edges", i.e. edges having only one endvertex. Then the matrix C contains only one non-zero entry, equal to 1, in a column corresponding to a semi-edge. The eigenspace of the eigenvalue -2 in generalized line graphs is then constructed by finding vectors $\mathbf{x}$ which satisfy $C\mathbf{x} = \mathbf{0}$ for this matrix C. $\qquad\square$

Remark 5.2.10 shows that the theory of graphs with least eigenvalue -2, a well-developed mathematical theory, has some applications beyond mathematics. For a recent application of the theory to convex quadratic programming see [Car].

Remark 5.2.11. It can easily be seen that the eigenspace of an integer eigenvalue of a graph has a basis consisting of vectors with integer coordinates. Suppose that each vector in such a basis has the smallest possible norm. The results of this section show that for generalized line graphs, the vectors in such a basis for the eigenspace of -2 have coordinates with absolute value at most 2. However, Fig. 3.2 and Table A5 show that in exceptional graphs, vectors with coordinates of larger absolute value can appear. This type of problem has been considered in [Haz] in relation to so-called Hodge cycles on certain abelian varieties of CM-type. $\qquad\square$

5.3 Exceptional graphs

Let G be a connected graph with least eigenvalue -2. By Theorem 5.1.6, G has a star complement H for -2 which is connected. By interlacing, the least eigenvalue of H is greater than -2. Now the connected graphs with least eigenvalue greater than -2 have been determined by M. Doob and D. Cvetković [DoCv], whose result appears as Theorem 2.3.20. We have seen in the previous section the role of graphs of types (i) and (ii) from this theorem as star complements of generalized line graphs. The exceptional graphs with least eigenvalue greater than -2 are those appearing in parts (iii)–(v) of the theorem. The main result of this section is the following.

Theorem 5.3.1 [CvRS4]. *Let G be a graph with least eigenvalue -2. Then G is exceptional if and only if it has an exceptional star complement for -2.*

Proof. If G has an exceptional star complement H then by Theorem 2.3.19, H contains as an induced subgraph F one of the graphs $F_1, \ldots, F_{20}$ from Table A2. Then F is an induced subgraph of G. By Proposition 5.1.4(i), G is connected because H is connected. Hence G is exceptional. Conversely, suppose that G is exceptional. By Theorem 2.3.19, G contains an induced subgraph F isomorphic to some F_j $(1 \le j \le 20)$. By Theorem 5.1.6, G has a connected star complement H for -2 which contains F as an induced subgraph. Since H is exceptional, the theorem follows.   □

A consequence of Theorem 5.3.1 and our earlier observations is that one can prove without recourse to root systems that there are only finitely many exceptional graphs: by Proposition 5.1.10, such graphs have order at most 36, and this confirms the bound of Theorem 3.6.7. Moreover we can use Theorem 5.3.1 to construct all of the maximal exceptional graphs. Details are given in Chapter 6, where we work also with representations in E_8; a construction independent of root systems is entirely possible but less efficient.

Computational results, described in [CvLRS1], establish the following relevant facts. There are exactly 10 maximal graphs having one of $F_1, \ldots, F_{20}$ as a star complement for the eigenvalue -2. If we take the 110 exceptional graphs of type (iv) in the role of a star complement for -2 then we find exactly 39 maximal graphs. These comprise one graph on 14 vertices (non-regular with spectrum 8, 1^6, $(-2)^7$), 28 graphs on 17 vertices, five graphs on 18 vertices, one on 19 vertices, two on 20, one on 22 vertices and one on 27 vertices (the Schläfli graph with spectrum 16, 4^6, $(-2)^{20}$). We show that the graphs which arise in this way are not maximal exceptional graphs because they have representations in E_7.

Proposition 5.3.2. *An exceptional graph with a representation in E_6 or E_7 is an induced subgraph of an exceptional graph of larger order.*

Proof. We take E_7 as the set of vectors in E_8 orthogonal to $\frac{1}{2}\mathbf{j}_8$, and E_6 as the set of vectors in E_8 orthogonal to $\frac{1}{2}\mathbf{j}_8$ and $\mathbf{e}_7 + \mathbf{e}_8$. We make use of the vectors mentioned in Remark 3.2.7, namely

$$\mathbf{w} = \frac{1}{4}(\mathbf{e}_1 + \mathbf{e}_2 + \mathbf{e}_3 + \mathbf{e}_4 + \mathbf{e}_5 + \mathbf{e}_6 - 3\mathbf{e}_7 - 3\mathbf{e}_8)$$

and

$$\mathbf{w'} = \frac{1}{3}(\mathbf{e}_1 + \mathbf{e}_2 + \mathbf{e}_3 + \mathbf{e}_4 - 2\mathbf{e}_5 - 2\mathbf{e}_6).$$

First consider a connected graph G with a representation $\mathcal{R}(G)$ in E_6. It is straightforward to check that $\mathbf{u}.\mathbf{w'} \in \{-1, 0, 1\}$ for all $\mathbf{u} \in E_6$; moreover, if $\mathbf{u}$ and $\mathbf{v}$ represent adjacent vertices of G then $\mathbf{u}.\mathbf{w'}$ and $\mathbf{v}.\mathbf{w'}$ cannot have opposite signs. Since G is connected, we may assume that $\mathbf{u}.\mathbf{w'} \in \{0, 1\}$ for all $\mathbf{u} \in \mathcal{R}(G)$. Since G is exceptional, we know that $\mathcal{R}(G)$ has dimension 6, and so $\mathbf{u}.\mathbf{w'} = 1$ for some $\mathbf{u} \in \mathcal{R}(G)$. Now let $\mathbf{n} = \frac{2}{3}\mathbf{w} + \mathbf{w'}$. Note that $\mathbf{n}.\mathbf{n} = 2$ and $\mathbf{n}.\mathbf{u} = \mathbf{w'}.\mathbf{u}$ for all $\mathbf{u} \in \mathcal{R}(G)$. It follows that $\{\mathbf{n}\} \cup \mathcal{R}(G)$ is the representation in E_7 of an exceptional graph containing G as a vertex-deleted subgraph.

Secondly, consider a connected graph with a representation in E_7, but not in E_6. Thus the representation has dimension 7. We may repeat the argument, representing the additional vertex by the vector $\frac{1}{4}\mathbf{j}_8 + \mathbf{w}$ in E_8. $\quad\square$

In view of Proposition 5.3.2, we can find all of the maximal exceptional graphs by taking as a star complement for -2 each of the 443 graphs arising in part (v) of Theorem 2.3.20. The results are described in Chapter 6.

5.4 Characterizations

In this section we discuss characterizations of graphs by star complements for -2 of the form $L(K)$, where K is a tree or an odd-unicyclic graph (cases (i) and (ii) of Theorem 2.3.20). By interlacing, -2 is the least eigenvalue of a graph G with such a star complement, and so it is straightforward to identify the non-exceptional graphs G that arise. In order to obtain characterizations among all graphs, rather than among the non-exceptional graphs, we can use the star complement technique (cf. [CvRS3]).

First, suppose that T is a tree with n vertices and that G is a non-exceptional graph with $L(T)$ as a star complement for -2. By Theorems 5.2.2 and 5.2.6, we know that G is the line graph of a bipartite graph H with T as a spanning tree. The bipartition of H is determined by the unique partition of $V(T)$ into two colour classes. Accordingly, we have the following result.

Theorem 5.4.1. *Let T be a tree on n vertices, and let G be a maximal graph with a star complement $L(T)$ for the eigenvalue -2. If G is not exceptional then $G = L(K_{p,q})$, where $p + q = n$ and $V(T)$ can be partitioned into colour classes of sizes p and q.*

We mention two special cases of Theorem 5.4.1, proved by F. K. Bell [Bel2] using the star complement technique; note that $L(P_{t+1}) = P_t$.

Theorem 5.4.2. *Let t be an integer ≥ 3, and let G be a maximal graph with the path P_t as a star complement for -2.*
(i) *If $t = 2m - 1 \neq 7$ then $G = L(K_{m,m})$.*
(ii) *If $t = 2m \neq 8$ then $G = L(K_{m,m+1})$.*

Turning from trees to orchid gardens as foundations, we begin with the following general observation.

Theorem 5.4.3. *Let G be a connected maximal graph with a star complement $L(K)$ for the eigenvalue -2, where K is an orchid garden on n vertices $(n > 3)$, with p pendant edges and q petals. If G is not exceptional then $G = L(K_s; a_1, \ldots, a_s)$, where $s + \sum_{i=1}^{s} a_i = n$ and $\sum_{i=1}^{s} a_i \leq p + q$.*

Proof. From Section 5.2, we know that G is a generalized line graph, say $L(H; a_1, \ldots, a_s)$, where H has s vertices and $s + \sum_{i=1}^{s} a_i = n$. The inequality $\sum_{i=1}^{s} a_i \leq p + q$ follows from the fact that any pendant edge of K may be one of a pair of double edges in the root multigraph of G. The edges of K not paired in this way belong to H. Lastly, $H = K_s$ because G is maximal. $\qquad\square$

Now let us consider the special case in which K is just an odd cycle C_t ($= L(C_t)$). Again we can find with minimal effort the non-exceptional graphs that arise. If G is a generalized line graph with root multigraph H, then by Theorem 5.4.3, C_t can be a spanning orchid garden of H only when H is a graph with t vertices. It follows that if G is a maximal graph with C_t (odd $t > 3$) as a star complement for -2 then either $G = L(K_t)$ or G is exceptional. The latter possibility is excluded by the following result of F. K. Bell [Bel1] mentioned in Section 5.1.

Theorem 5.4.4. *Let t be an odd integer ≥ 5, and let G be a maximal graph with C_t as a star complement for -2. Then G is the line graph $L(K_t)$.*

When t is even, C_t has -2 as an eigenvalue, and therefore cannot be taken as a star complement for -2. However, Bell also proved the following.

Theorem 5.4.5. *Let t be an even integer ≥ 6, and let G be a maximal graph with $C_r \cup C_s$ as a star complement for -2, where r and s are odd integers ≥ 3 with $r + s = t$. If $t \neq 8$ then G is the line graph $L(K_t)$.*

Here, if $t = 8$ then the star complement consists of a 3-cycle and a 5-cycle. In this case, there are many solutions to equation (5.3) in Section 5.1 (indeed, one yields a Chang graph), and M. Lepović has used a computer implementation of the star complement technique to find all the maximal exceptional graphs that arise; there are 434 such graphs. The conclusion of Theorem 5.4.2 remains true when we take for a star complement any union of odd cycles which does not include $C_3 \,\dot\cup\, C_5$ (see [Bel1, Theorem 3.4]).

Among the odd-unicyclic graphs on t vertices (odd $t > 3$), C_t has least index, whereas the graph with largest index is the graph R_t consisting of a triangle with $t - 3$ pendant edges attached at one vertex. Now $L(R_t)$ is also a star complement for -2 in $L(K_t)$, and this statement is true also for even $t > 3$. However, for no $t > 3$ is $L(K_t)$ the unique maximal graph with $L(R_t)$ as a star complement for -2: it is straightforward to verify (using Theorem 5.2.5(ii)) that if $0 \le p \le t - 3$ then the generalized line graph $L(K_{t-p}; p, 0, \ldots, 0)$ has $L(R_t)$ as a star complement for -2. We shall see that these graphs are maximal, and are the only maximal graphs that arise. We use the star complement technique, which is simpler to apply here than in the case of $L(P_{t+1})$ or $L(C_t)$ because $L(R_t)$ has a small divisor and only 4 distinct eigenvalues.

We work with $L(R_t)$ as a star complement H for -2, with $V(H) = V_1 \,\dot\cup\, V_2 \,\dot\cup\, V_3$, where V_1 consists of the vertex of degree 2, V_2 consists of the 2 vertices of degree $t - 1$, and V_3 consists of the $t - 3$ vertices of degree $t - 2$. Fig. 5.8 shows the allocation of the edges of R_t to the sets V_1, V_2, V_3 (and resolves any ambiguity in the cases $t = 4$, $t = 5$). We say that a vertex is of *type* (a, b, c) if it has a neigbours in V_1, b in V_2 and c in V_3.

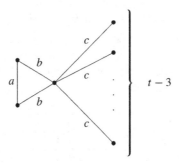

Figure 5.8: A partition of the edges of R_t.

Since $\overline{H}$ has 0 as a main eigenvalue of multiplicity $t - 2$, we know that H has -1 as an eigenvalue of multiplicity $t - 3$. In particular, H has at most four distinct eigenvalues. On the other hand, H has a divisor with matrix

$$D = \begin{pmatrix} 0 & 2 & 0 \\ 1 & 1 & t-3 \\ 0 & 2 & t-4 \end{pmatrix}$$

determined by V_1, V_2, V_3. It follows that if $m(x)$ is the minimal polynomial of H, then $m(x) = (x+1)\det(xI - D)$, that is,

(5.4) $m(x) = x^4 - (t-4)x^3 - (2t-3)x^2 + (t-8)x + 2t - 8$.

Note that $m(-2) = 4$, and that, by Proposition 5.1.11,

$$4(-2I - C)^{-1} = (t-6)I - C - (t-2)C^2 + C^3.$$

In equation (5.3) we take $\mu = -2$ and

$$C = \begin{pmatrix} 0 & J_{1,2} & O_{1,t-3} \\ J_{2,1} & J_{2,2} - I_2 & J_{2,t-3} \\ O_{t-3,1} & J_{t-3,2} & J_{t-3,t-3} - I_{t-3} \end{pmatrix}$$

to obtain

$$m(\mu)(\mu I - C)^{-1} = -4I + \begin{pmatrix} -t+4 & (t-2)J_{1,2} & -2J_{1,t-3} \\ (t-2)J_{2,1} & (-t+4)J_{2,2} & 2J_{2,t-3} \\ -2J_{1,2} & 2J_{t-3,2} & O_{t-3,t-3} \end{pmatrix}.$$

(5.5)

Now let $H + u$ be the graph obtained from H by adding a vertex u of type (a, b, c). We apply Corollary 5.1.9(i) with $X = \{u\}$ to find the values of (a, b, c) for which $H + u$ has -2 as an eigenvalue. From equation (5.3) we have

(5.6)
$$-8 = -4(a+b+c) + (-t+4)(a^2 + b^2)$$
$$- 4ac + 4bc + 2(t-2)ab,$$

and we consider in turn the six possibilities for (a, b). We find that u is one of the following types listed in Table 1.

all $t \geq 5$	$t = 8$
$(0, 0, 2)$	$(1, 0, 0)$
$(1, 1, 1)$	$(0, 1, b)\ (0 \leq b \leq t - 3)$
$(0, 2, t-4)$	$(1, 2, b)\ (0 \leq b \leq t - 3)$

Table 1

Thus when $t \neq 8$, $H + u$ has -2 as an eigenvalue if and only if u is of type $(0, 0, 2)$, $(1, 1, 1)$ or $(0, 2, t-4)$. Moreover we know that all $\binom{t-3}{2}$ possible vertices of type $(0, 0, 2)$ and all $2(t-3)$ possible vertices of type $(1, 1, 1)$ occur simultaneously in $L(K_t)$. To determine the vertices v which can occur simultaneously with a vertex u of type $(0, 2, t-4)$ we equate (u, v)-entries in equation

(5.3). Let $\rho_{uv} = |\overline{\Gamma}(u) \cap \overline{\Gamma}(v)|$, where $\overline{\Gamma}(u)$ denotes the $\overline{X}$-neighbourhood of u. We obtain

(5.7)
$$-4a_{uv} = -4\rho_{uv} + (-t+4)(a_u a_v + b_u b_v) - 2(a_u c_v + a_v c_u)$$
$$+ 2(b_u c_v + b_v c_u) + (t-2)(a_u b_v + a_v b_u).$$

It follows that if v is of type $(0, 0, 2)$ a necessary and sufficient condition for $H + u + v$ to have -2 as a double eigenvalue is that $a_{uv} = \rho_{uv} - 2$. This equation is satisfied if and only if $u \not\sim v$ and $\overline{\Gamma}(v) \subseteq \overline{\Gamma}(u)$. If v is of type $(1, 1, 1)$ we again obtain $a_{uv} = \rho_{uv} - 2$, an equation which is satisfied if and only if $u \not\sim v$ and $\overline{\Gamma}(v) \cap V_2 \subseteq \overline{\Gamma}(u)$. If v is of type $(0, 2, t-4)$ then we obtain $a_{uv} = \rho_{uv} - (t-4)$, that is, $a_{uv} = 1$, and so there is no constraint on the $\overline{X}$-neighbourhoods of u and v in this case. Now suppose that G is maximal, with precisely p vertices of type $(0, 2, t-4)$ in X, constituting the subset X' $(0 \leq p \leq t-3)$. In view of the constraints identified above, $X \setminus X'$ consists of all vertices v of type $(1, 1, 1)$ or $(0, 0, 2)$ such that $\overline{\Gamma}(v) \cap V_2 \subseteq \overline{\Gamma}(u)$ for all $u \in X'$.

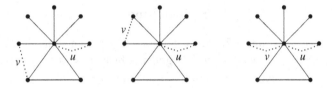

Figure 5.9: Three types of edge v

We can now show that G is the generalized line graph $L(K_{t-p}; p, 0, 0, \ldots, 0)$, that is, $G = L(H_{t-p,p})$ where $H_{t-p,p}$ is the multigraph obtained from K_{t-p} by adding p petals at a single vertex. Fig. 5.9 shows the three possible ways in which $H + u + v$ arises as a line graph when u is of type $(0, 2, t-4)$. It is now clear that when there are exactly p vertices of type $(0, 2, t-4)$ all other possible vertices occur simultaneously in $L(H_{t-p,p})$. Thus we have the following result.

Theorem 5.4.6. *If G is a maximal graph with $L(R_t)$ $(4 \leq t \neq 8)$ as a star complement for -2, then $G = L(K_{t-p}; p, 0, 0, \ldots, 0)$ $(0 \leq p \leq t-3)$.*

Again the value $t = 8$ is anomalous, and in Section 5.5 we describe how exceptional graphs can be constructed from $L(R_8)$ as a star complement for -2.

For odd $t > 3$, the graphs C_t and R_t are extremal with respect to the lexicographical ordering by spectral moments of odd-unicyclic graphs with t vertices. It is interesting to observe what happens when we replace R_t with the next graph in the ordering. This is the graph Q_t $(t \geq 4)$, obtained from a triangle by adding $t - 4$ pendant edges at one vertex and one pendant edge at another. The following result is proved in [CvRS3] by the star complement technique.

Theorem 5.4.7. *If G is a maximal graph with $L(Q_t)$ $(4 \leq t \neq 8)$ as a star complement for -2 then $G = L(K_{t-p-q}; p, q, 0, \ldots, 0)$ $(0 \leq p \leq t - 4, 0 \leq q \leq 1)$.*

It follows from Remark 1.2.11 that if -2 is a non-main eigenvalue of G then 1 is an eigenvalue of $\overline{G}$ with the same multiplicity. This fact is exploited in [CvRS3] to deal with the graphs $\overline{L(R_t)}$ $(t \geq 5)$ and $\overline{L(Q_t)}$ $(t \geq 6)$ as star complements for 1. Further results concerning 1 as an eigenvalue of $\overline{G}$ may be found in Chapter 7.

Finally we note that the graphs which arise as a star complement for -2, in some graph with least eigenvalue -2, are characterized in a forthcoming paper by F. K. Bell, E. M. Li Marzi and S. K. Simić [BLMS].

5.5 Switching

We have seen in Example 3.6.5, by considering representations in E_8, how exceptional graphs arise as cones over graphs which are switching-equivalent to $L(K_8)$. In this section we discuss this phenomenon in the context of star complements. We see how the eigenvalue -2 is related to cones, how $L(R_8)$ arises as a star complement for a large class of graphs, and how $L(R_8)$ can itself be switched to obtain an exceptional star complement common to many exceptional graphs.

We start with the general situation: suppose that X is a star set for the eigenvalue μ of the graph G, and let H be the star complement $G - X$. We say that the vertex u of X is *amenable to switching* if μ is an eigenvalue of the graph obtained from $H + u$ by switching with respect to $\{u\}$. In the notation of Section 5.1, we have $\langle\!\langle \mathbf{b}_u, \mathbf{b}_u \rangle\!\rangle = \mu$, and u is amenable to switching if and only if also $\langle\!\langle \mathbf{j} - \mathbf{b}_u, \mathbf{j} - \mathbf{b}_u \rangle\!\rangle = \mu$, equivalently,

$$(5.8) \qquad\qquad \langle\!\langle \mathbf{j}, \mathbf{j} \rangle\!\rangle = 2\langle\!\langle \mathbf{j}, \mathbf{b}_u \rangle\!\rangle.$$

Now suppose that μ is not a main eigenvalue of G. By Theorem 5.1.7, $\langle\!\langle \mathbf{j}, \mathbf{b}_u \rangle\!\rangle = -1$ for all $u \in X$. From equation (5.8), we see that every vertex in X is amenable if and only if $\langle\!\langle \mathbf{j}, \mathbf{j} \rangle\!\rangle = -2$. In this situation, we say that X is *amenable to switching*. We can now prove the following result.

Theorem 5.5.1 [RoJa]. *Let μ be a non-main eigenvalue of G, let X be a star set for μ in G, and suppose that X is amenable to switching. If G' is obtained from*

G by switching with respect to a subset of X, then μ is a non-main eigenvalue of G' and X is a star set for μ in G'.

Proof. Suppose that G' is obtained from G by switching with respect to the subset Y of X. We know already that $\langle\!\langle \mathbf{j} - \mathbf{b}_u, \mathbf{j} - \mathbf{b}_u \rangle\!\rangle = \mu$ for all $u \in Y$. It is straightforward to check that if $u, v \in Y$ then $\langle\!\langle \mathbf{j} - \mathbf{b}_u, \mathbf{j} - \mathbf{b}_v \rangle\!\rangle = \langle\!\langle \mathbf{b}_u, \mathbf{b}_v \rangle\!\rangle$, while if $u \in Y$ and $v \in X \setminus Y$ then $\langle\!\langle \mathbf{j} - \mathbf{b}_u, \mathbf{b}_v \rangle\!\rangle = -1 - \langle\!\langle \mathbf{b}_u, \mathbf{b}_v \rangle\!\rangle$. It now follows from Corollary 5.1.9 that X is a star set for μ in G'. Finally, by Theorem 5.1.7, μ is a non-main eigenvalue of G' because $\langle\!\langle \mathbf{j}, \mathbf{j} - \mathbf{b}_u \rangle\!\rangle = -1$ for all $u \in Y$ and $\langle\!\langle \mathbf{j}, \mathbf{b}_v \rangle\!\rangle = -1$ for all $v \in X \setminus Y$. $\qquad\square$

Now consider the special case $\mu = -2$, where -2 is a non-main eigenvalue of G. In this case, we have the following result.

Proposition 5.5.2 [RoJa]. *Let -2 be a non-main eigenvalue of G, let X be a star set for -2 in G, and let $H = G - X$. Then X is amenable to switching if and only if -2 is an eigenvalue of the cone over H. In this situation, H is a star complement for -2 in the cone over any graph G' obtained from G by switching with respect to a subset of X.*

Proof. It follows from equation (5.8) that X is amenable to switching if and only if $\langle\!\langle \mathbf{j}, \mathbf{j} \rangle\!\rangle = -2$. By Corollary 5.1.9(i), this holds if and only if -2 is an eigenvalue of the cone over H. In this situation, we have $\langle\!\langle \mathbf{j}, \mathbf{b}_u \rangle\!\rangle = \langle\!\langle \mathbf{j}, \mathbf{j} - \mathbf{b}_u \rangle\!\rangle = -1$ for all $u \in X$ and so, if G' obtained from G by switching with respect to a subset of X, then -2 is an eigenvalue of both the cone over G and the cone over G'. $\qquad\square$

Note that in the situation of Proposition 5.5.2, X cannot contain vertices with complementary H-neighbourhoods because $\langle\!\langle \mathbf{b}_u, \mathbf{j} - \mathbf{b}_u \rangle\!\rangle = 1$.

Since -2 is a non-main eigenvalue of $L(K_8)$ we can now see in a different way how exceptional graphs arise as cones over graphs switching-equivalent to $L(K_8)$. In particular, the following example shows how exceptional graphs can arise in the exceptional case $t = 8$ of Theorem 5.4.6.

Example 5.5.3. Consider $L(R_8)$ as a star complement H for -2, with corresponding star set X. From Table 1 of Section 5.4 we know that a vertex of type $(1, 2, 5)$ can be added to H, equivalently $\langle\!\langle \mathbf{j}, \mathbf{j} \rangle\!\rangle = -2$. Thus X is amenable to switching if and only if $\langle\!\langle \mathbf{j}, \mathbf{b}_u \rangle\!\rangle = -1$ for all $u \in X$. By equation (5.4), a vertex u of type (a, b, c) satisfies this condition if and only if $-6a + 4b - 2c = -4$, that is, if and only if (a, b, c) is one of $(0, 0, 2), (1, 1, 1), (0, 1, 4), (1, 2, 3)$. We may add the 20 vertices of type $(0, 0, 2)$ or $(1, 1, 1)$ to obtain $L(K_8)$. Any vertex of type $(0, 0, 2)$ can be switched to a vertex of type $(1, 2, 3)$, and any vertex of type $(1, 1, 1)$ can be switched to a vertex of type $(0, 1, 4)$. Finally, by

Proposition 5.5.2 (or in virtue of equations (5.6) and (5.7)), we may add the vertex of type $(1, 2, 5)$ to obtain a cone over a graph switching-equivalent to $L(K_8)$. In this way we can construct the cones over many graphs switching-equivalent to $L(K_8)$. We shall see in Chapter 6 that a large proportion of these exceptional graphs of order 29 are maximal exceptional graphs.   □

Example 5.5.3 illustrates a procedure for constructing exceptional graphs as extensions of a non-exceptional star complement for -2. They are graphs of the form $K_1 \triangledown G'$, where G' is obtained from $L(K_8)$ by switching with respect to a subset of X. By Theorem 5.3.1, each such graph has a star complement for -2 which is exceptional. If G' is not $L(K_8)$ itself then we may take H' to be the graph U_8 obtained from K_7 by adding a vertex of degree 5. To see this, note that G' contains a vertex v of type $(0, 1, 4)$ or $(1, 2, 3)$, and then the vertices in $\{v\} \cup V_2 \cup V_3$ induce a subgraph isomorphic to U_8. (The graph U_8 is one of the exceptional graphs arising in part (v) of Theorem 2.3.20.) We can obtain a further graph (G_{28}, say) with U_8 as a star complement for -2 by switching $L(K_8)$ with respect to $V_2 \cup V_3$. (Then $L(R_8)$ is switched to U_8.) We may repeat the procedure of Example 5.5.3 with G_{28} and U_8 in place of $L(K_8)$ and $L(R_8)$, respectively, but this merely replicates the graphs G' above. However, we shall see in Chapter 6 that every maximal exceptional graph which is not a cone has U_8 as a star complement for -2.

6

The maximal exceptional graphs

This chapter is based on the recent publications [CvLRS2],[CvLRS3],[CvRS5] and [CvRS6]. In Section 6.1 we describe the outcome of the exhaustive computer search for the maximal exceptional graphs, obtained as extensions of 443 exceptional star complements. Inspection of these graphs motivates the results proved in Section 6.2 using representations in E_8. It is convenient to partition the maximal exceptional graphs into three types: (a) those which are cones of order 29, (b) those with maximal degree 28 and more than 29 vertices, (c) those with maximal degree less than 28. (We know from Chapter 3 that no vertex has degree greater than 28.) Our results reveal how the computing requirements can be drastically reduced. In particular, we show in Section 6.3 that the graphs of type (b) can be found as extensions of just one star complement, denoted by U_8. In Sections 6.4 and 6.5 we determine the graphs of type (c) without recourse to a computer, and in Section 6.6 we see that it follows that all graphs of type (b) or (c) have both U_8 and K_8 as a star complement for -2.

6.1 The computer search

We know from Theorem 5.3.1 that every maximal exceptional graph has an exceptional star complement for -2, and from Proposition 5.3.2 that such a star complement is one of the 443 graphs of order 8 described in part (v) of Theorem 2.3.20. Here, these star complements are labelled $H001$ to $H443$, ordered lexicographically by spectral moments; the graphs are listed in Table A2. A computer implementation of the star complement technique was used by M. Lepović in 1998–99 to find all the maximal exceptional graphs as extensions of one of $H001, \ldots, H443$. The same graph can arise as extensions of different star complements, or as different extensions of the same star complement, and so the computer program necessarily includes an isomorphism check. (As an example of the effort required to deal with the difficult case of one particular star complement, it took a PC-586 computer about 24 hours to produce all of the 1 048 580 maximal graphs that arise, and to sort them into 457 isomorphism classes.) For each maximal exceptional graph, the output consists of an adjacency matrix, the spectrum, and the list of those graphs among $H001, \ldots, H443$ which appear as a star complement for -2.

It transpires that there are only 473 non-isomorphic maximal exceptional graphs, and they are listed in Table A6 as $G001, \ldots, G473$. They are ordered first by the number of vertices, secondly by the largest eigenvalue, and thirdly by the sequence of vertex-degrees. The distribution over the number of vertices is as follows:

number of vertices	22	28	29	30	31	32	33	34	36
number of graphs	1	1	432	25	7	3	1	2	1

Note that in this book the graphs $H001, \ldots, H443$ and $G001, \ldots, G473$ have an ordering and a labelling different from that in the original paper [CvLRS2]; the correspondence between old and new labels is given in the technical report [CvLRS3].

We comment in turn on (i) vertex degrees, (ii) particular graphs that arise, (iii) spectra, (iv) the star complements involved. First, none of the maximal exceptional graphs is regular. Of those with 29 vertices, all are cones except for two, namely $G425$ and $G430$. There are 467 maximal exceptional graphs with maximal degree 28, and in all of these graphs, the vertices of degree 28 induce a clique. The graphs $G001$, $G002$, $G425$, $G430$, $G455$ and $G463$ have maximal degree less than 28, and their vertex-degree sequences are as follows:

$$
\begin{array}{ll}
G001 & 16^{14}, 7^8 \\
G002 & 22^7, 16^{14}, 10^7 \\
G425 & 26, 24^2, 18^{16}, 12^8, 10^2 \\
G430 & 26^2, 22, 18^{16}, 14^6, 10^4 \\
G455 & 26^2, 24^1, 20^8, 17^8, 16^1, 14^2, 13^4, 11^4 \\
G463 & 26^3, 22^4, 19^8, 16^4, 15^6, 12^6
\end{array}
$$

Note that there are 430 maximal exceptional graphs of type (a), 37 of type (b), and 6 of type (c). Representations in E_8 of the six graphs of type (c) are given in Section 6.4.

The graph $G006$ is the cone over $L(K_8)$ while $G003$, $G004$ and $G005$ are cones over the Chang graphs (see Example 1.1.12 or Table A3). The graph $G434$ is the double cone $K_2 \triangledown G$ where G is the Schläfli graph. The graph $G473$

is obtained from $L(K_9)$ by switching with respect to a clique of order 8; for some time, it was the only known example of a maximal exceptional graph (see [CaGSS] and Section 3.5).

In all cases, maximal exceptional graphs with the same index are cospectral. Table A7 shows the index of each graph and so the cospectral maximal exceptional graphs are easily identified.

The graph $G001$ was found by W. A. Bridges and R. A. Mena [BrMe] as an example of a non-regular graph with just three distinct eigenvalues. The exceptional graphs with this property were found by E. R. van Dam [Dam2], and the maximal ones are as follows, with the spectra shown:

$G001$	$14, \ 2^7, (-2)^{14}$
$G003 - G005, \ G434$	$14, \ 4^7, (-2)^{21}$.
$G473$	$21, \ 5^7, (-2)^{28}$

Among the maximal exceptional graphs of order $\neq 29$, the integral graphs other than G001 and G473 are as follows, with the spectra shown:

$G002$	$17, \ 5, \ 3^6, (-2)^{20}$
$G466$	$19, \ 5^2, \ 4^4, \ 1, (-2)^{23}$
$G471, \ G472$	$20, \ 5^4, \ 4^3, (-2)^{26}$

Among the maximal exceptional graphs of order 29, all except $G425$ and $G430$ have a spectrum of the form $x, \ 4^6, \ y, (-2)^{21}$, where $x + y = 18$; it follows that the number of edges is given by $252 - xy$. For a computer-free proof of this result, using the Seidel spectrum, see [Cve13, Theorem 3.3]. The graphs $G425$ and $G430$ are the only maximal exceptional graphs of order 29 which are not cones, and their spectra (with non-integer eigenvalues correct to 4 decimal places) are:

$G425$	$17.3899, \ 4^5, \ 3, \ 1.6101, (-2)^{21}$
$G430$	$17.5887, \ 4^5, \ 3.4113, \ 1, (-2)^{21}$.

It follows that, among the maximal exceptional graphs of order 29, the integral graphs are the 67 graphs with integral index. Five different spectra occur here, and the graphs are readily identified from Table A7.

The 25 maximal graphs on 30 vertices have spectra of the form $x, \ y, \ 4^5, \ z, \ (-2)^{22}$, where $x + y + z = 24$. (For a computer-free proof of this result, see [Cve13, Theorem 3.4].) We have noted that the two 34-vertex graphs $G471$ and $G472$ share the spectrum $20, \ 5^4, \ 4^3, \ (-2)^{26}$. Since they also have the same main angles, we know from Proposition 1.2.13 that their complements are cospectral with eigenvalues $15, \ 1^{25}, \ -1, \ (-5)^3, \ (-6)^4$. However, $G471$ and $G472$ have different angles because they have different vertex-degrees. In both graphs the diameter is 2, and all vertices have eccentricity 2.

Finally, we note some facts about star complements. Each of the graphs $H424, H425, H431, H433, H435, H436, H437, H439, H440$ appears as a star complement for -2 in every maximal exceptional graph other than $G006$. Other star complements generate fewer maximal graphs. Thirty-nine of the graphs $H001, \ldots, H443$ appear as star complements in $G006$ (the cone over the non-exceptional graph $L(K_8)$). Each of the graphs $H001, H009$ and $H023$ generates 14 maximal graphs, while other exceptional star complements generate a larger number of maximal graphs. Each of the maximal graphs which is not a cone of order 29 has $H443$ as a star complement.

Further details arising from the computer investigation of maximal exceptional graphs can be found in the technical report [CvLRS3]. The Appendix to this book includes many of the results, but omits data on the extendability graphs and the maximal exceptional graphs that arise from each exceptional star complement.

6.2 Representations in E_8

We begin this section by using representations of maximal exceptional graphs in E_8 to establish, without recourse to a computer, some of the properties evident from (or suggested by) the computer results of the previous section. We also develop a first means of reducing the computing effort needed to determine the maximal exceptional graphs of type (b).

We know from Chapter 3 that if G is a maximal exceptional graph with adjacency matrix A then $8I + 4A$ is the Gram matrix of of a subset of the following set of 240 vectors:

> *type a*: 28 vectors of the form $\mathbf{a}_{ij} = 2\mathbf{e}_i + 2\mathbf{e}_j$; $i, j = 1, \ldots, 8$, $i < j$;
> *type a'*: 28 vectors opposite to those of type a;
> *type b*: 28 vectors of the form $\mathbf{b}_{ij} = -2\mathbf{e}_i - 2\mathbf{e}_j + \sum_{s=1}^{8} \mathbf{e}_s$;
> *type b'*: 28 vectors opposite to those of type b;
> *type c*: 56 vectors of the form $\mathbf{c}_{ij} = 2\mathbf{e}_i - 2\mathbf{e}_j$; $i, j = 1, \ldots, 8$, $i \neq j$;
> *type d*: 70 vectors of the form $\mathbf{d}_{ijkl} = -2\mathbf{e}_i - 2\mathbf{e}_j - 2\mathbf{e}_k - 2\mathbf{e}_l + \sum_{s=1}^{8} \mathbf{e}_s$,
>   for distinct $i, j, k, l \in \{1, \ldots, 8\}$;
> *type e*: 2 vectors $\mathbf{e}$ and $-\mathbf{e}$, where $\mathbf{e} = \sum_{s=1}^{8} \mathbf{e}_s$.

Note that if $\mathcal{R}(G)$ is a representation of G in E_8 then so is $-\mathcal{R}(G) = \{-\mathbf{u} : \mathbf{u} \in \mathcal{R}(G)\}$. Since also the graph $G(\overline{E}_8)$ is transitive (cf. Proposition 3.2.4), we may take $\mathbf{e}$ to represent any chosen vertex of G. When $\mathbf{e} \in \mathcal{R}(G)$, no vector of type a', b' features in the representation; moreover a second representation is given by $\phi(\mathcal{R}(G))$ where the involutory map ϕ is defined by: $\phi(\mathbf{e}) = \mathbf{e}$, $\phi(\mathbf{a}_{ij}) = \mathbf{b}_{ij}$, $\phi(\mathbf{b}_{ij}) = \mathbf{a}_{ij}$, $\phi(\mathbf{c}_{ij}) = \mathbf{c}_{ji}$ $(= -\mathbf{c}_{ij})$, $\phi(\mathbf{d}_{ijkl}) = \mathbf{d}_{\overline{ijkl}}$ $(= -\mathbf{d}_{ijkl})$. In fact, $\phi(\mathcal{R}(G)) = -\mathcal{R}'(G)$, where $\mathcal{R}'(G)$ is obtained from $\mathcal{R}(G)$ by a reflection in the hyperplane orthogonal to $\mathbf{e}$. We refer to $\mathcal{R}(G)$ and $\phi(\mathcal{R}(G))$ as *dual* representations. (One benefit of this duality is that we may assume if necessary that the number of vectors of type b in $\mathcal{R}(G)$ does not exceed the number of vectors of type a.)

In the next three propositions we provide a computer-free proof of some of the properties noted in the previous section. The first observation was anticipated in the proof of Theorem 3.6.7.

Proposition 6.2.1. *If v is any vertex of a graph which can be represented in E_8, then $\deg(v) \leq 28$; and if $\deg(v) = 28$ then the neighbours of v induce a subgraph switching-equivalent to $L(K_8)$.*

Proof. Since the graph $G(\overline{E}_8)$ is transitive, we may assume that v is represented by the vector $\mathbf{e}$. Every neighbour of v is represented by a vector whose inner product with $\mathbf{e}$ is 4, hence by a vector of type a or b. For each pair $\{i, j\}$ the vectors $\mathbf{a}_{ij}$ and $\mathbf{b}_{ij}$ are incompatible because they have inner product -4. Hence there are at most 28 neighbours of v. Moreover, if there are exactly 28 neighbours then for each pair $\{i, j\}$, either $\mathbf{a}_{ij}$ or $\mathbf{b}_{ij}$ is present. If we now switch with respect to the vertices represented by the vectors of type b, we obtain $L(K_8)$, represented by the 28 vectors of type a. $\square$

We shall frequently identify vertices with the vectors in E_8 which represent them, as in the following proof.

Proposition 6.2.2. *In an exceptional graph, the vertices of degree 28, if any, induce a clique.*

Proof. If G is an exceptional graph with two non-adjacent vertices of degree 28, we may take one to be $\mathbf{e}$ and the other to be a vector $\mathbf{v}$ of type c or d. Then $\mathbf{v}$ has inner product 4 with just six vectors of type a and just six vectors of type b. Since G has at most 36 vertices (Theorem 3.6.7), $\mathbf{v}$ has degree at most $(36 - 29) + 12$, a contradiction. $\square$

Proposition 6.2.3. *In a maximal exceptional graph, no vertex has degree 27.*

Proof. We may take $\mathbf{e}$ to be a vertex of degree < 28. By maximality there is at least one pair $\{i, j\}$ such that the vectors $\mathbf{a}_{ij}$ and $\mathbf{b}_{ij}$ are excluded by the presence of certain vectors of type c or d. Now the vectors of type c or d which exclude $\mathbf{a}_{ij}$ are those in the set A_{ij} comprising $\mathbf{c}_{hi}$ $(h \neq i, j)$, $\mathbf{c}_{hj}$ $(h \neq i, j)$ and the vectors $\mathbf{d}_{pqrs}$ for which $\{i, j\} \subseteq \{p, q, r, s\}$. Those which exclude $\mathbf{b}_{ij}$ are those in the set B_{ij} comprising $\mathbf{c}_{ik}$ ($k \neq i, j$), $\mathbf{c}_{jk}$ ($k \neq i, j$) and the vectors $\mathbf{d}_{pqrs}$ for which $\{i, j\} \cap \{p, q, r, s\} = \emptyset$. Note that the inner product of any vector in A_{ij} with any vector in B_{ij} is non-positive; in particular, two non-adjacent vectors $\mathbf{v}$ and $\mathbf{w}$, each of type c or d, must be present. These vectors $\mathbf{v}$ and $\mathbf{w}$ exclude $\mathbf{a}_{ij}$ and $\mathbf{b}_{ij}$ for two pairs $\{i, j\}$. (Explicitly, $\mathbf{c}_{12}$ and $\mathbf{c}_{34}$ exclude $\mathbf{a}_{23}$, $\mathbf{b}_{23}$ and $\mathbf{a}_{14}$, $\mathbf{b}_{14}$; the vectors $\mathbf{c}_{12}$, $\mathbf{d}_{5678}$ exclude $\mathbf{a}_{23}$, $\mathbf{b}_{23}$ and $\mathbf{a}_{24}$, $\mathbf{b}_{24}$; the vectors $\mathbf{c}_{56}$, $\mathbf{d}_{5678}$ exclude $\mathbf{a}_{57}$, $\mathbf{b}_{57}$ and $\mathbf{a}_{58}$, $\mathbf{b}_{58}$; and the vectors $\mathbf{d}_{3456}$, $\mathbf{d}_{5678}$ exclude $\mathbf{a}_{34}$, $\mathbf{b}_{34}$ and $\mathbf{a}_{78}$, $\mathbf{b}_{78}$.) It follows that $\mathbf{e}$ has at most 26 neighbours. □

It follows from Proposition 6.2.1 that a maximal exceptional graph of type (a) is a cone over a graph switching-equivalent to $L(K_8)$. We have seen in Chapter 3 that all such cones are exceptional graphs, but it turns out that not all are maximal exceptional graphs, and so our next task is to establish a criterion for the maximality of such a cone. This is a step towards a description of the maximal graphs of type (b). Let $G(P)$ denote the cone over the graph obtained from $L(K_8)$ by switching with respect to the edge-set $E(P)$, where P is a spanning subgraph of K_8. Thus for each edge ij of P the vector $\mathbf{a}_{ij}$ is replaced by $\mathbf{b}_{ij}$. We define properties (I), (II) of P as follows:

(I) P has a 4-clique and a 4-coclique on disjoint sets of vertices,
(II) P has six vertices adjacent to a seventh and non-adjacent to the eighth.

These configurations are called *dissections* of P of type I or II. Clearly, a dissection of type I yields a partition of the vertex set into two subsets of cardinality 4, while a dissection of type II yields a partition into subsets of cardinalities 6 and 2. Notice that (i) if P has a dissection of type II then P or $\overline{P}$ has an isolated vertex, (ii) P can have at most 5 dissections of type I, and (iii) P can have at most 7 dissections in all. (Note that (iii) provides another proof of the fact that an exceptional graph of type (b) has at most 36 vertices.)

Now we can characterize the cones which are not maximal. Theorems 6.2.4 and 6.2.5 below are due to D. Cvetković, M. Lepović, P. Rowlinson and S. K. Simić.

Theorem 6.2.4 [CvLRS2]. *The graph $G(P)$ is maximal if and only if P cannot be dissected (that is, P has neither property (I) nor property (II)).*

Proof. The graph $G(P)$ is not maximal if and only if a vector of type c or d can be added to the 29 vectors of types a, b and e described in the proof of Proposition 6.2.1. The presence of a vector $\mathbf{v}$ of type c or d excludes 6 vectors of type a and 6 vectors of type b, and so forces the presence of 6 vectors of type b and 6 vectors of type a (along with one of type e and one of type c or d). The 6 vectors of type a that are included correspond to 6 edges of P, and the 6 vectors of type a that are excluded correspond to 6 edges of $\overline{P}$. These conditions are reflected in property (I) or (II) according as **v** is of type d or c.   □

Note that a maximal exceptional graph of type (a) can be represented in the form $G(P)$ for several graphs P. For each of the 430 graphs G of type (a), Table A6.1 gives a graph P which yields G in accordance with Theorem 6.2.4. Next we formulate a more general result.

Theorem 6.2.5 [CvLRS2]. *Let G be an exceptional graph with $29 + k$ vertices ($k \geq 0$), and suppose that G has a vertex u of degree 28. Let Y be the set of vertices not adjacent to u. Then G is a maximal exceptional graph if and only if $G - Y$ is isomorphic to a cone $G(P)$ in which P has exactly k dissections.*

Proof. As before we take u to be the vector **e**, so that u and its 28 neighbours of type a or b induce a subgraph $G(P)$. An exceptional graph may be obtained from $G(P)$ by adding a vertex of type c or d if and only if P admits the corresponding dissection. Accordingly, it suffices to show that any two vertices which may be added individually may also be added simultaneously. Suppose that the additional vectors are **v** and **w**, each of type c or d. There are three cases to consider: (i) **v** and **w** are both of type c, (ii) **v** and **w** are both of type d, (iii) **v** is of type c and **w** is of type d. In each case we suppose by way of contradiction that **v** and **w** cannot be added simultaneously, that is, the inner product of **v** and **w** is negative. We cannot have $\mathbf{w} = -\mathbf{v}$ because $\mathbf{v}.\mathbf{a}_{ij} = -4$ for some ij: then $\mathbf{b}_{ij}$ is present, and $-\mathbf{v}.\mathbf{b}_{ij} = -4$.

Accordingly, in case (i) we may take $\mathbf{v} = \mathbf{c}_{12}$, $\mathbf{w} = \mathbf{c}_{23}$ without loss of generality; but then both $\mathbf{a}_{13}$ and $\mathbf{b}_{13}$ are excluded, a contradiction. In case (ii) we may take $\mathbf{v} = \mathbf{d}_{1234}$, $\mathbf{w} = \mathbf{d}_{4567}$ without loss of generality; but then both $\mathbf{a}_{23}$ and $\mathbf{b}_{23}$ are excluded, a contradiction. In case (iii) we may take $\mathbf{v} = \mathbf{c}_{12}$, $\mathbf{w} = \mathbf{d}_{1345}$ without loss of generality; but then both $\mathbf{a}_{26}$ and $\mathbf{b}_{26}$ are excluded, a contradiction.   □

In view of Theorem 6.2.5, we can determine the maximal exceptional graphs of type (b) by first finding the graphs P which have a dissection. Note that (i) P and $\overline{P}$ have the same dissections, (ii) $G(P) = G(\overline{P})$. Therefore, it suffices to take just one representative from $\{P, \overline{P}\}$, and we say that maximal exceptional graphs of the type described in Theorem 6.2.5 are generated by sets $\{P, \overline{P}\}$. Since at least one of the graphs P, $\overline{P}$ is connected, we can always select a connected representative of this set.

A computer catalogue of the connected graphs on 8 vertices was used to establish that 350 of them have a dissection, and of these, 207 have a disconnected complement. Among the remaining 143 graphs, 3 are self-complementary, and so we obtain $207 + 3 + 140/2 = 280$ representative graphs. Now we choose representatives with at most 14 edges, as listed in Table A6.2. Taking these graphs in turn for P, we can construct the corresponding extensions of $G(P)$, thereby obtaining an independent computer construction of the 37 maximal exceptional graphs of type (b). The results, from [CvLRS3], are reproduced in Table A6.3.

Examples 6.2.6. (i) If $P = K_{1,7}$ then P has exactly 7 dissections, all of type II, and the graph $G473$ is obtained by adding 7 vectors of type c to $G(P)$. The graph $G473$ can also be obtained from $G(P)$ when $P = K_1 \nabla (K_4 \dot\cup 3K_1)$. No other set $\{P, \overline{P}\}$ generates $G473$.

(ii) There are four sets $\{P, \overline{P}\}$ that generate $G472$: we may take P to be one of $K_1 \dot\cup (K_1 \nabla (K_4 \dot\cup 2K_1))$, $K_1 \dot\cup (K_5 \nabla 2K_1)$, $K_1 \dot\cup ((K_{1,3} \dot\cup K_1) \nabla K_2)$, $K_1 \dot\cup ((2K_1 \dot\cup K_2) \nabla K_3)$. In each case, P has exactly 5 dissections, and we obtain $G472$ by adding the 5 corresponding vectors to $G(P)$.

(iii) If $P = K_5 \dot\cup 3K_1$ then P has exactly 5 dissections, all of type I, and $G471$ is obtained from $G(P)$ by adding 5 vectors of type d. Two other sets $\{P, \overline{P}\}$ generate $G471$, given by $P = K_1 \dot\cup (K_4 \nabla 3K_1)$, $P = K_1 \dot\cup ((K_3 \dot\cup 2K_1) \nabla 2K_1)$.

(iv) If $P = K_1 \nabla (P_3 \dot\cup 4K_1)$ then P has exactly 4 dissections and we obtain $G470$ by adding the 4 corresponding vectors to $G(P)$. Six other sets $\{P, \overline{P}\}$ generate $G470$. $\qquad\square$

6.3 A versatile star complement

In this section we take an alternative approach to the construction of the maximal exceptional graphs of type (b). We prove (independently of the computer search) that all of these graphs can be constructed from $H443$ as a common star complement for -2. We have already encountered the graph $H443$ in Section 5.5 as a graph U_8 switching-equivalent to $L(R_8)$; it can be constructed from K_7 by adding a vertex of degree 5. Thus $H443$ is the unique graph with one vertex of degree 5, two vertices of degree 6 and five vertices of degree 7. If G is a graph with a star complement H isomorphic to $H443$, and corresponding star set X, then we say that a vertex of X is of *type pqr* if it has p, q, r neighbours in H of degree 5, 6, 7 in H, respectively.

Theorem 6.3.1. *If the cone $G(P)$ is not maximal then it has $H443$ as a star complement for -2, and an induced subgraph $S(K_{1,7})$ obtained from $H443$ by adding all vertices of types 115 and 024.*

Proof. As before, we take $\mathbf{e}$ to be a vertex of degree 28. Recall from the previous section that $G(P)$ also contains 6 vectors of type a, 6 vectors of type b and a vector $\mathbf{v}$ of type c or d. Now consider a 15th vector $\mathbf{w}$ of type a or b, chosen from one of the remaining 16 such vectors in $G(P)$. By duality we may assume that $\mathbf{w}$ is of type a. We show that the 15 vertices in question can be partitioned into sets U and V such that U induces $H443$, $U \overset{.}{\cup} V$ induces $S(K_{1,7})$, and the vertices in V are of type 115 or 024.

Suppose first that $\mathbf{v}$ is of type c, say $\mathbf{v} = \mathbf{c}_{12}$. Then $G(P)$ contains the vectors $\mathbf{a}_{13}, \mathbf{a}_{14}, \mathbf{a}_{15}, \mathbf{a}_{16}, \mathbf{a}_{17}, \mathbf{a}_{18}, \mathbf{b}_{23}, \mathbf{b}_{24}, \mathbf{b}_{25}, \mathbf{b}_{26}, \mathbf{b}_{27}, \mathbf{b}_{28}$, and without loss of generality $\mathbf{w}$ is $\mathbf{a}_{12}$ or $\mathbf{a}_{34}$. If $\mathbf{w} = \mathbf{a}_{12}$ then we may take

$$U = \{\mathbf{a}_{12}, \mathbf{b}_{23}, \mathbf{b}_{24}, \mathbf{a}_{15}, \mathbf{a}_{16}, \mathbf{a}_{17}, \mathbf{a}_{18}, \mathbf{e}\}, \ V = \{\mathbf{a}_{13}, \mathbf{a}_{14}, \mathbf{b}_{25}, \mathbf{b}_{26}, \mathbf{b}_{27}, \mathbf{b}_{28}, \mathbf{c}_{12}\}.$$

If $\mathbf{w} = \mathbf{a}_{34}$ then we may take

$$U = \{\mathbf{a}_{34}, \mathbf{b}_{23}, \mathbf{b}_{24}, \mathbf{b}_{25}, \mathbf{b}_{26}, \mathbf{b}_{27}, \mathbf{b}_{28}, \mathbf{e}\}, \ V = \{\mathbf{a}_{13}, \mathbf{a}_{14}, \mathbf{a}_{15}, \mathbf{a}_{16}, \mathbf{a}_{17}, \mathbf{a}_{18}, \mathbf{c}_{12}\}.$$

If $\mathbf{v}$ is of type d, say $\mathbf{v} = \mathbf{d}_{5678}$, then $G(P)$ contains the vectors $\mathbf{a}_{56}, \mathbf{a}_{57}, \mathbf{a}_{58}, \mathbf{a}_{67}, \mathbf{a}_{68}, \mathbf{a}_{78}, \mathbf{b}_{12}, \mathbf{b}_{13}, \mathbf{b}_{14}, \mathbf{b}_{23}, \mathbf{b}_{24}, \mathbf{b}_{34}$. Without loss of generality, $\mathbf{w} = \mathbf{a}_{45}$ and we may take

$$U = \{\mathbf{a}_{45}, \mathbf{b}_{57}, \mathbf{b}_{58}, \mathbf{b}_{78}, \mathbf{a}_{14}, \mathbf{a}_{24}, \mathbf{a}_{34}, \mathbf{e}\},$$
$$V = \{\mathbf{b}_{67}, \mathbf{b}_{68}, \mathbf{b}_{56}, \mathbf{a}_{12}, \mathbf{a}_{13}, \mathbf{a}_{23}, \mathbf{d}_{5678}\}.$$

In each case the vectors in U are listed in non-decreasing order of their degrees as vertices of $H443$. It is straightforward to check that, in each case, V consists of two vertices of type 115 (the first two vectors listed) and five vertices of type 024. They are necessarily all of the possible vertices of these types, and so this completes the proof of the theorem. □

It follows from Theorem 6.3.1 that to describe the maximal exceptional graphs of type (b), it suffices to determine the maximal star sets X with at least 22 vertices, including the seven vertices of the types prescribed. It follows from Corollary 5.1.9(i) that the possibilities for vertices in X are: 1 of type 125, 2 of type 115, 10 of type 114, 10 of type 023, 5 of type 024, 10 of type 102, 10 of type 103, 20 of type 012, 10 of type 011 and 5 of type 001. Necessary and sufficient conditions for the simultaneous addition of vertices are obtained using Corollary 5.1.9(ii), and these are listed in Table 1. (For example, all vertices of type 023 may be added simultaneously; any vertex of type 023 is compatible with any vertex of type 024; a vertex of type 023 and a vertex of type 102 are compatible if and only if they have at least one common neighbour; no vertex of type 001 can be added when the vertex of type 125 is present.)

In Table 2 we describe just some of the ways in which vertices of various types may be added to $H443$ to obtain the graphs of type (b) with more than 31 vertices. These examples were constructed by inspection of adjacency matrices in the computer output.

Number	1	2	10	10	5	10	10	20	10	5
Type	125	115	114	023	024	102	103	012	011	001
125	–	all	all	all	all	all	all	all	all	none
115		all	all	all	all	all	all	all	(b)	all
114			all	all	all	all	all	(b)	(a)	(a)
023				all	all	(a)	(b)	(b)	all	(a)
024					all	(a)	all	all	(a)	all
102						all	(b)	(a)	all	(a)
103							all	(a)	(a)	all
012								(a)	(a)	all
011									all	(a)
001										all

(a) at least one common neighbour

(b) at least two common neighbours

Table 1

Type	125	115	114	023	024	102	103	012	011	001
$G473$	1	2	10	10	5	0	0	0	0	0
	1	2	10	0	5	0	10	0	0	0
$G472$	1	2	10	1	5	0	7	0	0	0
	1	2	10	7	5	0	1	0	0	0
	1	2	10	9	3	1	0	0	0	0
	1	2	8	9	5	0	0	1	0	0
$G471$	1	2	10	4	5	0	4	0	0	0
$G470$	1	2	10	8	2	2	0	0	0	0
	0	2	4	3	5	0	3	6	0	2
$G469$	1	2	10	3	5	0	3	0	0	0
$G468$	1	2	10	3	5	0	3	0	0	0
$G467$	1	2	10	2	5	0	4	0	0	0

Table 2

Note that it may or may not be necessary to describe the H-neighbourhoods explicitly. For example, the graph $G472$ may be obtained by adding all vertices of the types 125, 115, 114 and 024, together with one vertex of type 023 and seven vertices of type 103. From the restrictions in Table 1, the H-neighbourhoods are determined uniquely to within a labelling of vertices (and the corresponding vertices in X constitute a maximal star set). On the other hand, to distinguish $G468$ and $G469$ it is necessary to identify essentially different families of H-neighbourhoods.

6.4 Graphs with maximal degree less than 28

In Sections 6.4 and 6.5 we provide a computer-free proof that there are precisely six maximal exceptional graphs with maximal degree less than 28 (cf. [CvRS5, CvRS6]). We represent such a graph G by a subset $\mathcal{R}(G)$ of the 240 vectors of length $2\sqrt{2}$ listed in Section 6.2. In view of the transitivity of $G(\overline{E}_8)$, we can assume that $\mathbf{e}$ represents a vertex of maximal degree, and in this case we call $\mathcal{R}(G)$ a *standard* representation. Note that the dual of a standard representation is again a standard representation. The six graphs that arise are $G001$, $G002$, $G425$, $G430$, $G455$ and $G463$, and we give standard representations of these graphs below.

- $G001$ (22 vertices, with degrees 16^{14}, 7^8; the vertices of degree 16 induce the cocktail party graph $\overline{7K_2}$, while those of degree 7 form a coclique)
  $\mathbf{a}_{ij}$ ($ij = 12, 13, 14, 15, 23, 24, 26, 34, 37, 48$);
  $\mathbf{b}_{ij}$ ($ij = 56, 57, 58, 67, 68, 78$);
  $\mathbf{c}_{ij}$ ($ij = 15, 26, 37, 48$); $\mathbf{d}_{5678}$; $\mathbf{e}$.
- $G002$ (28 vertices, with degrees 22^7, 16^{14}, 10^7; the vertices of degree 10 form a coclique)
  $\mathbf{a}_{ij}$ ($ij = 12, 13, 14, 17, 18, 23, 25, 27, 28, 36, 37, 38, 78$);
  $\mathbf{b}_{ij}$ ($ij = 45, 46, 47, 48, 56, 57, 58, 67, 68$);
  $\mathbf{c}_{ij}$ ($ij = 14, 25, 36$); $\mathbf{d}_{4567}$, $\mathbf{d}_{4568}$; $\mathbf{e}$.
- $G425$ (29 vertices, with degrees 26^1, 24^2, 18^{16}, 12^8, 10^2)
  $\mathbf{a}_{ij}$ ($ij = 12, 15, 16, 17, 18, 25, 26, 27, 28, 57, 68$);
  $\mathbf{b}_{ij}$ ($ij = 13, 24, 34, 35, 36, 37, 38, 45, 46, 47, 48, 56, 58, 67, 78$);
  $\mathbf{c}_{12}$, $\mathbf{c}_{24}$; $\mathbf{e}$.
- $G430$ (29 vertices, with degrees 26^2, 22^1, 18^{16}, 14^6, 10^4)
  $\mathbf{a}_{ij}$ ($ij = 12, 15, 16, 17, 18, 25, 26, 27, 28$);
  $\mathbf{b}_{ij}$ ($ij = 13, 24, 34, 35, 36, 37, 38, 45, 46, 47, 48, 56, 57, 58, 67, 68, 78$);
  $\mathbf{c}_{13}$, $\mathbf{c}_{24}$; $\mathbf{e}$.
- $G455$ (30 vertices, with degrees 26^2, 24^1, 20^8, 17^8, 16^1, 14^2, 13^4, 11^4)
  $\mathbf{a}_{ij}$ ($ij = 12, 15, 16, 17, 18, 25, 26, 27, 28, 56$);
  $\mathbf{b}_{ij}$ ($ij = 13, 24, 34, 35, 36, 37, 38, 45, 46, 47, 48, 57, 58, 67, 68, 78$);
  $\mathbf{c}_{13}$, $\mathbf{c}_{24}$; $\mathbf{d}_{3478}$; $\mathbf{e}$.
- $G463$ (31 vertices, with degrees 26^3, 22^4, 19^8, 16^4, 15^6, 12^6)
  $\mathbf{a}_{ij}$ ($ij = 12, 15, 16, 17, 18, 25, 26, 27, 28, 56, 67$);
  $\mathbf{b}_{ij}$ ($ij = 13, 24, 34, 35, 36, 37, 38, 45, 46, 47, 48, 57, 58, 68, 78$);
  $\mathbf{c}_{13}$, $\mathbf{c}_{24}$; $\mathbf{d}_{3458}$, $\mathbf{d}_{3478}$; $\mathbf{e}$.

These standard representations are not unique; indeed others arise in the course of our arguments, and in such cases we specify an isomorphism with one of the above graphs, the isomorphism constructed using Corollary 5.1.8.

In a standard representation $\mathcal{R}(G)$ of an exceptional graph G, the following are the pairs of vectors which are incompatible because they have inner product -4.

(CC) $\mathbf{c}_{ij}$ and $\mathbf{c}_{jk}$,

(DD) $\mathbf{d}_{ijkl}$ and $\mathbf{d}_{i'j'k'l'}$ whenever $|\{i, j, k, l\} \cap \{i', j', k', l'\}| \leq 1$,

(AB) $\mathbf{a}_{ij}$ and $\mathbf{b}_{ij}$,

(AC) $\mathbf{a}_{ij}$ and $\mathbf{c}_{hj}$ $(h \neq i, j)$, $\mathbf{a}_{ij}$ and $\mathbf{c}_{hi}$ $(h \neq i, j)$,

(AD) $\mathbf{a}_{uv}$ and $\mathbf{d}_{ijkl}$ whenever $\{u, v\} \subseteq \{i, j, k, l\}$,

(BC) $\mathbf{b}_{ij}$ and $\mathbf{c}_{ik}$ $(k \neq i, j)$, $\mathbf{b}_{ij}$ and $\mathbf{c}_{jk}$ $(k \neq i, j)$,

(BD) $\mathbf{b}_{uv}$ and $\mathbf{d}_{ijkl}$ whenever $\{u, v\} \cap \{i, j, k, l\} = \emptyset$,

(CD) $\mathbf{c}_{uv}$ and $\mathbf{d}_{ijkl}$ whenever $\{u, v\} \cap \{i, j, k, l\} = \{u\}$.

The following consequence is a reformulation of Proposition 6.2.3.

Lemma 6.4.1. *If for some pair i, j the vectors $\mathbf{a}_{ij}$ and $\mathbf{b}_{ij}$ are absent from a standard representation $\mathcal{R}(G)$ of a maximal exceptional graph G then $\mathcal{R}(G)$ includes vectors $\mathbf{v}$ and $\mathbf{w}$ such that $\mathbf{e}$, $\mathbf{v}$, $\mathbf{w}$ are pairwise orthogonal.*

Proof. By the maximality of G, $\mathbf{a}_{ij}$ and $\mathbf{b}_{ij}$ are excluded by the presence of certain vectors, which in view of the complete list of incompatibilities above, are of type c or d. Now the vectors of type c or d which exclude $\mathbf{a}_{ij}$ are those in the set A_{ij} comprising $\mathbf{c}_{hi}$ $(h \neq i, j)$, $\mathbf{c}_{hj}$ $(h \neq i, j)$ and the vectors $\mathbf{d}_{pqrs}$ for which $\{i, j\} \subseteq \{p, q, r, s\}$. Those which exclude $\mathbf{b}_{ij}$ are those in the set B_{ij} comprising $\mathbf{c}_{ik}$ $(k \neq i, j)$, $\mathbf{c}_{jk}$ $(k \neq i, j)$ and the vectors $\mathbf{d}_{pqrs}$ for which $\{i, j\} \cap \{p, q, r, s\} = \emptyset$. Note that the inner product of any vector in A_{ij} with any vector in B_{ij} is non-positive; in particular, two orthogonal vectors $\mathbf{v}$ and $\mathbf{w}$, each of type c or d, must be present. Since these vectors are orthogonal to $\mathbf{e}$ the lemma is proved. $\qquad\square$

Henceforth we consider a standard representation $\mathcal{R}(G)$ of a maximal exceptional graph G with maximal degree less than 28. By Proposition 3.2.4, $W(E_8)$ acts as a rank 3 group on $G(\overline{E}_8)$, and so in Lemma 6.4.1 we may assume that $\mathbf{v}$ is a vector of type c; moreover, $W(E_7)$ acts as a rank 3 group on $G(\overline{E}_7)$ and so we may further assume that $\mathbf{w}$ is of type c. Now let θ be the maximum number of pairwise orthogonal vectors of type c in $\mathcal{R}(G)$, and note that $2 \leq \theta \leq 4$. We analyse the cases $\theta = 4, \theta = 3$ in this section, and the case $\theta = 2$ in Section 6.5.

When $\theta = 4$ we find that G is $G001$; when $\theta = 3$ we find that G is $G002$; and when $\theta = 2$ we find that G is one of $G001$, $G002$, $G425$, $G430$, $G455$, $G463$. As before, we identify vertices of G with the corresponding vectors in $\mathcal{R}(G)$.

Lemma 6.4.2. *If $\theta = 4$ then G is the graph $G001$.*

Proof. Without loss of generality, $\mathcal{R}(G)$ contains the vectors $c_{15}, c_{26}, c_{37}, c_{48}$.

In view of the incompatibilities (AC), (BC), (CC) the further possible vectors of types a, b, c in $\mathcal{R}(G)$ are

$$a_{ij} \ (ij = 12, 13, 14, 23, 24, 34, 15, 26, 37, 48);$$

$$b_{ij} \ (ij = 15, 26, 37, 48, 56, 57, 58, 67, 68, 78);$$

and

$$c_{ij} \ (ij = 16, 17, 18, 25, 27, 28, 35, 36, 38, 45, 46, 47).$$

Moreover if d_{ijkl} is present then $|\{i, j, k, l\} \cap \{5, 6, 7, 8\}| \geq 2$. (For example, neither d_{1234} nor d_{2345} is compatible with c_{26}.) It follows that d_{5678} is compatible with all possible vectors, hence is present by maximality. Now d_{5678} is adjacent to each of $c_{15}, c_{26}, c_{37}, c_{48}$, and is adjacent to all possible neighbours of e except a_{ij} and b_{ij} ($ij = 15, 26, 37, 48$).

Recall now that $\deg(e) \geq \deg(d_{5678})$, while for given ij at most one of a_{ij}, b_{ij} is present. It follows that (i) $\deg(e) = \deg(d_{5678})$; (ii) one of a_{ij}, b_{ij} is present for each $ij = 15, 26, 37, 48$; (iii) no further vectors of type c are present (for any such vector would be adjacent to d_{5678}); (iv) similarly, if another vector d_{ijkl} is present then $|\{i, j, k, l\} \cap \{5, 6, 7, 8\}| = 2$.

It follows from (iv) by (CD) that the only possible vectors of type d are d_{ijkl} for $ijkl = 1256, 1357, 1458, 2367, 2468, 3478$.

Next we show that either all a_{ij} ($ij = 15, 26, 37, 48$) are present or all b_{ij} ($ij = 15, 26, 37, 48$) are present. Without loss of generality, suppose by way of contradiction that a_{15} and b_{26} are present. Then the vectors d_{ijkl} ($ijkl = 1256, 1357, 1458, 3478$) are excluded and d_{2678} is compatible with all of the possible vectors remaining; but then by maximality d_{2678} is present, a contradiction.

The presence of a_{ij} ($ij = 15, 26, 37, 48$) or b_{ij} ($ij = 15, 26, 37, 48$) now excludes all possible vectors of type d other than d_{5678}, and so there remain just two possible maximal sets of 22 pairwise compatible vectors. By duality we may assume that the number of vectors of type b does not exceed the number of vectors of type a. Accordingly just one graph arises, namely the graph $G001$ defined above. $\qquad\square$

Lemma 6.4.3. *If $\theta = 3$ then G is the graph G002.*

Proof. Without loss of generality, suppose that $\{c_{14}, c_{25}, c_{36}\}$ is a largest set of pairwise orthogonal vectors of type c. In view of the incompatibilities (AC), (BC), (CC) the further possible vectors of types a, b, c in $\mathcal{R}(G)$ are

$$a_{ij} \ (ij = 12, 13, 17, 18, 23, 27, 28, 37, 38, 78, 14, 25, 36);$$
$$b_{ij} \ (ij = 45, 46, 47, 48, 56, 57, 58, 67, 68, 78, 14, 25, 36);$$

and

$$c_{ij} \ (ij = 15, 16, 17, 18, 24, 26, 27, 28, 34, 35, 37, 38, 74, 75, 76, 84, 85, 86).$$

Moreover, if d_{ijkl} is present then by (CD) either $|\{i, j, k, l\} \cap \{4, 5, 6\}| \geq 2$ or $ijkl \in \{1478, 2578, 3678\}$.

Now the compatible vectors d_{4567}, d_{4568} are compatible with all possible vectors, and are therefore present by maximality.

We show next that the vectors a_{12}, a_{13}, a_{23} and b_{45}, b_{46}, b_{56} are all present. If a_{12} is absent it must be excluded by d_{1245}, and if b_{45} is absent it must be excluded by d_{3678}. (The reasons are that a_{12}, b_{45} are compatible with all possible vectors of type c, while any vector of type d which is present must be compatible with c_{14} and c_{25}.) If d_{1245} is present then d_{3678} is absent and so a_{45} is present; but then $\deg(b_{45}) > \deg(e)$ since $a_{14}, a_{25}, b_{67}, b_{68}, b_{78}, b_{36}$ are excluded. Similarly, if d_{3678} is present then b_{12} is present and $\deg(a_{12}) > \deg(e)$. In either case we have a contradiction and so a_{12}, b_{45} are present. Similarly, a_{13}, b_{46} are present and a_{23}, b_{56} are present.

Let

$$S = \{a_{ij} : ij = 37, 38, 78, 14, 25, 36\} \cup \{b_{ij} : ij = 67, 68, 78, 14, 25, 36\},$$
$$T = \{a_{ij} : ij = 13, 17, 18, 23, 27, 28\} \cup \{b_{ij} : ij = 46, 47, 48, 56, 57, 58\},$$

and let α be the number of adjacencies between $\{a_{12}, b_{45}\}$ and vectors of type c or d.

Note that the elements of $T \cap \Delta(e)$, together with $e, c_{14}, c_{25}, d_{4567}, d_{4568}$, are adjacent to both a_{12} and b_{45}; while those of $S \cap \Delta(e)$, together with $\{a_{12}, b_{45}\}$, are adjacent to just one of a_{12}, b_{45}. It follows that:

$$\deg(a_{12}) + \deg(b_{45}) = |S \cap \Delta(e)| + 2|T \cap \Delta(e)| + 4 + \alpha.$$

Now

$$2\deg(e) = 2|S \cap \Delta(e)| + 2|T \cap \Delta(e)| + 4 \geq \deg(a_{12}) + \deg(b_{45})$$

and so $|S \cap \Delta(e)| \geq \alpha$. On the other hand, $\alpha \geq 8$ and $|S \cap \Delta(e)| \leq 8$, whence $|S \cap \Delta(e)| = \alpha = 8$.

It follows that (i) $\deg(\mathbf{e}) = \deg(\mathbf{a}_{12}) = \deg(\mathbf{b}_{45})$; (ii) $\mathbf{a}_{37}$, $\mathbf{a}_{38}$, $\mathbf{b}_{67}$, $\mathbf{b}_{68}$ are present, and either $\mathbf{a}_{ij}$ or $\mathbf{b}_{ij}$ is present for each $ij = 14, 25, 36, 78$.

If both $\mathbf{a}_{14}$ and $\mathbf{a}_{25}$ are present then so are $\mathbf{a}_{36}$ and $\mathbf{a}_{78}$, because $\deg(\mathbf{e}) = \deg(\mathbf{a}_{12}) = \deg(\mathbf{b}_{45})$. Similarly, if both $\mathbf{b}_{14}$ and $\mathbf{b}_{25}$ are present then so are $\mathbf{b}_{36}$ and $\mathbf{b}_{78}$. Identical arguments hold when we apply the permutation (123)(456) to subscripts, and we conclude that either $\mathbf{a}_{ij}$ ($ij = 14, 25, 36, 78$) are present or $\mathbf{b}_{ij}$ ($ij = 14, 25, 36, 78$) are present. Moreover all of $\mathbf{a}_{12}$, $\mathbf{a}_{13}$, $\mathbf{a}_{23}$, $\mathbf{b}_{45}$, $\mathbf{b}_{46}$, $\mathbf{b}_{56}$ have the same degree as $\mathbf{e}$. It follows that there are no further vectors of type c, and no vectors of type d other than $\mathbf{d}_{4567}$, $\mathbf{d}_{4568}$. The 28 vectors which remain are $\mathbf{a}_{12}$, $\mathbf{b}_{45}$, $\mathbf{c}_{14}$, $\mathbf{c}_{25}$, $\mathbf{c}_{36}$, $\mathbf{d}_{4567}$, $\mathbf{d}_{4568}$, $\mathbf{e}$, the 8 vectors in $S \cap \Delta(\mathbf{e})$ and the 12 vectors in T. By duality we may assume that the number of vectors of type b does not exceed the number of vectors of type a. Accordingly just one graph arises, namely the graph $G002$ defined above. $\square$

We may summarize the results of this section as follows.

Theorem 6.4.4. *Let G be a maximal exceptional graph with maximal degree less than 28. If G has a standard representation in E_8 which includes three pairwise orthogonal vectors of type c then G is one of the graphs $G001$ or $G002$.*

Proof. Here $\theta = 3$ or 4, and so the result follows immediately from Lemmas 6.4.2 and 6.4.3. $\square$

6.5 The last subcase

In this section we complete the proof of the following result.

Theorem 6.5.1. *If G is a maximal exceptional graph in which every vertex has degree less than 28 then G is one of the six graphs $G001, G002, G425, G430, G455$ and $G463$.*

As before, we consider a standard representation $\mathcal{R}(G)$ of G. We know from Theorem 6.4.4 that if $\mathcal{R}(G)$ contains three pairwise orthogonal vectors of type c then G is one of the graphs $G001$, $G002$. Accordingly, we may suppose, without loss of generality, that $\{\mathbf{c}_{13}, \mathbf{c}_{24}\}$ is a largest set of pairwise orthogonal vectors of type c. In view of the incompatibilities (AC), (BC), (CC) the further possible vectors of types a, b, c in $\mathcal{R}(G)$ are

$\mathbf{a}_{ij}$ ($ij = 13, 24; 12; 15, 16, 17, 18, 25, 26, 27, 28; 56, 57, 58, 67, 68, 78$);

$\mathbf{b}_{ij}$ ($ij = 13, 24; 34; 35, 36, 37, 38, 45, 46, 47, 48; 56, 57, 58, 67, 68, 78$);

and

$$\mathbf{c}_{ij} \quad (ij = 14, 15, 16, 17, 18, 23, 25, 26, 27, 28, 53, 54, 63, 64, 73, 74, 83, 84).$$

Lemma 6.5.2. *The vectors $\mathbf{a}_{12}$ and $\mathbf{b}_{34}$ are present.*

Proof. If $\mathbf{a}_{12}$ is absent then it is excluded by a vector of type d compatible with $\mathbf{c}_{13}$ and $\mathbf{c}_{24}$, and this is necessarily $\mathbf{d}_{1234}$. Similarly, if $\mathbf{b}_{34}$ is absent then it is excluded by $\mathbf{d}_{5678}$. Since $\mathbf{d}_{1234}$ and $\mathbf{d}_{5678}$ are incompatible at least one of $\mathbf{a}_{12}$, $\mathbf{b}_{34}$ is present. If only $\mathbf{a}_{12}$ is present then $\mathbf{d}_{5678}$ is present and so the further possible vectors of type a or b are:

$$\mathbf{a}_{ij} \quad (ij = 13, 24, 12, 15, 16, 17, 18, 26, 26, 27, 28)$$

and

$$\mathbf{b}_{ij} \quad (ij = 35, 36, 37, 38, 45, 46, 47, 48; 56, 57, 58, 67, 68, 78).$$

Thus $\mathbf{a}_{12}$ is adjacent to all other neighbours of $\mathbf{e}$, as well as to $\mathbf{d}_{5678}$ and $\mathbf{c}_{13}$. Then $\deg(\mathbf{a}_{12}) > \deg(\mathbf{e})$, a contradiction. If only $\mathbf{b}_{34}$ is present then we obtain similarly the contradiction $\deg(\mathbf{b}_{34}) > \deg(\mathbf{e})$. Consequently both $\mathbf{a}_{12}$ and $\mathbf{b}_{34}$ are present.   □

Let us now introduce some more notation: γ_1 (resp. γ_2) is the number of vectors of type c adjacent to one (resp. both) of $\mathbf{a}_{12}$, $\mathbf{b}_{34}$, while δ_1 (resp. δ_2) is the number of vectors of type d adjacent to one (resp. both) of $\mathbf{a}_{12}$, $\mathbf{b}_{34}$. Note that $\gamma_2 \geq 2$. Let

$$S = \{\mathbf{a}_{ij} : ij = 13, 24, 56, 57, 58, 67, 68, 78\}$$
$$\cup \{\mathbf{b}_{ij} : ij = 13, 24, 56, 57, 58, 67, 68, 78\},$$
$$T = \{\mathbf{a}_{ij} : ij = 15, 16, 17, 18, 25, 26, 27, 28\}$$
$$\cup \{\mathbf{b}_{ij} : ij = 35, 36, 37, 38, 45, 46, 47, 48\}.$$

Lemma 6.5.3. *With the above notation, the following holds:*

(6.1) $$\gamma_1 + 2\gamma_2 + \delta_1 + 2\delta_2 \leq |S \cap \Delta(\mathbf{e})|.$$

Proof. We have

$$\deg(\mathbf{a}_{12}) + \deg(\mathbf{b}_{45}) = 4 + 2|S \cap \Delta(\mathbf{e})| + |T \cap \Delta(\mathbf{e})| + \gamma_1 + 2\gamma_2 + \delta_1 + 2\delta_2$$

and

$$\deg(\mathbf{e}) = 2 + |S \cap \Delta(\mathbf{e})| + |T \cap \Delta(\mathbf{e})|.$$

The lemma follows because $\deg(\mathbf{a}_{12}) + \deg(\mathbf{b}_{45}) \leq 2 \deg(\mathbf{e})$.   □

Lemma 6.5.4. *At most one of the vectors* c_{14} *and* c_{23} *is present.*

Proof. If both c_{14} and c_{23} are present then $\delta_2 \geq 4$ and so $|S \cap \Delta(e)| \geq 8$ by Lemma 6.5.3. On the other hand, a_{24} and b_{13} are excluded by c_{14}, while b_{24} and a_{13} are excluded by c_{23}. Thus $|S \cap \Delta(e)| \leq 6$, a contradiction. $\square$

The next three lemmas are symmetric in 5,6,7,8.

Lemma 6.5.5. *If* a_{78} *and* b_{56} *are absent then either* (a) d_{3478} *is present or* (b) b_{78} *and* a_{56} *are absent. In particular, if* b_{78} *and* a_{56} *are present then* d_{3478} *is present.*

Proof. The vector d_{3478} is compatible with all remaining possible vectors of type a, b or c. Accordingly if (a) does not hold then d_{3478} is excluded by some vector d_{ijkl}. Since b_{34} is present (by Lemma 6.5.2), it follows from (DD) and (BD) that $\{i, j, k, l\} \cap \{3, 4, 7, 8\}$ is $\{3\}$ or $\{4\}$. In the former case, $ijkl = 1356$ since c_{24} excludes d_{2356}: and in the latter case, $ijkl = 2456$ since c_{13} excludes d_{1456}. In both cases, b_{78} and a_{56} are excluded. $\square$

Note that the assertions of Lemmas 6.5.2 to 6.5.5 remain true of $\phi(\mathcal{R}(G))$ when we apply the permutation (13)(24) to subscripts, and this justifies the duality arguments used in the sequel.

Lemma 6.5.6. *If* a_{56} *and* b_{56} *are absent then there exist vectors* d_{ijkl} *and* $d_{i'j'k'l'}$ *with* $\{5, 6\} \subseteq \{i, j, k, l\}$ *and* $\{5, 6\} \cap \{i', j', k', l'\} = \emptyset$.

Proof. The vector a_{56} can be excluded by c_{15}, c_{16}, c_{25}, c_{26} or d_{ijkl} where $\{5, 6\} \subseteq \{i, j, k, l\}$; and b_{56} can be excluded by c_{53}, c_{54}, c_{63}, c_{64} or $d_{i'j'k'l'}$ where $\{i', j', k', l'\}$ and $\{5, 6\}$ are disjoint. By duality it suffices to exclude two possibilities: (i) each of a_{56}, b_{56} is excluded by a vector of type c, (ii) a_{56} is excluded by a vector of type c and b_{56} is excluded by a vector of type d.

In case (i) we may assume without loss of generality first that a_{56} is excluded by c_{15}, and then that b_{56} is excluded by c_{63} (since c_{15} excludes c_{53} and c_{54}). In view of (AC) and (BC) the possible vectors in $S \cap \Delta(e)$ are a_{ij} ($ij = 67, 68, 78; 13, 24$) and b_{ij} ($ij = 57, 58, 78; 13, 24$). Note that by (AB), $|S \cap \Delta(e)| \leq 7$. Since $\gamma_1 \geq 2$ and $\gamma_2 \geq 2$ we have $|S \cap \Delta(e)| \geq 6$ by Lemma 6.5.3. Hence at most one of the vectors b_{57}, b_{58}, a_{67}, a_{68} is absent. Without loss of generality, a_{67} and b_{58} are present. By Lemma 6.5.5, d_{3467} is present, and so $\delta_2 \geq 1$. Now Lemma 6.5.3 yields the contradiction $|S \cap \Delta(e)| \geq 8$.

In case (ii) we may suppose without loss of generality that a_{56} is excluded by c_{15} and b_{56} is excluded by $d_{i'j'k'l'}$. This last vector must be compatible with c_{13}, c_{24}, c_{15}, and hence is one of d_{3478}, d_{2478}, d_{2347}, d_{2348}.

Since $\mathbf{a}_{12}$ is adjacent to $\mathbf{e}$, $\mathbf{b}_{34}$, $\mathbf{c}_{13}$, $\mathbf{c}_{24}$ and $\mathbf{c}_{15}$, we know that

$$\deg(\mathbf{a}_{12}) \geq 5 + |S \cap \Delta(\mathbf{a}_{12})| + |T \cap \Delta(\mathbf{e})|,$$

while

$$\deg(\mathbf{e}) = 2 + |S \cap \Delta(\mathbf{a}_{12})| + |S \cap \Delta(\mathbf{b}_{34})| + |T \cap \Delta(\mathbf{e})|.$$

Since $\deg(\mathbf{e}) \geq \deg(\mathbf{a}_{12})$, it follows that $|S \cap \Delta(\mathbf{b}_{34})| \geq 3$. Since $S \cap \Delta(\mathbf{b}_{34})$ is a subset of $\{\mathbf{b}_{24}, \mathbf{a}_{67}, \mathbf{a}_{68}, \mathbf{a}_{78}\}$ we conclude that not both $\mathbf{b}_{67}$ and $\mathbf{b}_{68}$ are present. If say $\mathbf{b}_{67}$ is absent then we may apply Lemma 6.5.5 to $\mathbf{a}_{58}$, $\mathbf{b}_{67}$ to deduce that either (a) $\mathbf{d}_{3458}$ is present or (b) $\mathbf{b}_{58}$, $\mathbf{a}_{67}$ are absent.

In subcase (a), $|S \cap \Delta(\mathbf{e})| = 7$ and either $\mathbf{a}_{58}$, $\mathbf{b}_{57}$ are present or $\mathbf{a}_{68}$, $\mathbf{b}_{57}$ are present. By Lemma 6.5.5 either $\mathbf{d}_{3467}$ or $\mathbf{d}_{3468}$ is present, and so $\delta_2 \geq 2$. By Lemma 6.5.3, $|S \cap \Delta(\mathbf{e})| \geq 8$, a contradiction. In subcase (b), $|S \cap \Delta(\mathbf{e})| = 5$, $S \cap \Delta(\mathbf{b}_{34}) = \{\mathbf{b}_{24}, \mathbf{a}_{68}, \mathbf{a}_{78}\}$, $\delta_1 = 0$ and $\delta_2 = 0$. Then $\mathbf{d}_{i'j'k'l'} = \mathbf{d}_{2478}$, a contradiction because this vector is not compatible with $\mathbf{a}_{78}$. $\qquad\square$

Lemma 6.5.7. *If $\mathbf{a}_{56}$ and $\mathbf{b}_{56}$ are absent then so are $\mathbf{a}_{78}$ and $\mathbf{b}_{78}$.*

Proof. We suppose that the conclusion does not hold, and obtain a contradiction. By Lemma 6.5.5, either $\mathbf{a}_{78}$ and $\mathbf{d}_{3456}$ are present or $\mathbf{b}_{78}$ and $\mathbf{d}_{3478}$ are present. By duality, we may assume that the former is the case. By Lemma 6.5.6, a vector $\mathbf{d}_{ijkl}$ is present, with $\{5,6\} \cap \{i,j,k,l\} = \emptyset$. This vector must be compatible with $\mathbf{a}_{12}$ and $\mathbf{a}_{78}$, and so $ijkl$ is one of 1347,1348,2347,2348. Without loss of generality, suppose that $\mathbf{d}_{1347}$ is present. Now

$$S \cap \Delta(\mathbf{e}) \subseteq \{\mathbf{b}_{13}, \mathbf{a}_{24}, \mathbf{b}_{24}, \mathbf{a}_{57}, \mathbf{b}_{57}, \mathbf{a}_{58}, \mathbf{a}_{67}, \mathbf{b}_{67}, \mathbf{a}_{68}, \mathbf{a}_{78}\}$$

and so $|S \cap \Delta(\mathbf{a}_{12})| \leq 3$. Also, in view of (AB), we have $|S \cap \Delta(\mathbf{e})| \leq 7$. On the other hand, $\gamma_2 \geq 2$, $\delta_1 \geq 1$ and $\delta_2 \geq 1$, whence $|S \cap \Delta(\mathbf{e})| \geq 7$ by Lemma 6.5.3.

Next, $\mathbf{b}_{34}$ is adjacent to $\mathbf{a}_{12}$, $\mathbf{c}_{13}$, $\mathbf{c}_{24}$, $\mathbf{d}_{3456}$, $\mathbf{d}_{1347}$ and $\mathbf{e}$ and so

$$\deg(\mathbf{b}_{34}) \geq 6 + |S \cap \Delta(\mathbf{b}_{34})| + |T \cap \Delta(\mathbf{e})|.$$

Now, arguing as in Lemma 6.5.6, we obtain the contradiction $|S \cap \Delta(\mathbf{a}_{12})| \geq 4$. $\qquad\square$

We are now in a position to determine the graphs which can arise. It is convenient to discuss the various possibilities in terms of the graph Q on $\{5,6,7,8\}$ in which i and j are joined by a red edge if $\mathbf{a}_{ij}$ is present, and by a blue edge if $\mathbf{b}_{ij}$ is present. By duality we may assume that $n_b(Q)$, the number of blue edges of Q, is not less than $n_r(Q)$, the number of red edges. We distinguish five cases:

(1) Q is incomplete, (2) $(n_b(Q), n_r(Q)) = (6, 0)$, (3) $(n_b(Q), n_r(Q)) = (5, 1)$, (4) $(n_b(Q), n_r(Q)) = (4, 2)$, (5) $(n_b(Q), n_r(Q)) = (3, 3)$.

Case 1: Q is incomplete.

Without loss of generality, suppose that $\mathbf{a}_{56}$ and $\mathbf{b}_{56}$ are absent. By Lemma 6.5.6, a vector $\mathbf{d}_{ijkl}$ is present, with $\{5, 6\} \cap \{i, j, k, l\} = \emptyset$. If also $\{1, 2\} \cap \{i, j, k, l\} = \emptyset$ then $\mathbf{d}_{ijkl} = \mathbf{d}_{3478}$ and $\delta_2 \geq 2$. Since also $\gamma_2 \geq 2$, Lemma 6.5.3 yields $|S \cap \Delta(\mathbf{e})| \geq 8$, a contradiction. Since $\mathbf{d}_{ijkl}$ must be compatible with $\mathbf{c}_{13}$ and $\mathbf{c}_{24}$ the possibilities for $ijkl$ are 1378, 2478.

By Lemma 6.5.7, $\mathbf{a}_{78}$ and $\mathbf{b}_{78}$ are absent, and so similarly either $\mathbf{d}_{1356}$ or $\mathbf{d}_{2456}$ is present. Since $\mathbf{d}_{1356}$ and $\mathbf{d}_{2478}$ are incompatible, and $\mathbf{d}_{2456}$ and $\mathbf{d}_{1378}$ are incompatible, we may assume without loss of generality that $\mathbf{d}_{2456}$ and $\mathbf{d}_{2478}$ are present. Then $\mathbf{a}_{24}$ and $\mathbf{b}_{13}$ are excluded and we note that $\mathbf{c}_{14}$ is compatible with all possible vectors of type a, b or c. It follows that $\mathbf{c}_{14}$ is present, for otherwise it is excluded by a vector of type d compatible with $\mathbf{a}_{12}$ and $\mathbf{b}_{34}$: such a vector has the form $\mathbf{d}_{13uv}$ where $\{u, v\} \subseteq \{5, 6, 7, 8\}$, and is therefore not compatible with both $\mathbf{d}_{2456}$ and $\mathbf{d}_{2478}$.

Now $\mathbf{a}_{24}$, $\mathbf{b}_{13}$ are excluded, and we have $\gamma_2 \geq 3$. Since $|S \cap \Delta(\mathbf{e})| \leq 6$ it follows from Lemma 6.5.3 that $|S \cap \Delta(\mathbf{e})| = 6$, $\gamma_2 = 3$, $\gamma_1 = \delta_1 = \delta_2 = 0$ and $\deg(\mathbf{a}_{12}) = \deg(\mathbf{b}_{34}) = \deg(\mathbf{e})$. In view of Lemma 6.5.5 (applied to non-adjacent edges of H), either $S \cap \Delta(\mathbf{e}) = \{\mathbf{a}_{13}, \mathbf{a}_{57}, \mathbf{a}_{68}, \mathbf{b}_{24}, \mathbf{b}_{67}, \mathbf{b}_{58}\}$ or $S \cap \Delta(\mathbf{e}) = \{\mathbf{a}_{13}, \mathbf{b}_{57}, \mathbf{b}_{68}, \mathbf{b}_{24}, \mathbf{a}_{67}, \mathbf{a}_{58}\}$.

Since $\gamma_2 = 3$ and $\gamma_1 = 0$ there can be no vectors of type c other than $\mathbf{c}_{13}$, $\mathbf{c}_{24}$, $\mathbf{c}_{14}$. Moreover there are no vectors of type d other than $\mathbf{d}_{2456}$, $\mathbf{d}_{2478}$: the only possible vectors of type d compatible with $\mathbf{a}_{12}$, $\mathbf{b}_{34}$, $\mathbf{c}_{24}$, $\mathbf{d}_{2456}$, $\mathbf{d}_{2478}$ are $\mathbf{d}_{i457}$, $\mathbf{d}_{i458}$, $\mathbf{d}_{i467}$, $\mathbf{d}_{i468}$ ($i = 1, 2$), but each of these is incompatible with both candidates for $S \cap \Delta(\mathbf{e})$. (For example, $\mathbf{d}_{i467}$ is incompatible with both $\mathbf{b}_{58}$ and $\mathbf{a}_{67}$.)

Since $\mathbf{d}_{2456}$ excludes $\mathbf{a}_{25}$, $\mathbf{a}_{26}$, $\mathbf{b}_{37}$, $\mathbf{b}_{38}$ and $\mathbf{d}_{2478}$ excludes $\mathbf{b}_{35}$, $\mathbf{b}_{36}$, $\mathbf{a}_{27}$, $\mathbf{a}_{28}$, we have

$$T \cap \Delta(\mathbf{e}) = \{\mathbf{a}_{15}, \mathbf{a}_{16}, \mathbf{a}_{17}, \mathbf{a}_{18}, \mathbf{b}_{45}, \mathbf{b}_{46}, \mathbf{b}_{47}, \mathbf{b}_{48}\}.$$

By applying the permutation (56) if necessary we may assume that

$$S \cap \Delta(\mathbf{e}) = \{\mathbf{a}_{13}, \mathbf{a}_{57}, \mathbf{a}_{68}, \mathbf{b}_{24}, \mathbf{b}_{58}, \mathbf{b}_{67}\}.$$

The vectors which remain are $\mathbf{a}_{12}$, $\mathbf{b}_{34}$, $\mathbf{c}_{13}$, $\mathbf{c}_{24}$, $\mathbf{c}_{14}$, $\mathbf{d}_{2456}$, $\mathbf{d}_{2478}$, $\mathbf{e}$ together with those in $S \cap \Delta(\mathbf{e})$ and $T \cap \Delta(\mathbf{e})$; they are pairwise compatible and determine a maximal graph with 22 vertices. An isomorphism ψ from the resulting graph to $G001$ is given by:

$$\begin{array}{c|cccccccc} \mathbf{u} & \mathbf{a}_{13} & \mathbf{c}_{24} & \mathbf{b}_{34} & \mathbf{a}_{12} & \mathbf{b}_{46} & \mathbf{b}_{47} & \mathbf{b}_{45} & \mathbf{b}_{48} \\ \psi(\mathbf{u}) & \mathbf{d}_{5678} & \mathbf{a}_{15} & \mathbf{e} & \mathbf{a}_{12} & \mathbf{a}_{13} & \mathbf{a}_{14} & \mathbf{b}_{67} & \mathbf{b}_{68} \end{array}$$

$$\begin{array}{c|cccccccccccc} \mathbf{u} & \mathbf{a}_{17} & \mathbf{a}_{16} & \mathbf{b}_{24} & \mathbf{c}_{13} & \mathbf{a}_{57} & \mathbf{a}_{68} & \mathbf{e} & \mathbf{a}_{18} & \mathbf{a}_{15} & \mathbf{c}_{14} & \mathbf{b}_{67} & \mathbf{b}_{58} & \mathbf{d}_{2456} & \mathbf{d}_{2478} \\ \psi(\mathbf{u}) & \mathbf{a}_{23} & \mathbf{a}_{24} & \mathbf{a}_{34} & \mathbf{a}_{26} & \mathbf{a}_{37} & \mathbf{a}_{48} & \mathbf{b}_{56} & \mathbf{b}_{57} & \mathbf{b}_{58} & \mathbf{b}_{78} & \mathbf{c}_{15} & \mathbf{c}_{26} & \mathbf{c}_{37} & \mathbf{c}_{48} \end{array}.$$

Here, and in isomorphisms exhibited subsequently, the first eight vectors induce a subgraph H isomorphic to $H443$, with degrees in H equal to 5,6,6,7,7,7,7,7. In each case, the isomorphism is constructed using Corollary 5.1.8.

Case 2: $n_b(Q) = 6$, $n_r(Q) = 0$.

Arguing as in Lemma 6.5.6, we find that $|S \cap \Delta(\mathbf{b}_{34})| \geq 2$. Since $S \cap \Delta(\mathbf{e})$ contains the six vectors $\mathbf{b}_{ij}$ ($\{i, j\} \subseteq \{5, 6, 7, 8\}$) it follows that

$$S \cap \Delta(\mathbf{e}) = \{\mathbf{b}_{ij} : ij = 13, 24, 56, 57, 58, 67, 68, 78\}.$$

In view of (BC) the only vectors of type c which are present are $\mathbf{c}_{13}$ $\mathbf{c}_{24}$; and in view of (BD) there are no vectors of type d. The 29 vectors which remain are $\mathbf{a}_{12}$, $\mathbf{b}_{34}$, $\mathbf{c}_{13}$, $\mathbf{c}_{24}$, $\mathbf{e}$ together with those in $S \cap \Delta(\mathbf{e})$ and T. They are pairwise compatible and determine a maximal graph which is the graph $G430$ defined in Section 6.4.

Case 3: $n_b(Q) = 5$, $n_r(Q) = 1$.

We may suppose that the red edge of Q is 56. Thus $\mathbf{a}_{56}$ and $\mathbf{b}_{78}$ are present, while $\mathbf{b}_{56}$ and $\mathbf{a}_{78}$ are absent. By Lemma 6.5.5, $\mathbf{d}_{3478}$ is present, and so $\deg(\mathbf{a}_{12}) \geq 5 + |S \cap \Delta(\mathbf{a}_{12})| + |T \cap \Delta(\mathbf{e})|$. Since $\deg(\mathbf{e}) \geq \deg(\mathbf{a}_{12})$, it follows that $|S \cap \Delta(\mathbf{b}_{34})| \geq 3$, and hence that $\mathbf{b}_{13}$ and $\mathbf{b}_{24}$ are present. Since also the vectors $\mathbf{b}_{ij}$ ($ij = 57, 58, 67, 68, 78$) are present, we conclude from (BC) that $\mathbf{c}_{13}$, $\mathbf{c}_{24}$ are the only vectors of type c present. If another vector of type d is present then it must have the form $\mathbf{d}_{ij78}$ ($ij \neq 34$) for compatiblity with $\mathbf{a}_{56}$ and $\mathbf{b}_{ij}$ ($ij = 56, 57, 67, 68, 78$). However, no such vector is compatible with $\mathbf{a}_{12}$, $\mathbf{c}_{13}$, $\mathbf{c}_{24}$, $\mathbf{b}_{13}$, $\mathbf{b}_{24}$. The 30 vectors which remain are $\mathbf{a}_{12}$, $\mathbf{b}_{34}$, $\mathbf{c}_{13}$, $\mathbf{c}_{24}$, $\mathbf{d}_{3478}$, $\mathbf{e}$ together with those in $S \cap \Delta(\mathbf{e})$ and T. They are pairwise compatible and determine a maximal graph which is the graph $G455$ defined in Section 6.4.

Case 4: $n_b(Q) = 4$, $n_r(Q) = 2$.

We distinguish two subcases depending on the factorization of Q induced by the edge-colouring.

Subcase 4a: The two red edges are non-adjacent.

We assume, without loss of generality, that edges 57 and 68 are red, so that $S \cap \Delta(\mathbf{e})$ contains $\mathbf{a}_{57}$, $\mathbf{a}_{68}$, $\mathbf{b}_{56}$, $\mathbf{b}_{58}$, $\mathbf{b}_{67}$, $\mathbf{b}_{78}$. These vectors exclude all vectors of type d, and all further vectors of type c other than $\mathbf{c}_{14}$, $\mathbf{c}_{23}$.

By Lemma 6.5.4, at most one of $\mathbf{c}_{14}$, $\mathbf{c}_{23}$ is present. If say $\mathbf{c}_{14}$ is present then $\mathbf{b}_{13}$, $\mathbf{a}_{24}$ are excluded, and by maximality $\mathbf{a}_{13}$, $\mathbf{b}_{24}$ are present. In this case the vectors which remain are $\mathbf{a}_{12}$, $\mathbf{b}_{34}$, $\mathbf{c}_{13}$, $\mathbf{c}_{24}$, $\mathbf{c}_{14}$, $\mathbf{e}$ together with those in $S \cap \Delta(\mathbf{e})$ and T. They are pairwise compatible and determine a maximal graph with 30 vertices. An isomorphism ψ from this graph to $G455$ is given by:

$$
\begin{array}{c|ccccccc|cccccccc}
\mathbf{u} & \mathbf{b}_{35} & \mathbf{b}_{46} & \mathbf{a}_{15} & \mathbf{a}_{12} & \mathbf{a}_{18} & \mathbf{b}_{36} & \mathbf{a}_{17} & \mathbf{e} & \mathbf{a}_{16} & \mathbf{b}_{37} & \mathbf{b}_{48} & \mathbf{a}_{25} & \mathbf{b}_{38} & \mathbf{b}_{24} & \mathbf{a}_{57} \\
\psi(\mathbf{u}) & \mathbf{a}_{18} & \mathbf{a}_{25} & \mathbf{b}_{38} & \mathbf{a}_{12} & \mathbf{a}_{15} & \mathbf{b}_{36} & \mathbf{b}_{37} & \mathbf{e} & \mathbf{a}_{16} & \mathbf{a}_{17} & \mathbf{a}_{26} & \mathbf{a}_{27} & \mathbf{a}_{28} & \mathbf{a}_{56} & \mathbf{b}_{13}
\end{array}
$$

$$
\begin{array}{c|cccccccccccccc}
\mathbf{u} & \mathbf{a}_{68} & \mathbf{b}_{34} & \mathbf{a}_{27} & \mathbf{a}_{26} & \mathbf{a}_{28} & \mathbf{b}_{45} & \mathbf{b}_{47} & \mathbf{b}_{58} & \mathbf{b}_{78} & \mathbf{b}_{56} & \mathbf{b}_{67} & \mathbf{a}_{13} & \mathbf{c}_{13} & \mathbf{c}_{24} & \mathbf{c}_{14} \\
\psi(\mathbf{u}) & \mathbf{b}_{24} & \mathbf{b}_{34} & \mathbf{b}_{35} & \mathbf{b}_{45} & \mathbf{b}_{46} & \mathbf{b}_{47} & \mathbf{b}_{48} & \mathbf{b}_{57} & \mathbf{b}_{58} & \mathbf{b}_{67} & \mathbf{b}_{68} & \mathbf{b}_{78} & \mathbf{c}_{13} & \mathbf{c}_{24} & \mathbf{d}_{3478}
\end{array}.
$$

If $\mathbf{c}_{14}$ and $\mathbf{c}_{23}$ are absent then $\mathbf{c}_{14}$ is excluded by $\mathbf{b}_{13}$ or $\mathbf{a}_{24}$, while $\mathbf{c}_{23}$ is excluded by $\mathbf{a}_{13}$ or $\mathbf{b}_{24}$. Thus either $\mathbf{b}_{13}$, $\mathbf{b}_{24}$ are present and we obtain the graph $G425$ defined in Section 6.4; or $\mathbf{a}_{13}$, $\mathbf{a}_{24}$ are present and an isomorphism ψ from the resulting graph to $G430$ is given by:

$$
\begin{array}{c|cccccc|cccccc}
\mathbf{u} & \mathbf{a}_{15} & \mathbf{b}_{35} & \mathbf{b}_{45} & \mathbf{a}_{12} & \mathbf{a}_{16} & \mathbf{a}_{17} & \mathbf{a}_{18} & \mathbf{e} & \mathbf{b}_{36} & \mathbf{b}_{37} & \mathbf{b}_{38} & \mathbf{b}_{46} & \mathbf{b}_{47} & \mathbf{b}_{48} \\
\psi(\mathbf{u}) & \mathbf{b}_{35} & \mathbf{a}_{15} & \mathbf{a}_{25} & \mathbf{a}_{12} & \mathbf{b}_{36} & \mathbf{b}_{37} & \mathbf{b}_{38} & \mathbf{e} & \mathbf{a}_{16} & \mathbf{a}_{17} & \mathbf{a}_{18} & \mathbf{a}_{26} & \mathbf{a}_{27} & \mathbf{a}_{28}
\end{array}
$$

$$
\begin{array}{c|cccccccccccc}
\mathbf{u} & \mathbf{b}_{34} & \mathbf{a}_{25} & \mathbf{a}_{26} & \mathbf{a}_{27} & \mathbf{a}_{28} & \mathbf{b}_{24} & \mathbf{b}_{13} & \mathbf{b}_{78} & \mathbf{b}_{68} & \mathbf{b}_{67} & \mathbf{b}_{58} & \mathbf{b}_{57} & \mathbf{b}_{56} & \mathbf{c}_{13} & \mathbf{c}_{24} \\
\psi(\mathbf{u}) & \mathbf{b}_{34} & \mathbf{b}_{45} & \mathbf{b}_{46} & \mathbf{b}_{47} & \mathbf{b}_{48} & \mathbf{b}_{13} & \mathbf{b}_{24} & \mathbf{b}_{56} & \mathbf{b}_{57} & \mathbf{b}_{58} & \mathbf{b}_{67} & \mathbf{b}_{68} & \mathbf{b}_{78} & \mathbf{c}_{13} & \mathbf{c}_{24}
\end{array}.
$$

Subcase 4b: Two red edges are adjacent.

We assume, without loss of generality, that edges 56 and 67 are red, so that $S \cap \Delta(\mathbf{e})$ contains $\mathbf{a}_{56}$, $\mathbf{a}_{67}$, $\mathbf{b}_{57}$, $\mathbf{b}_{58}$, $\mathbf{b}_{68}$, $\mathbf{b}_{78}$. By Lemma 6.5.8 (applied to $\mathbf{a}_{78}$, $\mathbf{b}_{56}$ and to $\mathbf{a}_{58}$, $\mathbf{b}_{67}$), we know that $\mathbf{d}_{3478}$, $\mathbf{d}_{3458}$ are present. It follows that $\delta_2 \geq 2$. Since also $\gamma_2 \geq 2$, it follows from Lemma 6.5.3 that $\gamma_1 = \delta_1 = 0$, $\gamma_2 = \delta_2 = 2$, $|S \cap \Delta(\mathbf{e})| = 8$ and $\deg(\mathbf{a}_{12}) = \deg(\mathbf{b}_{34}) = \deg(\mathbf{e})$. Since $\deg(\mathbf{a}_{12}) = 6 + |S \cap \Delta(\mathbf{a}_{12})| + |T \cap \Delta(\mathbf{e})|$, it follows that $|S \cap \Delta(\mathbf{b}_{34})| \geq 4$, and hence that $\mathbf{b}_{13}$ and $\mathbf{b}_{24}$ is present. In view of (BD) there are no further vectors of type d. The 31 vectors which remain are $\mathbf{a}_{12}$, $\mathbf{b}_{34}$, $\mathbf{c}_{13}$, $\mathbf{c}_{24}$, $\mathbf{d}_{3458}$, $\mathbf{d}_{3478}$, $\mathbf{e}$ together with those in $S \cap \Delta(\mathbf{e})$ and T. They are pairwise compatible and determine a maximal graph which is the graph $G463$ defined in Section 6.4.

Case 5: $n_b(Q) = 3$, $n_r(Q) = 3$.

We show first that the three red edges form a path. Otherwise by duality we may assume that they form a star, say with edges 56, 57 and 58. By

Lemma 6.5.5, the vectors $\mathbf{d}_{3478}$, $\mathbf{d}_{3468}$, $\mathbf{d}_{3467}$ are present. Now we have $\gamma_2 \geq 2$ and $\delta_2 \geq 3$, contradicting Lemma 6.5.3.

Accordingly we assume, without loss of generality, that edges 56, 58 and 67 are red, so that $S \cap \Delta(\mathbf{e})$ contains $\mathbf{a}_{56}$, $\mathbf{a}_{58}$, $\mathbf{a}_{67}$, $\mathbf{b}_{57}$, $\mathbf{b}_{68}$, $\mathbf{b}_{78}$. By Lemma 6.5.5 (applied to $\mathbf{a}_{78}$, $\mathbf{b}_{56}$), we know that $\mathbf{d}_{3478}$ is present, and so $\delta_2 \geq 1$. The vectors in $S \cap \Delta(\mathbf{e})$ exclude all further possible vectors of type c other than $\mathbf{c}_{14}$ and $\mathbf{c}_{23}$.

By Lemma 6.5.4, at most one of $\mathbf{c}_{14}$, $\mathbf{c}_{23}$ is present. If say $\mathbf{c}_{14}$ is present then $\mathbf{b}_{13}$, $\mathbf{a}_{24}$ are excluded, and $\gamma_2 \geq 3$. By Lemma 6.5.3, we have $\gamma_1 = \delta_1 = 0, \gamma_2 = 3, \delta_2 = 1, |S \cap \Delta(\mathbf{e})| = 8$ and $\deg(\mathbf{a}_{12}) = \deg(\mathbf{b}_{34}) = \deg(\mathbf{e})$. In particular, $\mathbf{a}_{13}$ and $\mathbf{b}_{24}$ are present, but no further vectors of type c are present.

Now the only further vector of type d compatible with $S \cap \Delta(\mathbf{e})$, $\mathbf{a}_{12}$, $\mathbf{c}_{13}$, $\mathbf{c}_{24}$ and $\mathbf{c}_{14}$ is $\mathbf{d}_{2478}$. If this vector is present then $\mathbf{a}_{27}, \mathbf{a}_{28}, \mathbf{b}_{35}$, $\mathbf{b}_{36}$ are excluded. The 28 vectors which remain are $\mathbf{a}_{12}$, $\mathbf{b}_{34}$, $\mathbf{c}_{13}$, $\mathbf{c}_{24}$, $\mathbf{c}_{14}$, $\mathbf{d}_{2478}$, $\mathbf{d}_{3478}$, $\mathbf{e}$ together with the 8 vectors in $S \cap \Delta(\mathbf{e})$ and the 12 vectors in $T \cap \Delta(\mathbf{e})$. They are pairwise compatible and determine a maximal graph with 28 vertices. An isomorphism ψ from this graph to $G002$ is given by:

$$
\begin{array}{c|cccccccc|ccccccc}
\mathbf{u} & \mathbf{d}_{2478} & \mathbf{a}_{12} & \mathbf{e} & \mathbf{b}_{48} & \mathbf{a}_{15} & \mathbf{b}_{78} & \mathbf{a}_{13} & \mathbf{a}_{16} & \mathbf{b}_{34} & \mathbf{a}_{67} & \mathbf{a}_{26} & \mathbf{b}_{45} & \mathbf{c}_{13} & \mathbf{b}_{37} \\
\psi(\mathbf{u}) & \mathbf{a}_{14} & \mathbf{b}_{45} & \mathbf{b}_{46} & \mathbf{a}_{12} & \mathbf{a}_{13} & \mathbf{a}_{17} & \mathbf{a}_{18} & \mathbf{e} & \mathbf{a}_{23} & \mathbf{a}_{25} & \mathbf{a}_{27} & \mathbf{a}_{28} & \mathbf{a}_{36} & \mathbf{a}_{37}
\end{array}
$$

$$
\begin{array}{c|cccccccccccc}
\mathbf{u} & \mathbf{a}_{18} & \mathbf{b}_{57} & \mathbf{a}_{17} & \mathbf{b}_{38} & \mathbf{b}_{47} & \mathbf{c}_{14} & \mathbf{d}_{3478} & \mathbf{b}_{24} & \mathbf{a}_{56} & \mathbf{b}_{68} & \mathbf{c}_{24} & \mathbf{a}_{58} & \mathbf{b}_{46} & \mathbf{a}_{25} \\
\psi(\mathbf{u}) & \mathbf{a}_{38} & \mathbf{a}_{78} & \mathbf{b}_{47} & \mathbf{b}_{48} & \mathbf{b}_{56} & \mathbf{b}_{57} & \mathbf{b}_{58} & \mathbf{b}_{67} & \mathbf{b}_{68} & \mathbf{c}_{14} & \mathbf{c}_{25} & \mathbf{c}_{36} & \mathbf{d}_{4567} & \mathbf{d}_{4568}
\end{array}.
$$

If $\mathbf{d}_{2478}$ is absent then we obtain a maximal graph with 31 vertices, and an isomorphism ψ from this graph to $G463$ is given by:

$$
\begin{array}{c|cccccccc|ccccccc}
\mathbf{u} & \mathbf{b}_{35} & \mathbf{a}_{15} & \mathbf{b}_{47} & \mathbf{b}_{34} & \mathbf{a}_{18} & \mathbf{a}_{16} & \mathbf{b}_{37} & \mathbf{e} & \mathbf{b}_{46} & \mathbf{b}_{48} & \mathbf{a}_{25} & \mathbf{a}_{27} & \mathbf{a}_{13} & \mathbf{b}_{78} & \mathbf{b}_{68} & \mathbf{b}_{57} \\
\psi(\mathbf{u}) & \mathbf{a}_{18} & \mathbf{b}_{38} & \mathbf{b}_{48} & \mathbf{a}_{12} & \mathbf{a}_{15} & \mathbf{a}_{16} & \mathbf{a}_{17} & \mathbf{e} & \mathbf{a}_{25} & \mathbf{a}_{26} & \mathbf{a}_{27} & \mathbf{a}_{28} & \mathbf{a}_{56} & \mathbf{a}_{67} & \mathbf{b}_{13} & \mathbf{b}_{24}
\end{array}
$$

$$
\begin{array}{c|cccccccccccccc}
\mathbf{u} & \mathbf{a}_{12} & \mathbf{b}_{38} & \mathbf{b}_{36} & \mathbf{a}_{17} & \mathbf{a}_{26} & \mathbf{a}_{28} & \mathbf{b}_{45} & \mathbf{a}_{67} & \mathbf{a}_{56} & \mathbf{a}_{58} & \mathbf{b}_{24} & \mathbf{c}_{13} & \mathbf{c}_{24} & \mathbf{c}_{14} & \mathbf{d}_{3478} \\
\psi(\mathbf{u}) & \mathbf{b}_{34} & \mathbf{b}_{35} & \mathbf{b}_{36} & \mathbf{b}_{37} & \mathbf{b}_{45} & \mathbf{b}_{46} & \mathbf{b}_{47} & \mathbf{b}_{57} & \mathbf{b}_{58} & \mathbf{b}_{68} & \mathbf{b}_{78} & \mathbf{c}_{13} & \mathbf{c}_{24} & \mathbf{d}_{3478} & \mathbf{d}_{3458}
\end{array}.
$$

If $\mathbf{c}_{14}$ and $\mathbf{c}_{23}$ are absent then (arguing as above) we find that the only possible vectors of type d in addition to $\mathbf{d}_{3478}$ are $\mathbf{d}_{1378}$, $\mathbf{d}_{2478}$. If both are present then the vectors $\mathbf{a}_{ij}$ ($ij = 27, 28, 17, 18, 24, 13$) and $\mathbf{b}_{ij}$ ($ij = 35, 36, 45, 46, 24, 13$) are excluded. The vectors which remain are $\mathbf{a}_{12}$, $\mathbf{b}_{34}$, $\mathbf{c}_{13}$, $\mathbf{c}_{24}$, $\mathbf{d}_{1378}$ $\mathbf{d}_{2478}$, $\mathbf{d}_{3478}$, $\mathbf{e}$ together with the 6 vectors in $S \cap \Delta(\mathbf{e})$ and the 8 vectors in $T \cap \Delta(\mathbf{e})$. They are pairwise compatible and determine a maximal graph with 22 vertices. An isomorphism ψ from this graph to $G001$ is given by:

$$
\begin{array}{c|cccccccc}
\mathbf{u} & \mathbf{d}_{3478} & \mathbf{e} & \mathbf{b}_{57} & \mathbf{a}_{16} & \mathbf{b}_{78} & \mathbf{b}_{47} & \mathbf{b}_{37} & \mathbf{a}_{12} \\
\psi(\mathbf{u}) & \mathbf{d}_{5678} & \mathbf{a}_{15} & \mathbf{e} & \mathbf{a}_{12} & \mathbf{a}_{13} & \mathbf{a}_{14} & \mathbf{b}_{67} & \mathbf{b}_{68}
\end{array}
$$

$$
\begin{array}{c|cccccccccccccc}
\mathbf{u} & \mathbf{b}_{38} & \mathbf{b}_{34} & \mathbf{a}_{67} & \mathbf{a}_{25} & \mathbf{b}_{68} & \mathbf{a}_{58} & \mathbf{a}_{15} & \mathbf{a}_{56} & \mathbf{b}_{48} & \mathbf{a}_{26} & \mathbf{d}_{2478} & \mathbf{c}_{13} & \mathbf{d}_{1378} & \mathbf{c}_{24} \\
\psi(\mathbf{u}) & \mathbf{a}_{23} & \mathbf{a}_{24} & \mathbf{a}_{26} & \mathbf{a}_{34} & \mathbf{a}_{37} & \mathbf{a}_{48} & \mathbf{b}_{56} & \mathbf{b}_{57} & \mathbf{b}_{58} & \mathbf{b}_{78} & \mathbf{c}_{15} & \mathbf{c}_{26} & \mathbf{c}_{37} & \mathbf{c}_{48}
\end{array}
$$

If just one of $\mathbf{d}_{1378}$, $\mathbf{d}_{2478}$ is present we obtain a contradiction because either $\mathbf{c}_{23}$ or $\mathbf{c}_{24}$ is then not excluded. If neither $\mathbf{d}_{1378}$ nor $\mathbf{d}_{2478}$ is present then either $\mathbf{a}_{13}$, $\mathbf{a}_{24}$ or $\mathbf{b}_{13}$, $\mathbf{b}_{24}$ must be present to exclude $\mathbf{c}_{14}$ and $\mathbf{c}_{23}$. By duality we may assume that $\mathbf{b}_{13}$, $\mathbf{b}_{24}$ are present. The vectors which remain are $\mathbf{a}_{12}$, $\mathbf{b}_{34}$, $\mathbf{c}_{13}$, $\mathbf{c}_{24}$, $\mathbf{d}_{3478}$, $\mathbf{e}$ together with the 8 vectors in $S \cap \Delta(\mathbf{e})$ and the 16 vectors in T. They are pairwise compatible and determine a maximal graph with 30 vertices. An isomorphism ψ from this graph to $G455$ is given by:

$$
\begin{array}{c|ccccccc|ccccccc}
\mathbf{u} & \mathbf{b}_{35} & \mathbf{a}_{25} & \mathbf{a}_{15} & \mathbf{b}_{34} & \mathbf{b}_{37} & \mathbf{b}_{36} & \mathbf{b}_{38} & \mathbf{e} & \mathbf{a}_{16} & \mathbf{a}_{18} & \mathbf{b}_{48} & \mathbf{b}_{46} & \mathbf{a}_{27} & \mathbf{b}_{78} & \mathbf{b}_{68} \\
\psi(\mathbf{u}) & \mathbf{a}_{18} & \mathbf{a}_{25} & \mathbf{b}_{38} & \mathbf{a}_{12} & \mathbf{a}_{15} & \mathbf{b}_{36} & \mathbf{b}_{37} & \mathbf{e} & \mathbf{a}_{16} & \mathbf{a}_{17} & \mathbf{a}_{26} & \mathbf{a}_{27} & \mathbf{a}_{28} & \mathbf{a}_{56} & \mathbf{b}_{13}
\end{array}
$$

$$
\begin{array}{c|ccccccccccccccc}
\mathbf{u} & \mathbf{b}_{57} & \mathbf{a}_{12} & \mathbf{a}_{17} & \mathbf{b}_{45} & \mathbf{a}_{28} & \mathbf{a}_{26} & \mathbf{b}_{47} & \mathbf{a}_{67} & \mathbf{b}_{24} & \mathbf{b}_{13} & \mathbf{a}_{58} & \mathbf{a}_{56} & \mathbf{c}_{13} & \mathbf{c}_{24} & \mathbf{d}_{3478} \\
\psi(\mathbf{u}) & \mathbf{b}_{24} & \mathbf{b}_{34} & \mathbf{b}_{35} & \mathbf{b}_{45} & \mathbf{b}_{46} & \mathbf{b}_{47} & \mathbf{b}_{48} & \mathbf{b}_{57} & \mathbf{b}_{58} & \mathbf{b}_{67} & \mathbf{b}_{68} & \mathbf{b}_{78} & \mathbf{c}_{13} & \mathbf{c}_{24} & \mathbf{d}_{3478}
\end{array}
$$

This completes the proof of Theorem 6.5.1. $\qquad\qquad\square$

6.6 Concluding remarks

In view of the results in the previous section, we have a computer-free proof of the following result, which identifies two tractable star complements that we can use to construct all of the maximal exceptional graphs.

Theorem 6.6.1. *A maximal exceptional graph has as a star complement for* -2 *either the graph* $L(R_8)$ *or the graph* U_8 *obtained from* $L(R_8)$ *by switching with respect to the vertex of degree 2.*

Proof. Since a maximal exceptional graph of type (a) is a cone over a graph switching-equivalent to $L(K_8)$, it follows from Section 5.5 that the graphs of type (a) have $L(R_8)$ as a star complement. By Theorem 6.3.1, graphs of type (b) have $H443$ as a star complement for -2. Since $H443$ is the graph obtained from $L(R_8)$ by switching with respect to the vertex of degree 2, it now suffices to show that graphs of type (c) have $H443$ as a star complement. By Theorem 6.5.1, the graphs of type (c) are the six graphs defined in Section 6.4. In the course of proving Theorem 6.5.1, we identified an induced subgraph $H443$ in each of $G001$, $G002$, $G430$, $G455$, $G463$. Finally, in terms of the representation of $G425$ given in Section 6.4, the vertices $\mathbf{a}_{27}$, $\mathbf{b}_{24}$, $\mathbf{b}_{47}$, $\mathbf{b}_{34}$, $\mathbf{b}_{45}$, $\mathbf{b}_{46}$, $\mathbf{b}_{48}$, $\mathbf{e}$ induce a subgraph of $G425$ isomorphic to $H443$. $\square$

We know from the computer results described in Section 6.1 that we cannot construct every maximal exceptional graph from just one exceptional star complement for -2. On the other hand, there are nine exceptional graphs with the property that they serve as a star complement for -2 in every maximal exceptional graph except the cone over $L(K_8)$. One such graph is the graph $H440$ (another induced subgraph of $\overline{S(K_{1,7})}$), and following [Cve13] we strengthen this observation. First we introduce some further terminology: a graph G is said to be *star covered* by the graph K for the eigenvalue μ if G is an induced subgraph of a graph which has K as a star complement for eigenvalue μ.

Theorem 6.6.2. *An exceptional graph is either star covered by the graph $H440$ for the eigenvalue -2 or isomorphic to the cone over $L(K_8)$.*

Proof. Let $G = L(K_8)\nabla K_1$, the graph $G006$ of Table A6; it has a vertex of degree 28 and 28 vertices of degree 13. All exceptional graphs which are proper induced subgraphs of G are induced subgraphs of $G - x$, where x is any vertex of degree 13. Consider the graph H obtained from $L(K_8)$ by switching with respect to the single vertex x. Since the graph $K_2 \,\dot\cup\, 6K_1$ has no dissections, the cone $H\nabla K_1$ is a maximal exceptional graph by Theorem 6.2.4. (It can be identified in Table A6 as the graph $G013$, with vertex degrees 12^{12}, 14^{15}, 16, 28.) Since $G - x$ is an induced subgraph of $G013$, all exceptional graphs contained in G as proper induced subgraphs are contained also in $G013$. Hence all exceptional graphs other than $G006$ are star covered by $H440$ for the eigenvalue -2. This completes the proof. $\square$

Our results can be reformulated as follows.

Theorem 6.6.3. *A connected graph G is exceptional if and only if either $G = L(K_8)\nabla K_1$ or the following two conditions hold:*
(i) G contains one of the graphs $F_1, F_2, \ldots, F_{20}$ as an induced subgraph;
(ii) G is star covered by the graph $H440$ for the eigenvalue -2.

Since the maximal exceptional graphs of type (a) have an elegant description in terms of $L(K_8)$, it is reasonable to concentrate our attention on the construction of the graphs of type (b) and (c). Here we can use the 'simplest' exceptional star complement $H443$, but so far we have not considered non-exceptional star complements other than $L(R_8)$. It can be shown (using

Theorem 2.3.20(i)(ii)) that no non-exceptional star complement for -2 is common to all the maximal exceptional graphs. However all the graphs of type (b) and (c) have K_8 as a non-exceptional star complement. For graphs of type (b) this follows from the presence of an 8-clique in $\overline{S(K_{1,7})}$. For the graph $G001$, we may take as the vertices of an 8-clique the vectors $\mathbf{a}_{12}, \mathbf{a}_{13}, \mathbf{a}_{14}, \mathbf{a}_{15}, \mathbf{b}_{67}, \mathbf{b}_{68}, \mathbf{b}_{78}, \mathbf{e}$ (cf. Example 5.1.14). For $G002$, we use $\mathbf{a}_{18}, \mathbf{a}_{28}, \mathbf{a}_{38}, \mathbf{a}_{78}, \mathbf{b}_{45}, \mathbf{b}_{46}, \mathbf{b}_{56}, \mathbf{e}$, and for $G425, G430, G455, G463$, we use $\mathbf{a}_{12}, \mathbf{a}_{15}, \mathbf{a}_{16}, \mathbf{a}_{17}, \mathbf{b}_{34}, \mathbf{b}_{38}, \mathbf{b}_{48}, \mathbf{e}$. Accordingly, we have a computer-free proof that every maximal exceptional graph can be constructed from K_8 in (at least) one of two ways. M. Lepović has shown by computer that exactly 320 graphs of type (a) have K_8 as a star complement. We summarize our observations as follows.

Theorem 6.6.4 [Row4]. *If G is a maximal exceptional graph then either* (a) *G is the cone over a graph switching-equivalent to $L(K_8)$ or* (b) *G has K_8 as a star complement for -2, or both.*

The construction of G from K_8 in case (b) of Theorem 6.6.4 is described in Theorem 5.1.15.

7

Miscellaneous results

In this chapter we survey some results which are closely related to the subject of the book but which do not fit readily into earlier chapters.

7.1 Graphs with second largest eigenvalue not exceeding 1

The graphs with $\lambda_2(G) \leq 1$ were investigated in 1982 by D. Cvetković [Cve8]. In many cases, such graphs are the complements of graphs whose least eigenvalue is greater than or equal to -2, but the general situation is best described in the context of the well-known Courant-Weyl inequalities for the eigenvalues of the sum of two Hermitian $n \times n$ matrices A and B (see [Pra, Theorem 34.2.1]):

$$\lambda_i(A+B) \leq \lambda_j(A) + \lambda_{i-j+1}(B) \ (i \geq j),$$

$$\lambda_i(A+B) \geq \lambda_j(A) + \lambda_{i-j+n}(B) \ (i \leq j).$$

In particular, if A and B are the adjacency matrices of G and $\overline{G}$ then $A + B = J - I$ and so $\lambda_2(G) + \lambda_{n-1}(\overline{G}) \geq -1$, $\lambda_2(G) + \lambda_n(\overline{G}) \leq -1$. We deduce the following result.

Theorem 7.1.1 [Cve8]. *Suppose that G is a graph with $\lambda_2(G) \leq 1$. Then either*

(a) $\lambda(\overline{G}) \geq -2$, *or*
(b) $\overline{G}$ *has exactly one eigenvalue less than* -2.

Conversely if $\lambda(\overline{G}) \geq -2$ then $\lambda_2(G) \leq 1$. A graph which satisfies condition (b) may or may not have $\lambda_2(G) \leq 1$: see the graphs in Fig. 7.1, where in each case $\lambda(\overline{G})$ and $\lambda_2(G)$ are as shown. Note that if a graph has $\lambda(G) = -2$ with multiplicity at least 2, then $\lambda_n(G) = \lambda_{n-1}(G)$, and so necessarily $\lambda_2(\overline{G}) = 1$.

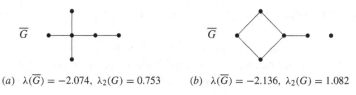

(a) $\lambda(\overline{G}) = -2.074$, $\lambda_2(G) = 0.753$ (b) $\lambda(\overline{G}) = -2.136$, $\lambda_2(G) = 1.082$

Figure 7.1: Examples related to Theorem 7.1.1.

Since $\mathcal{E}_G(\lambda) \cap \langle \mathbf{j} \rangle^{\perp} = \mathcal{E}_{\overline{G}}(-1 - \lambda) \cap \langle \mathbf{j} \rangle^{\perp}$ for any eigenvalue λ of G, it is clear that the regular graphs with $\lambda_2 \leq 1$ are just the complements of regular graphs with least eigenvalue $\lambda(G) \geq -2$. Moreover we can state a partial converse of Theorem 7.1.1 as follows.

Theorem 7.1.2 [Cve8]. (i) *If G is a graph with $\lambda(G) > -2$ then $\lambda_2(\overline{G}) < 1$.*
(ii) *If G is a graph with $\lambda(G) = -2$ then $\lambda_2(\overline{G}) \leq 1$, with strict inequality if and only if -2 is a simple main eigenvalue of G.*

The generalized line graphs G with $\lambda(G) = -2$ and $\lambda_2(\overline{G}) < 1$ were determined in [Cve8]; they can also be found using Theorem 2.2.8 and Corollary 5.2.9.

By Proposition 2.2.2, an eigenvalue -2 in a line graph is always a non-main eigenvalue. It follows from Theorem 7.1.2(ii) that if G is the complement of a line graph having least eigenvalue -2, then $\lambda_2(G) = 1$. Such graphs without triangles are characterized in the following theorem.

Theorem 7.1.3 [CvSi1]. *A graph G without triangles has a line graph for its complement if and only if G is an induced subgraph of the Petersen graph, or $K_{3,n} - E(C_6)$, or a graph of the form $K_1 \dot{\cup} (K_{m,n} - E(pK_2))$.*

The proof uses the forbidden subgraph technique; in particular, the complements of the graphs $H_1, \ldots, H_9$ shown in Fig. 2.1 are forbidden as induced subgraphs of G (see Theorem 2.1.3).

By interlacing, a graph G with $\lambda_2(G) \leq 1$ either has girth at most 6 or is a union of trees with diameter at most 4 [Cve8]: the reason is that a cycle of length greater than 6, and a path of length greater than 4, have at least two eigenvalues greater than 1. The bipartite graphs G with $\lambda_2(G) \leq 1$ were characterized in 1991 by M. Petrović [Pet2]. He showed that such graphs are precisely the induced subgraphs of a graph in one of four families of bipartite graphs (three infinite families and one family of four particular graphs). Some nine years earlier, the particular case of trees with second largest eigenvalue less than 1 had been treated by A. Neumaier [Neu1], who proposed an algorithm for deciding if the second largest eigenvalue of an arbitrary tree is less than some given bound.

The line graphs and generalized line graphs with second largest eigenvalue at most 1 were determined in [PeMi1] and [PeMi2], respectively. Some related results are given in the next section. The characterization of all graphs G with $\lambda_2(G) \leq 1$ (or $\lambda_2(G) = 1$) remains an interesting open question in spectral graph theory. A representation of graphs with $\lambda_2(G) = 1$ in Lorentz space was given by A. Neumaier and J. J Seidel in 1983 [NeSe].

We now turn our attention to upper bounds for λ_2 which are less than 1; in particular we discuss the graphs G for which $\lambda_2 \leq \sigma$, where σ denotes the golden ratio $(\sqrt{5} - 1)/2 \approx 0.618033989$.

It is an elementary fact (see, for example, [CvDSa, p. 163]) that $\lambda_2(K_n) = -1$ $(n > 1)$ and $\lambda_2(K_{n_1,n_2,...,n_k}) = 0$ $(n_1 n_2 \cdots n_k > 1)$, while $\lambda_2(G) > 0$ for all other non-trivial connected graphs G.

In 1993, D. Cao and Y. Hong [CaHo] showed that the second largest eigenvalue of a graph G on n vertices lies between 0 and $\frac{1}{3}$ if and only if $G = (n-3)K_1 \bigtriangledown (K_1 \dot\cup K_2)$. They posed the problem of characterizing the graphs G with $\frac{1}{3} < \lambda_2(G) < \sigma$.

The graphs G with $\lambda_2(G) < \sqrt{2} - 1$ (≈ 0.41421356) were determined by M. Petrović [Pet3]. Such graphs were characterized independently by J. Li [Li], who described a corresponding family of minimal forbidden subgraphs.

It was proved by S. K. Simić [Sim6] that the family of minimal forbidden subgraphs for the property $\lambda_2(G) < \sigma$ is finite. The structure of graphs G with $\lambda_2(G) \leq \sigma$ was investigated by D. Cvetković and S. K. Simić [CvSi7], and some of their results were announced in [CvSi6].

We introduce some terminology. Graphs having the property $\lambda_2(G) \leq \sigma$ (the σ-*property*) will be called σ-*graphs*. For convenience, graphs G for which $\lambda_2(G) < \sigma$, $\lambda_2(G) = \sigma$, $\lambda_2(G) > \sigma$ will be called σ^--*graphs*, σ^0-*graphs*, σ^+-*graphs*, respectively.

The next observation, taken from [Wol] (see also [BeHa]), enables us to define a class of graphs to which every σ^--graph belongs.

Proposition 7.1.4. *If $\overline{G}$ is a connected graph and if G has no isolated vertices, then G contains $2K_2$ or P_4 as an induced subgraph.*

Now assume that G is a σ^--graph. If $\overline{G}$ is a connected graph, then G must have at least one isolated vertex (otherwise G contains $2K_2$ or P_4 as an induced subgraph, and hence is not a σ^--graph). On the other hand, if $\overline{G}$ is a disconnected graph, then G itself is a join of at least two graphs. Since the σ^--property is hereditary, it follows that G belongs to the class $\mathcal{C}$ of graphs defined as the smallest family of graphs that contains K_1 and is closed under adding isolated

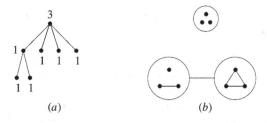

Figure 7.2: An expression tree and associated diagram.

vertices and taking joins. Thus if $G \in \mathcal{C}$, then $G \, \dot{\cup} \, K_1 \in \mathcal{C}$; and if $G_1, G_2 \in \mathcal{C}$, then $G_1 \, \triangledown \, G_2 \in \mathcal{C}$. The class $\mathcal{C}$ may be described alternatively as the class of graphs without $2K_2$ or P_4 as an induced subgraph. Clearly, any σ^--graph belongs to $\mathcal{C}$, but not vice versa. The class $\mathcal{C}$ was introduced in [Sim6], where a graph G from $\mathcal{C}$ is represented by a weighted rooted tree T_G (also called an *expression tree* of G) defined inductively as follows:

> any subgraph $H = (((H_1 \, \triangledown \, H_2) \, \triangledown \cdots) \, \triangledown \, H_m) \, \dot{\cup} \, nK_1$ $(m \geq 0, \ n > 0)$ of G is represented by a subtree T_H with a root v of weight n whose neighbours in T_H are the roots $v_1, v_2, \ldots, v_m$ of the subtrees representing $H_1, H_2, \ldots, H_m$ respectively.

Example 7.1.5. If $G = (((((K_1 \, \triangledown \, K_1) \, \dot{\cup} \, K_1) \, \triangledown \, K_1) \, \triangledown \, K_1) \, \triangledown \, K_1) \, \dot{\cup} \, 3K_1$, then the corresponding expression tree is depicted in Fig. 7.2(a). In Fig. 7.2(b) we represent the same graph by a diagram in which a line between two circled sets of vertices denotes that each vertex inside one set is adjacent to every vertex inside the other set. □

It turns out that weighted trees can be used to categorize σ^--graphs: the weighted tree of any such graph is of one of the nine types illustrated in Fig. 7.3.

It was proved in [Sim6] by similar means that the family of minimal forbidden subgraphs for the σ^--property is finite. Except for $2K_2$ and P_4, they all belong to $\mathcal{C}$. The complete list of minimal forbidden subgraphs is described in [CvSi9]: the list is extensive and contains graphs of much larger order than those encountered in the characterizations from Chapter 2.

We now present the main results of [CvSi7].

Theorem 7.1.6. *A σ-graph has at most one non-trivial component; and for such a component G, one of the following holds:*

(a) *G is a complete multipartite graph,*
(b) *G is an induced subgraph of C_5,*
(c) *G contains a triangle.*

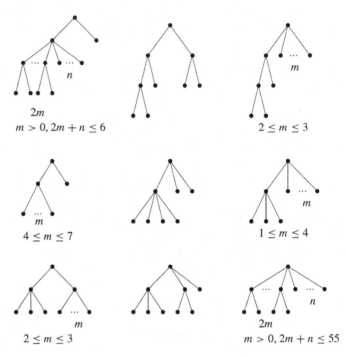

Figure 7.3: Expression trees for σ^--graphs.

To describe σ-graphs containing a triangle we introduce some notation. Let G be a σ-graph with vertex set V, and let T be a triangle in G induced by the vertices x, y, z. Next, let $A(= A(G, T))$, $B(= B(G, T))$, $C(= C(G, T))$ be the sets of vertices outside T which are adjacent to exactly one, two, three vertices from T, respectively. Also, let G_A, G_B, G_C be the component (containing T) of the subgraph of G induced by the vertex set $V - B - C$, $V - A - C$, $V - A - B$, respectively. Let $d(u, T)$ denote the distance of the vertex u from the triangle T, that is, the length of the shortest path between u and a vertex from T.

The σ-graphs containing triangles can now described in more detail in terms of the induced subgraphs G_A, G_B, G_C.

Theorem 7.1.7. *Let G be a connected σ-graph which contains a triangle. For any triangle T of G the following hold for subgraphs G_A, G_B, G_C.*

(1) *G_A is an induced subgraph of a graph from Fig. 7.4.*
(2) *For G_B one of the following holds:*
    (i) *G_B is an induced subgraph of a graph from Fig. 7.5;*
    (ii) *$G_B = P_4 \triangledown (H \,\dot\cup\, K_1)$ for some σ-graph H;*
    (iii) *$G_B = H_1 \triangledown H_2 \triangledown H_3$ for some σ-graphs H_1, H_2, H_3.*

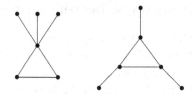

Figure 7.4: Some graphs from Theorem 7.1.7.

Figure 7.5: Further graphs from Theorem 7.1.7.

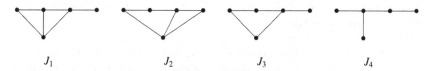

Figure 7.6: Some graphs from Theorem 7.1.8.

(3) *For G_C one of the following holds:*

    (i) *G_C is an induced subgraph of $(K_3 \ \dot\cup \ K_1) \ \triangledown \ H$ for some σ-graph H;*

    (ii) *G_C is obtained from $K_n \ \triangledown \ K_3 \ \triangledown \ H$ by adding a pendant edge to each vertex of K_n, where $n \geq 2$ and H is a σ-graph containing no induced subgraph isomorphic to one of the graphs $K_3 \ \dot\cup \ K_1$, $K_2 \ \dot\cup \ 3K_1$, $K_{1,2} \ \dot\cup \ 2K_1$, $K_{2,4} \ \dot\cup \ K_1$, $K_{3,3} \ \dot\cup \ K_1$.*

It is also proved in [CvSi7] that the set of minimal forbidden subgraphs for the σ-property is finite. The next theorem provides more details.

Theorem 7.1.8 [CvSi7]. *If H is a minimal forbidden subgraph for the σ-property, then either*

(a) *H is one of the graphs $2K_2$, J_1, J_2, J_3, J_4 (see Fig. 7.6), or*

(b) *H belongs to the class $\mathcal{C}$.*

A complete set of minimal forbidden subgraphs for the σ-property is not yet known. On the other hand, more can be said if we require that both a graph and its complement are σ-graphs; see Theorem 7.2.8 below.

7.2 Graphs sharing properties with their complements

A natural type of problem in graph theory, which has featured frequently in the literature, is to determine all the graphs that share a given property with their complements. In this section we mention some results in this area relevant to the topic of the book. Further results can be found in a series of papers by J. Akiyama and F. Harary, who systematically investigated this kind of problem (see, for example, [AkHa]).

Suppose that $\mathcal{C}$ is the class of graphs determined by some hereditary property $\mathcal{P}$, and let $\mathcal{F}$ be a family of forbidden subgraphs (not necessarily minimal) which characterizes $\mathcal{C}$. Let $\overline{\mathcal{C}} = \{\overline{G} : G \in \mathcal{C}\}$ and $\overline{\mathcal{F}} = \{\overline{H} : H \in \mathcal{F}\}$. Then $\overline{\mathcal{F}}$ is a family of forbidden subgraphs which characterizes $\overline{\mathcal{C}}$. Moreover, if $\mathcal{F}^*$ is a family of minimal forbidden subgraphs which characterizes $\mathcal{C} \cap \overline{\mathcal{C}}$ then $\mathcal{F}^* \subseteq \mathcal{F} \cup \overline{\mathcal{F}}$. We may have $\mathcal{F}^* \neq \mathcal{F} \cup \overline{\mathcal{F}}$ even when $\mathcal{F}$ and $\overline{\mathcal{F}}$ consist of minimal forbidden subgraphs: this is the case when, for example, $\mathcal{P}$ is the property of being a line graph because (in the notation of Fig. 2.1) H_1 is a proper induced subgraph of $\overline{H_4}$. In fact, in this situation, $\mathcal{F}^*$ consists of 10 graphs rather than 18, as the next result shows.

Theorem 7.2.1. *G and $\overline{G}$ are both line graphs if and only if G (or $\overline{G}$) does not contain as an induced subgraph any of the following graphs: H_1, H_2, H_3, H_6, H_9 (see Fig. 2.1), $\overline{H_1}$, $\overline{H_2}$, $\overline{H_3}$, $\overline{H_6}$ and $\overline{H_9}$.*

Graphs G with the property that both G and $\overline{G}$ are line graphs were named *coderived graphs* by L. W Beineke, who showed that, apart from cliques and cocliques, there are just 37 pairs of graphs of this type [Bei2]. Since the coderived property is hereditary it suffices to describe the maximal graphs that arise.

Theorem 7.2.2. *The graphs G and $\overline{G}$ are both line graphs if and only if one of them is either*

(a) *a clique, or*

(b) *an induced subgraph of one of the following graphs: C_5, $3K_2$, $K_3 \circ K_1$, $C_6 \overset{.}{\cup} K_1$ and $K_3 + K_3$.*

The above result was subsequently extended as follows. Let $\mathcal{L}$ denote the family of all line graphs, $\mathcal{L}^*$ the family of all generalized line graphs and $\mathcal{A}$ the family of all graphs whose least eigenvalue is not less than -2. Then $\mathcal{L} \subset \mathcal{L}^* \subset \mathcal{A}$. Thus Theorem 7.2.1 specifies some, but not all, graphs G such that both G and $\overline{G}$ belongs to $\mathcal{L}^*$ (or $\mathcal{A}$). Accordingly, the next two natural steps are to consider the analogous problem for the sets $\mathcal{L}^*$ and $\mathcal{A}$.

For generalized line graphs, an analogue of Theorem 7.2.1 was obtained by Z. Radosavljević, S. K. Simić, M. Syslo and J. Topp [RaSST]. (The result was proved independently by the first two and the second two authors, initially in two separate submissions to the same journal at the same time.) In this case Theorem 2.3.18 was used to obtain a characterization by a family $\mathcal{F}^*$ of 50 minimal forbidden subgraphs.

Theorem 7.2.3. *The graphs G and $\overline{G}$ are both generalized line graphs if and only if G (or $\overline{G}$) does not contain as an induced subgraph any of the graphs $G^{(i)}$ ($i \in \{1, 2, \ldots, 31\} \setminus \{14, 18, 21, 23\}$) and $\overline{G_i}$ ($i \in \{1, 2, \ldots, 31\} \setminus \{4, 14, 17, 18, 19, 21, 23, 27\}$) (see Fig. 2.4).*

It is clear that all GCPs, as well as their complements, belong to $\mathcal{L}^*$. In order to determine the remaining graphs G in $\mathcal{L}^*$, the following structural details were established in [RaSST]:

  (i) if G is disconnected then G contains one non-trivial component and at most two isolated vertices;

 (ii) if G contains a GCP with at least two edges removed then G is an induced subgraph of $L(K_3; 2, 0, 0)$;

(iii) if G contains two disjoint GCPs, each of them having just one edge removed, then G is an induced subgraph of $L(K_{1,2}; 1, 1, 1)$ or $L(K_{1,4}; 0, 1, 1, 0, 0)$ (where petals are added at two endvertices);

 (iv) if G contains two GCPs with a common vertex, each of them having just one edge removed, then G is an induced subgraph of $L(K_3; 1, 1, 1)$;

  (v) if G contains just one GCP with only one edge removed, then G is an induced subgraph of a generalized line graph with a root graph from Fig. 7.7.

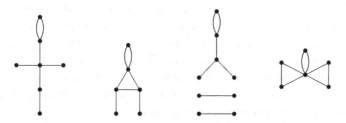

Figure 7.7: Some root multigraphs.

Gathering together these five facts, we arrive at the following result.

Theorem 7.2.4. *G and $\overline{G}$ are both generalized line graphs if and only if one of them is either*

(a) *a generalized cocktail party graph, or*

(b) *an induced subgraph of (at least one) of the graphs $Y_1, \ldots, Y_6$ from Fig. 7.8.*

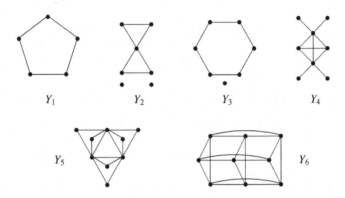

Figure 7.8: The graphs $Y_1, \ldots, Y_6$ from Theorem 7.2.4.

Remark 7.2.5. Note that Y_1 and Y_5 are self-complementary graphs (and that they both feature in Theorem 7.2.2); each of the remaining graphs in Fig. 7.8 is an extension (not necessarily proper) of a GCP or a maximal graph from Theorem 7.2.2. □

The final case (in which both G and $\overline{G}$ belong to $\mathcal{A}$) was settled in [RaSi2]. In order to determine the remaining graphs G such that both $\lambda(G) \geq -2$ and $\lambda(\overline{G}) \geq -2$, we start from the assumption that at least one graph of the pair $(G, \overline{G})$ is not a generalized line graph, i.e. it belongs to the set $\mathcal{A} \setminus \mathcal{L}^*$. Such a graph G must then contain as an induced subgraph one of the 50 graphs mentioned in Theorem 7.2.3. But on inspecting the spectra of all these graphs, one finds that their number is reduced to the fourteen graphs on six vertices shown in Fig. 7.9. Since the spectra of these graphs and their complements are bounded from below by -2, we know from Theorem 7.1.2 that the eigenvalues of all these graphs, other than their indices, lie in the interval $[-2, 1]$. The graphs of Fig. 7.9 are chosen so that, from a complementary pair, the graph with the smaller number of edges is illustrated.

Since the property '$\lambda(G) \geq -2$ and $\lambda(\overline{G}) \geq -2$' is a hereditary property, we can proceed in the following way. We start with an 'initial layer' (layer 0) consisting of all fourteen graphs from Fig. 7.9. In any layer, a graph with

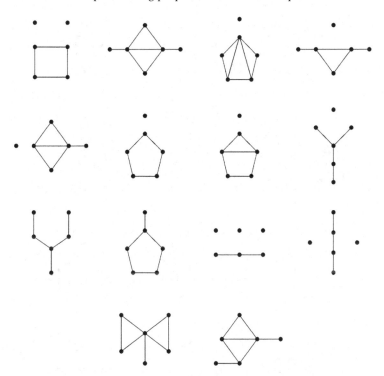

Figure 7.9: Graphs in the 'initial layer'.

no proper extension possessing the property is maximal, while the next layer consists of all one-vertex extensions which do have the property. We find two maximal graphs on 9 vertices (in the third layer), eight on 11 vertices, two on 12 vertices, and two on 14 vertices (in the eighth layer). These 14 maximal graphs $M_1, \ldots, M_{14}$ are displayed in Fig. 7.10.

Theorem 7.2.6 [RaSi2]. *If at least one graph of the pair $(G, \overline{G})$ is not a generalized line graph, then $\lambda(G) \geq -2$, and $\lambda(\overline{G}) \geq -2$ if and only if G or $\overline{G}$ is an induced subgraph of (at least) one of the graphs $M_1, \ldots, M_{14}$ from Fig. 7.10.*

In fact, the graphs M_1-M_{14} are interesting in themselves. For instance, M_7 and M_9 are cospectral integral graphs with cospectral integral complements, as discussed in Section 7.6. Also, denoting by $G - K_1$ the deletion of an isolated vertex from G, we see that for the eight graphs on 11 vertices we have

$$M_{i+1} = \overline{(M_i - K_1)} \,\dot\cup\, K_1, \quad i = 3, 5, 7, 9.$$

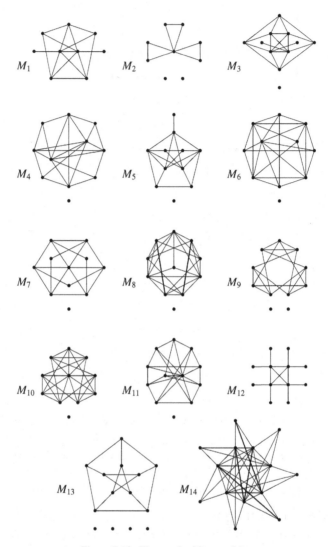

Figure 7.10: The graphs $M_1, \ldots, M_{14}$.

(Hence M_8 and M_{10} are also cospectral.) Further, some of the graphs can be expressed in terms of graph operations on small graphs: $M_2 = (3K_2 \bigtriangledown K_1) \dot\cup K_1$, $M_{11} = (P \bigtriangledown K_1) \dot\cup K_1$, $M_{12} = K_4 \circ 2K_1$, $M_{13} = P \dot\cup 4K_1$, where P is the Petersen graph. The graph M_{14} can be described as follows. It has seven vertices of degree 9 which induce the subgraph K_7 and seven vertices of degree 3 which induce $\overline{K_7}$. Every vertex of degree 3 is adjacent to the vertices of a triangle of the induced subgraph K_7. If we consider these seven triangles as blocks in the set of vertices of degree 9, then every vertex is contained in three blocks and, since the triangles are edge-disjoint, every pair of vertices is contained in one block; thus M_{14} can be described in terms of a Steiner triple system (a BIBD with parameters $k = 3$, $\lambda = 1$).

Now, looking again at the graphs of Figs. 7.8 and 7.9, we see that Y_1 is contained in M_1 as an induced subgraph, Y_2 is contained in M_2, Y_3 in M_7, Y_4 in M_{12}, Y_5 in $\overline{M_{14}}$, and Y_6 in M_9. Accordingly, we can summarize our conclusions as follows.

Theorem 7.2.7 [RaSi2]. *The least eigenvalue of each of the graphs G and $\overline{G}$ is less than or equal to -2 if and only if (at least) one of them is either*
(a) *a GCP (generalized cocktail party graph), or*
(b) *an induced subgraph of (at least) one of the graphs $M_1, \ldots, M_{14}$ (see Fig. 7.10).*

Recall from Section 7.1 that G is said to be a σ-graph if $\lambda_2(G) < \sigma$, where $\sigma = (\sqrt{5} - 1)/2$. In [Sim7] S. K. Simić showed that there are exactly 27 minimal forbidden subgraphs for graphs G with the property '$\lambda_2(G) < \sigma$ and $\lambda_2(\overline{G}) < \sigma$'. Here we give an explicit description of the graphs in question.

Theorem 7.2.8 [Sim7]. *The graphs G and $\overline{G}$ are both σ-graphs if and only if (at least) one of them is one of the following graphs:*

$K_m \dot{\cup} n K_1$ $(m, n \geq 0)$, $K_{2,1,1} \dot{\cup} m K_1$ $(m \geq 0)$, $K_{2,1} \dot{\cup} m K_1$ $(m \geq 0)$,
$K_{3,1} \dot{\cup} m K_1$ $(m \leq 3)$, $K_{2,1,1,1} \dot{\cup} m K_1$ $(m \leq 2)$,
$((K_{2,1,1} \dot{\cup} K_1) \triangledown K_1) \dot{\cup} K_1$, $((K_{2,1} \dot{\cup} 2K_1) \triangledown K_1) \dot{\cup} K_1$, $((K_{2,1} \dot{\cup} K_1) \triangledown K_1) \dot{\cup} K_1$,
$(K_{m,1} \dot{\cup} K_1) \triangledown K_n$ $(m \geq 2, n \geq 0)$,
$(K_{2,1,1} \dot{\cup} K_1) \triangledown K_m$ $(m \leq 2)$, $(K_{2,1} \dot{\cup} 2K_1) \triangledown K_m$ $(m \leq 2)$,
$(K_{3,1} \dot{\cup} 2K_1) \triangledown K_1$, $(K_{2,1,1,1} \dot{\cup} K_1) \triangledown K_1$, $(((K_{2,1} \dot{\cup} K_1) \triangledown K_1) \dot{\cup} K_1) \triangledown K_1$.

We note that line graphs G for which both $\lambda_2(G) \leq 1$ and $\lambda_2(\overline{G}) \leq 1$ were determined explicitly in [PeMi1]. The analogous result for generalized line graphs can be found in [PeMi2]. Note that the sets of graphs obtained are wider than the sets of graphs described in Theorems 7.2.2 and 7.2.4.

Finally, we add (as a curiosity) the observation that the graphs G with the property that both G and $\overline{G}$ are orientable to line digraphs were determined in [RaST].

7.3 Spectrally bounded graphs

The graphs G with least eigenvalue $\lambda(G) \geq -2$ constitute just one (infinite) family of graphs whose spectrum is bounded from below. Here we discuss infinite families of graphs having some uniform, but unspecified, lower bound on their eigenvalues. The main contributions are due to A. J. Hoffman [Hof9].

Consider graphs G and H on the same vertex set V, so that $G \cup H$ is the graph on V with edge-set $E(G) \cup E(H)$. We shall now introduce the *distance* between G and H, denoted by $d(G, H)$. For each vertex $v \in V$, let $n(v, G) = \deg_{G \cup H}(v) - \deg_G(v)$, with $n(v, H)$ defined analogously. Let $n(v, G, H) = \max\{n(v, G), n(v, H)\}$. Then $d(G, H) = \max_v\{n(v, G, H)\}$. It is easy to show that d is a distance function on the set of all graphs with a fixed vertex set. Now we can state the following theorem.

Theorem 7.3.1 [Hof9]. *Let $\mathcal{G}$ be an infinite set of graphs. Then the following statements about $\mathcal{G}$ are equivalent:*

(i) *there exists a real number λ such that, for all $G \in \mathcal{G}$, $\lambda(G) \geq \lambda$;*
(ii) *there exists a positive integer t such that, for all $G \in \mathcal{G}$, neither $K_{1,t}$ nor $K_t \nabla(K_t \dot\cup K_1)$ is an induced subgraph of G;*
(iii) *there exists a positive integer u such that, for each $G \in \mathcal{G}$, there exists a graph H such that $d(G, H) \leq u$ and H contains a distinguished family of cliques K^i ($i \in I$) satisfying:*
    (a) *each edge of H is in at least one of the cliques K^i;*
    (b) *each vertex of H is in at most u of the cliques K^i;*
    (c) *$|V(K^i) \cap V(K^j)| \leq u$ whenever $i \neq j$.*

Remarks 7.3.2. Let (G_i) be an infinite sequence of graphs with $\lim_{i \to +\infty} \lambda(G_i) = -\infty$, and let (H_i) be the infinite sequence $K_{1,1}$, $K_1 \nabla(K_1 \dot\cup K_1)$, $K_{1,2}$, $K_2 \nabla(K_2 \dot\cup K_1)$, $K_{1,3}$, $K_3 \nabla(K_3 \dot\cup K_1)$, ... From the equivalence of statements (i) and (ii) in Theorem 7.3.1, it follows that there exist subsequences of these sequences, say (G_{i_k}) and (H_{j_k}), such that for each k, H_{i_k} is an induced subgraph of G_{j_k}.

In statement (iii) it would be desirable to assert (a), (b), (c) for G rather than H. To see that we cannot do this, take $\lambda = -2$, $G = CP(n)$ (for some fixed $n \geq 2$) and suppose that G satisfies (a), (b), (c). Let $k = \max\{|V(K^i)| : i \in I\}$, say $k = |V(K^1)|$. Fix $y \in V(K^1)$ and let x be the vertex non-adjacent to y. The number of pairs (z, K^j) such that $x \sim z$, $z \in V(K^1)$ and $xz \in E(K^j)$ is at least $k - 1$ by (a). On the other hand, by (b) there are at most u cliques K^j containing x, while $|V(K^1) \cap V(K^j)| \leq u$ by (c). It follows that $k - 1 \leq u^2$.

Since y lies in at most u cliques, we have $\deg(y) \le (k-1)u$; but this cannot be true if $2n-1 > u^3$. □

The next result, also due to A. J. Hoffman, tells us that there are no graphs with arbitrarily large minimal degree having least eigenvalue in the interval $(-1-\sqrt{2}, -2)$. Let Λ_{min} be the set of real numbers x for which there exists a graph whose least eigenvalue is x; and let $\delta(G)$ denote the minimal degree of the graph G.

Theorem 7.3.3 [Hof11]. *There exists an integer-valued function f, defined on the set $\Lambda_{min} \cap (-1 - \sqrt{2}, -1]$, such that for a connected graph G:*

(i) *if $-2 < \lambda \le -1$, $\lambda(G) = \lambda$ and $\delta(G) \ge f(\lambda)$, then G is a clique and $\lambda(G) = -1$ (in particular, G is a generalized line graph);*
(ii) *if $-1 - \sqrt{2} < \lambda \le -2$, $\lambda(G) = \lambda$ and $\delta(G) \ge f(\lambda)$, then G is a generalized line graph and $\lambda(G) = -2$.*

Remark 7.3.4. In the original proof (which we omit), the estimated values of the function f from Theorem 7.3.3 were astronomically large – due mainly to a crude use of Ramsey type theorems. From the theory of root systems we now know the best possible bounds, namely $f(-1) = 1$ and $f(-2) = 29$ (since vertex degrees in exceptional graphs are at most 28). □

The first detailed study of the second largest eigenvalue of a graph, in the context of the hereditary property $\lambda_2(G) \le a$ (for some constant a), was undertaken by L. Howes in the early 1970s – see [How1] and [How2]. The following characterization is taken from the second of these papers.

Theorem 7.3.5 [How2]. *For an infinite set of graphs $\mathcal{G}$, the following statements are equivalent.*

(i) *There exists a real number a such that $\lambda_2(G) \le a$ for every $G \in \mathcal{G}$.*
(ii) *There exists a positive integer s such that for each $G \in \mathcal{G}$, the graphs $(K_s \dot\cup K_1) \triangledown K_s$, $(sK_1 \dot\cup K_{1,s}) \triangledown K_1$, $(K_{s-1} \dot\cup sK_1) \triangledown K_1$, $K_s \dot\cup K_{1,s}$, $2K_{1,s}$, $2K_s$ and the graphs in Fig. 7.11 (each obtained from two copies of $K_{1,s}$ by adding extra edges) are not induced subgraphs of G.*

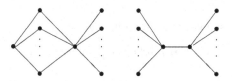

Figure 7.11: The graphs from Theorem 7.3.5.

7.4 Embedding a graph in a regular graph

This section is based on a paper of M. Doob and D. Cvetković [DoCv]. As usual, $\lambda(G)$ denotes the least eigenvalue of the graph G. We use the notation $H \subseteq G$ to indicate that H is embedded in G, i.e. H is an induced subgraph of G. In this situation, $\lambda(G) - \lambda(H) \leq 0$. Since any graph can be embedded in a regular graph, one may ask how this may be done with minimal perturbation of the least eigenvalue. This question was posed by A. J. Hoffman, who defined

$$\lambda_R(H) = \sup\{\lambda(G) : H \subseteq G; \; G \text{ regular}\}.$$

It is immediate that if H is regular then $\lambda_R(H) = \lambda(H)$. When H is not regular, $\lambda_R(H)$ is not easy to determine, and was originally found only for $H = K_{1,2}$ and $H = K_{1,3}$. Here we shall find $\lambda_R(H)$ for a large class of graphs which are not necessarily regular; in particular we determine λ_R for all line graphs. Without loss of generality we shall assume for the remainder of this section that H is connected.

Proposition 7.4.1. *If $s > 1$ then $\lambda_R(P_s) = \lambda(C_n) = -2\cos\frac{\pi}{n}$, where $n = 2\lfloor(s-1)/2\rfloor + 3$.*

Proof. By Corollary 2.3.22, $\lambda_R(H) > -2$ only if H is a clique or H is embedded in an odd cycle. If we take $H = P_s$ then in the former case, $s = 2$ and $\lambda_R(P_2) = -1 = \lambda(C_3)$. In the latter case, $s > 1$ and $\lambda_R(P_s)$ is determined by the odd cycles that contain P_s; in particular, we need to embed P_s in the odd cycle with the smallest possible number of vertices, namely $\lfloor(s-1)/2\rfloor + 3$. $\qquad\square$

Note that when $s = 3$, Proposition 7.4.1 asserts that $\lambda_R(K_{1,2}) = \lambda(C_5) = -2\cos\frac{\pi}{5}$. From the proof of the Proposition we know that if $\lambda_R(H) > -2$ then H is a clique, an odd cycle or a path. The converse statement is immediate, and so we have the following result.

Proposition 7.4.2. *$\lambda_R(H) \leq -2$ if and only if $H \neq C_{2n+1}$, $H \neq P_n$ and $H \neq K_n$.*

Corollary 7.4.3. *If H is not a clique and H has a vertex of degree greater than 2, then $\lambda_R(H) \leq -2$.*

The next result supplements Proposition 7.4.2.

Theorem 7.4.4 [DoCv]. *If H is a line graph other than C_{2n+1}, P_n or K_n, then $\lambda_R(H) = -2$.*

Proof. We know from the above corollary that $\lambda_R(H) \leq -2$. Now let $H = L(H')$, where H' has n vertices, and let $G = L(K_n)$. Then G is regular and $\lambda(G) \geq -2$. Since $H \subseteq G$, the result follows. $\qquad\square$

Lemma 7.4.5. *If H is the complete multipartite graph $K_{n_1, n_2, \ldots, n_t}$ with each $n_i \leq 2$, then*

$$
\lambda_R(H) = \begin{cases}
0 & \text{if } n_1 = 1, \, t = 1; \\
-1 & \text{if } n_1 = n_2 = \cdots = n_t = 1, \, t \geq 2; \\
\frac{1}{2}(-1 - \sqrt{5}) & \text{if } \{n_1, n_2\} = \{1, 2\} \text{ and } t = 2; \\
-2 & \text{otherwise.}
\end{cases}
$$

Proof. Since $H \subseteq CP(t)$, we have $\lambda_R(H) \geq -2$. Moreover, Proposition 7.4.2 shows that $\lambda_R(H) \leq -2$ except for the three special cases. □

Any regular graph H with $\lambda(H) = -2$ is either a generalized line graph or a graph with a representation in E_8. The maximal regular exceptional graphs are the three Chang graphs, the Schläfli graph, and five special graphs of degree 9 on 22 vertices (cf. Theorems 4.1.5 and 4.4.20). Each of these graphs has least eigenvalue equal to -2, and so we have the following observation.

Lemma 7.4.6. *If H is one of*
 (a) *a line graph,*
 (b) *$K_{n_1, n_2, \ldots, n_t}$ with each $n_i \leq 2$, or*
 (c) *an induced subgraph of one of: the Chang graphs, the Schläfli graph, five special graphs of degree 9 on 22 vertices (graph nos. 148–152 in Table A3),*

then either $\lambda_R(H) = -2$ or H is one of K_n, C_{2n+1}, P_n for some n.

The nine minimal graphs forbidden for line graphs are shown in Fig. 2.1. One of these has $\lambda(G) < -2$, but the others can be embedded in the Schläfli graph or a Chang graph, and so have $\lambda_R = -2$. In particular this implies that $\lambda_R(K_{1,3}) = -2$.

We conjecture that the converse of the Lemma 7.4.6 is true. In this direction, we have the following result.

Proposition 7.4.7. *Suppose that $H \subseteq G$, where G is regular and $\lambda(G) = -2$. Then H satisfies (a), (b), or (c) in Lemma 7.4.6.*

Proof. We know that either G is a generalized line graph or G has a representation in E_8. In the former case either G is $CP(t)$ for some t, and H satisfies (b), or G is a line graph and H satisfies (a). If G arises from E_8, then G satisfies (c), as is shown in Theorem 4.4.20. □

Hoffman [Hof7] has also investigated the embeddings of a graph in a regular graph of large degree. To this end he defined

$$\mu_R(H) = \limsup_{d \to \infty} \{\lambda(G) : G \text{ is } d\text{-regular}; \ H \subseteq G\}.$$

For example, $\mu_R(K_n) = -1$, while if H is not a clique, then $\mu_R(H) \leq -2$. The second assertion here follows from Corollary 2.3.22: any regular graph G with degree greater than two that is not a clique has $\lambda(G) \leq -2$. In view of Theorem 7.4.4 we have the following result.

Theorem 7.4.8 [Hof7]. *The graph H is such that $\mu_R(H) = -2$ if and only if H is a line graph other than K_n or an induced subgraph of a cocktail party graph.*

Finally we note that Hoffman also defined

$$\mu(H) = \limsup_{d \to \infty} \{\lambda(G) : H \subseteq G, \ \delta(G) > d\},$$

where $\delta(G)$ is the minimum degree of a vertex of G. Clearly $\lambda(H) \geq \mu(H) \geq \mu_R(H)$.

Theorem 7.4.9 [DoCv]. *Let H' be any connected graph that is neither a tree nor a unicyclic graph with an odd cycle, and let $H = L(H')$. Then $\mu(H) = -2$.*

Proof. From Theorem 2.3.20 we have $\lambda(H) = -2$, and from Theorem 7.4.8 we have $\mu_R(H) = -2$. The result follows.   □

7.5 Reconstructing the characteristic polynomial

Let G be a graph with vertex set $\{1, \ldots, n\}$. We refer to the collection (with repetitions) of all vertex-deleted subgraphs $G - i$ $(i = 1, \ldots, n)$ as the *g-deck* of G. We also consider the *p-deck* of G, which consists of the characteristic polynomials $P_{G-i}(x)$ $(i = 1, \ldots, n)$. We denote these decks by $\mathcal{G}(G)$ and $\mathcal{P}(G)$ repectively. A. J. Schwenk [Sch2] noted that the following four reconstruction problems arise (see also [CvDGT, Section 3.5]):

(i) the reconstruction of G from $\mathcal{G}(G)$;
(ii) the reconstruction of $P_G(x)$ from $\mathcal{G}(G)$;
(iii) the reconstruction of G from $\mathcal{P}(G)$;
(iv) the reconstruction of $P_G(x)$ from $\mathcal{P}(G)$.

In each case we ask whether G or $P_G(x)$ is determined uniquely by its g-deck or p-deck when $n > 2$.

Problem (i) is Ulam's famous reconstruction problem, and is not yet settled: except for a positive answer for certain classes of graphs, there are few general results. For line graphs the problem is equivalent to the reconstruction of a graph from the deck of edge-deleted subgraphs. A positive answer to this variant of Problem (i), under certain restrictions, appears in [Lov], and a generalization is given in [Schm]. Problem (ii) was settled, affirmatively, by W. T. Tutte [Tut1]. Clearly, Problem (iii) is much harder than Problem (iv), and in what follows we restrict ourselves to Problem (iv).

Problem (iv) was posed by D. Cvetković at the 18th International Scientific Colloquium in Ilmenau in 1973, and the first results were obtained by I. Gutman and D. Cvetković [GuCv] (see also [CvDSa, p. 267] and [CvDGT, pp. 68–70]). No example of non-unique reconstruction of the characteristic polynomials is known so far. Some relations between Problems (i) and (iv) are described in [CvRS2, Section 5.4].

Since $P'_G(x) = \sum_{i=1}^{n} P_{G-i}(x)$ (see [Clarke4] or [CvDSa, p. 60]), we can readily determine the characteristic polynomial $P_G(x)$ except for the constant term. If we know just one eigenvalue of G, then the constant term is determined. In particular, this is the case if we know a multiple root λ of some polynomial $P_{G-i}(x)$, for then (by the Interlacing Theorem) λ is an eigenvalue of G.

In [GuCv], Problem (iv) was solved affirmatively for regular graphs and for a large class of bipartite graphs including trees without a 1-factor. The result was extended to all trees in [Cve12] and [CvLe]:

Theorem 7.5.1. *If G is a tree, then its characteristic polynomial is determined uniquely by $\mathcal{P}(G)$.*

In the spirit of this book, we outline here a proof by S. K. Simić that the same conclusion holds when G is a connected graph for which no polynomial from the p-deck has a root less than -2 (see [Sim5]). For further results, see [Sci].

Lemma 7.5.2. *For a connected graph G on n vertices,*

  (i) $P_G(-2) = (-1)^n(n+1)$ *when* $G = L(T)$, *where* T *is a tree;*

  (ii) $P_G(-2) = (-1)^n 4$ *when* $G = L(T; 1, 0, \ldots, 0)$, *where* T *is a tree, or when* $G = L(U)$, *where* U *is an odd-unicyclic graph;*

(iii) $P_G(-2) = (-1)^n(9-n)$ *when* G *generates* E_n $(6 \le n \le 8)$.

For parts (i) and (ii) of Lemma 7.5.2, see [CvDo1]; part (iii) can be checked by computer (see [MiHu]).

Theorem 7.5.3 [Sim5]. *The characteristic polynomial of any connected graph whose vertex-deleted subgraphs have spectra bounded from below by* -2 *is reconstructible from its p-deck.*

Sketch proof. Suppose that $\mathcal{P}$ $(= \{P_1(x), \ldots, P_n(x)\})$ is the p-deck of some connected graph whose vertex-deleted subgraphs have no eigenvalue less than -2, and consider a connected graph G such that $\mathcal{P}(G) = \mathcal{P}$. We assume that $n > 10$ because the result can be checked by computer for graphs with 10 or fewer vertices. We need to prove that $P_G(x)$ is uniquely determined by $\mathcal{P}$. We consider two cases.

Case 1: $P_i(-2) \neq 0$ for all i.
We note first that $\lambda(G) \geq -2$, for otherwise by Theorem 2.4.6, G has a proper induced subgraph with least eigenvalue -2 and the Interlacing Theorem is contradicted. Hence G is either a (connected) generalized line graph or an exceptional graph. Moreover, the multiplicity of -2 as an eigenvalue of G cannot exceed 1.

Now G is a generalized line graph for otherwise one of the graphs $G - i$ is an exceptional graph with $\lambda(G - i) > -2$; and this contradicts Theorem 2.3.20 because $G - i$ has order greater than 8. It follows from Theorems 2.2.4 and 2.2.8, together with an inspection of the p-decks of graphs which arise there, that G belongs to one of the following classes of graphs:

$\mathcal{B}_1$ = { $L(\hat{H})$: $\hat{H}$ consists of two petals joined by a non-trivial path};

$\mathcal{B}_2$ = { C_n : n even };

$\mathcal{B}_3$ = { $L(\hat{H})$: $\hat{H}$ consists of a petal and an odd cycle together with a path (possibly of length zero) between them};

$\mathcal{B}_4$ = { $L(B)$: B is a bicyclic graph consisting of two odd cycles and a path (possibly of length zero) between them }

$\mathcal{C}_1$ = { $L(T)$: T is a tree };

$\mathcal{C}_2$ = { $L(\hat{H})$: $\hat{H}$ is a tree with one petal attached};

$\mathcal{C}_3$ = { $L(U)$: U is an odd-unicyclic graph }.

Note that the graphs in $\mathcal{C}_1 - \mathcal{C}_3$ are those in Theorem 2.3.20(i)(ii); they all have least eigenvalue greater than -2. In contrast, the graphs in $\mathcal{B}_1 - \mathcal{B}_4$ have -2 as a simple eigenvalue (see Theorems 2.2.4 and 2.2.8).

We next observe that by Lemma 7.5.2 the value of the characteristic polynomial at -2 is constant on each of the above classes. Therefore, to reconstruct $P_G(x)$ from $\mathcal{P}$, it suffices to determine only the class to which G belongs. Now graphs from $\mathcal{B}_1 - \mathcal{B}_4$ can be recognized from $\mathcal{P}$ by calculating, for each vertex i, the number of edges and triangles of G incident with i and by recalling (where necessary) some of the forbidden subgraphs for generalized line graphs shown in Fig. 2.4. In such cases we have $P_G(-2) = 0$, and we are done. Otherwise, G must be a graph from $\mathcal{C}_1 - \mathcal{C}_3$, and we have only to distinguish between $\mathcal{C}_1$ and $\mathcal{C}_2$ because $n > 10$. Here we need check only whether $P_i(-2) = \pm 4$ for some i. If so, then $P_G(-2) = (-1)^n 4$, and if not, then $P_G(-2) = (-1)^n(n+1)$.

Case 2: $P_i(-2) = 0$ for at least one i.
In this case the Interlacing Theorem shows that either $\lambda(G) = -2$, or $\lambda(G) < -2$. The second possibility is ruled out by Theorem 2.4.5, because $n > 10$. Since -2 is now known to be an eigenvalue of G, we are done. □

In the following theorems much more is said about the polynomial reconstruction of disconnected graphs.

Theorem 7.5.4 [CvLe]. *If G is a disconnected graph with at least three components, then the characteristic polynomial of G is determined uniquely by its p-deck.*

Theorem 7.5.5 [CvLe]. *Let G be a graph of order at least three with exactly two connected components. If these components have different orders, then the characteristic polynomial of G is determined uniquely by its p-deck.*

7.6 Integral graphs

The eigenvalues of a graph are not only real numbers but also algebraic integers. Thus any rational eigenvalue is an integer. Recall that a graph is called an *integral* graph if all its eigenvalues are integers. The first paper on integral graphs, by F. Harary and A. J. Schwenk [HaSc1], appeared in 1974 under the title: *Which graphs have integral spectra?* In its full generality, the problem here appears to be intractable, and the results that have appeared in the literature to date are roughly of the following types:

(a) a search for integral graphs within particular (possibly finite) classes of graphs;
(b) constructions of integral graphs by means of certain graph operations and/or transformations;
(c) computer generation of all integral graphs of small order (so far up to 13 vertices).

It is worth mentioning that problem (a) is unsolved even for trees: the known results concern integral trees of diameter less than 11. Concerning (b), observe that the family of integral graphs is closed under the NEPS operation (cf. Theorem 1.2.32); in the context of this book, note that line graphs of regular integral graphs are also integral (cf. Theorem 1.2.16). (For more details on the topic, see [PeRa] or the survey paper [BaCRSS].)

In what follows we shall be concerned with connected integral graphs whose least eigenvalue is equal to -2. (Recall that the non-trivial complete graphs are the only connected graphs whose least eigenvalue is equal to -1, and that all such graphs are integral.) Thus the graphs in question are either connected generalized line graphs or exceptional graphs. The former class of graphs was first studied in the context of graphs with maximum degree $\Delta \leq 4$. The latter class is finite, and in the light of the recent enumeration of all maximal exceptional graphs, the integral exceptional graphs can be found by a computer search.

Concerning integral graphs with bounded vertex degrees, we first mention a result of D. Cvetković [Cve5] who showed that any class of such graphs is finite. This is because the diameter is bounded also. Explicitly, if D is the diameter of G and m is the number of distinct eigenvalues of G then $D \leq m - 1$ [CvDSa, Theorem 3.13]. On the other hand, $m \leq 2\lambda_1 + 1 \leq 2\Delta + 1$, since the absolute value of each eigenvalue is at most Δ. A crude upper bound (cf. [PeRa]) for the order n of G is given by:

$$ n \leq \frac{\Delta(\Delta - 1)^D - 2}{\Delta - 2}. $$

One of the first results on integral graphs was the enumeration of all connected non-cubic integral graphs with $\Delta \leq 3$ [CvGT]. There are seven such graphs: three complete graphs (K_1, K_2 and K_3), and four graphs with least eigenvalue equal to -2 (the cycles C_4 and C_6, and the trees $K_2 \circ 2K_1$ and $S(K_{1,3})$). Note that the index of any such graph is at most 2 and so the last four graphs can be found as Smith graphs (see Fig. 3.2). The spectra of the Smith graphs are given in [CvGu]; see also Table A1.

The connected cubic integral graphs were enumerated by D. Cvetković and F. C. Bussemaker [Cve5, BuCv], and independently by A. J. Schwenk [Sch1]. The first two authors exploited the Hoffman polynomial (see Theorem 1.2.7) before embarking on a computer search, while Schwenk avoided the use of a computer by observing that only bipartite graphs need be considered. The trick is to transform a non-bipartite integral graph G to the bipartite integral graph $G \times K_2$ (a NEPS with basis $\{(1, 1)\}$ in the sense of Definition 1.2.31). Note that a vertex of degree d is transformed to two vertices of degree d. Thus

Figure 7.12a: Five integral graphs.

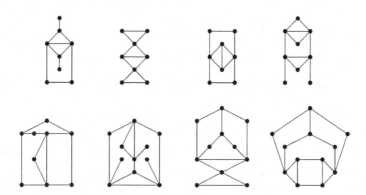

Figure 7.12b: Eight integral graphs.

once all the bipartite integral graphs from a given class are known, the non-bipartite ones can be extracted by reversing the construction. It turns out that there are exactly 13 connected cubic integral graphs. Of these, one is complete (K_4), four have least eigenvalue -2 (the Petersen graph, the three-sided prism $K_3 + K_2$, the line graph $L(S(K_4))$ and one exceptional graph), while the remaining eight are bipartite graphs (with least eigenvalue -3), the largest having order 30.

Z. Radosavljević and S. K. Simić [RaSi2, SiRa] determined all connected non-regular non-bipartite integral graphs with $\Delta = 4$ (thirteen in total). These graphs have least eigenvalue equal to -2, and so the topic clearly falls within the scope of this book. The graphs in question are generalized line graphs (the five graphs of Fig. 7.12a), or exceptional graphs (the eight graphs of Fig. 7.12b). We shall give here only an outline of the proof, since the complete proof is too involved. Some of the techniques are relevant to the wider class of integral graphs with $\Delta \leq 4$.

We first note that any graph of the required type has index 3, because the only candidate among the Smith graphs is the bipartite graph $K_{1,4}$. Moreover, the diameter is at most 5 since the number of different eigenvalues is at most six.

Consider first the family $\mathcal{S}_1$ of generalized line graphs that arise. If $G = L(H; a_1, \ldots, a_n) \in \mathcal{S}_1$, then $\Delta(H) \leq 3$; and if $\deg_H(u) = 3$ then $a_u = 0$, for

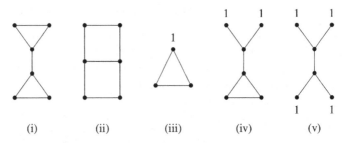

Figure 7.13: The root multigraphs of the graphs in Fig. 7.12a.

otherwise the index of G is greater than 3. Next, if u and v are adjacent vertices of H then

(7.1) $\deg_H(u) + \deg_H(v) + 2(a_u + a_v - 1) \leq 4,$

since $\Delta(G) = 4$. It follows that $a_u + a_v < 2$ and that each of $a_1, \ldots, a_n$ is 0 or 1; otherwise, only $K_{1,4}$ can arise. If $a_u = 1$ and $a_v = 0$, then $\deg_H(u) + \deg_H(v) \leq 4$ (and $\deg_H(u) \neq 3$, as noted above). Also, since $\Delta(G) = 4$, there exist adjacent vertices u and v in H for which equality holds in (7.1). These facts impose substantial restrictions on $H(a_1, \ldots, a_n)$. It turns out that the root multigraphs that arise are those appearing in Fig. 7.13, which shows the graphs H with 1 used as a label to identify the vertices i for which $a_i = 1$.

To prove that the graphs of Fig. 7.13 are all the root multigraphs for the graphs from $\mathcal{S}_1$, we proceed as in [SiRa]. We assume first that H contains at least two independent cycles. Since G has diameter at most five, and maximum degree at most 4, the number of edges in $H(a_1, \ldots, a_n)$ is constrained. Moreover, the edges must be such that the following hold:

- they are not too close to each other, i.e. there are no dense subgraphs (otherwise, the index of G exceeds three);
- the edges are not separated to the extent that $H(a_1, \ldots, a_n)$ contains disjoint induced subgraphs U, V such that $\lambda_1(L(U)) > 2$ and $\lambda_1(L(V)) \geq 2$ (otherwise, the second largest eigenvalue of G exceeds 2; cf. [RaSi3, Theorem 3.2]).

With some effort one can now show that the first two root multigraphs of Fig. 7.13 are the only graphs that not only survive these opposing requirements but also yield integral generalized line graphs.

We next assume that H is either a unicyclic graph or a tree. We can then take advantage of the simple structures of H; for example, the girth of a unicyclic graph appears as an important parameter. Also, it is now easy to relate the diameter of G and the diameter of H. We can also use the formulas (2.4) and

(2.6) from Section 2.2 to compute the multiplicities of the eigenvalue -2 in generalized line graphs. To fix some ideas, we discuss the implications for trees; for unicyclic graphs, the analysis is more involved and yields the third and fourth graphs of Fig. 7.13.

If H is a tree then we know from Theorems 2.2.4 and 2.2.8 that $\sum_{i=1}^{n} a_i \geq 2$. In view of our earlier remarks, it follows that H contains two non-adjacent vertices labelled by 1. There are two cases to consider.

Case 1. There are two vertices of H labelled by 1 at distance at least three.

We show that the multiplicity of -2 in G is exactly one. Suppose that u and v are the two vertices with $a_u = a_v = 1$ and $d(u, v) \geq 3$. Consider the subgraph consisting of the $u - v$ path in H together with the petals at u and v. Its generalized graph has two eigenvalues less than -1 (as can be verified by direct calculation), and so the multiplicity of -2 in G is at least two (by interlacing). This means that at least three vertices of H have labels equal to 1. Proceeding in the same way, we find that there exist at least four vertices labelled by 1. By considering the index and second largest eigenvalue of G, we find that the only labelled tree which meets all of our conditions is the one shown in Fig. 7.13(v).

Case 2. Any two vertices of H labelled by 1 are at distance two.

In this case H contains at most three vertices labelled by 1, and then we can easily check (as above) that there are no integral graphs in this situation.

Secondly we consider the family $\mathcal{S}_2$ of exceptional graphs that meet our conditions. The mean degree of any graph $G \in \mathcal{S}_2$ is less than 3, and so by Theorem 3.6.7, the order of G is at most 13. The graphs that arise have order at most 12: they can be constructed by adding vertices to the minimal forbidden subgraphs for generalized line graphs which have least eigenvalue greater than -2 and maximum degree at most four (see Fig. 2.4). Alternatively one can consult forthcoming tables of integral graphs with at most 13 vertices.   □

The corresponding problem for bipartite graphs is not yet solved; some details can be found in [BaSi1, BaSi2, BaSi3]. The 4-regular integral graphs were initially investigated in [CvSS]; further information may be found in [PeRa].

Finally, we add come comments on the generation of small integral graphs (results of type (c)). First, there are exactly 150 connected integral graphs with up to ten vertices (see [BaCLS]). Secondly, the integral graphs of order 11 and 12 have recently been generated (see [BaKSZ1, BaKSZ2]): the results were first obtained using genetic algorithms, and subsequently confirmed essentially by brute force at the Supercomputing Centre in Poznań. The integral graphs on

13 vertices have been determined by K.T. Balińska *et al* and publication of their results is awaited. The numbers i_n of connected integral graphs with n vertices ($n = 1, 2, \ldots, 12$) are given in the following table.

n	1	2	3	4	5	6	7	8	9	10	11	12
i_n	1	1	1	2	3	6	7	22	24	83	263	325

7.7 Graph equations

Graph equations are equations in which the unknowns are graphs. A classification and a survey of graph equations can be found in [Sim1], [CvSi3] and [CvSi5]. The equality relation in graph equations is usually taken as the isomorphism relation.

In this section we present several results on graph equations involving line graphs; in particular, we are interested in the question of when a line graph (or its complement) can arise from certain binary operations on graphs defined in Section 1.2. Spectral techniques are useful in solving such graph equations since the lower bound on eigenvalues of a line graph enables many potential solutions to be eliminated. Our first remarks concern the equation $L(G) = G_1 \nabla G_2$ in the case that $L(G)$ is regular.

A graph is said to be ∇-*prime* if it cannot be represented as the join of two graphs. Note that if an r-regular graph G_0 of order n is not ∇-prime then $\overline{G}_0$ is a disconnected regular graph of degree $n - r - 1$. Thus the index of $\overline{G}_0$ is $n - r - 1$ with multiplicity at least 2. By Corollary 1.2.14, $r - n$ is an eigenvalue of G_0 and so $r - n \geq \lambda(G_0)$, where as before $\lambda(G_0)$ denotes the least eigenvalue of G_0. Since $\lambda(G_0) \geq -r$, we have $r \geq \frac{1}{2}n$. However, if $G_0 = L(G)$ then $\lambda(G_0) \geq -2$ and we have $r \geq n - 2$; in this situation, either $r = n - 1$ and $G_0 = K_n = L(K_{1,n})$ or $r = n - 2$ and G_0 is a cocktail party graph. In the latter case, $n \leq 6$ because G_0 is a line graph, and we can state our first result.

Proposition 7.7.1 [Cve2]. *If $L(G)$ is a regular line graph such that $L(G) = G_1 \nabla G_2$ then one of the following holds:*

(a) $L(G)$, G_1, G_2 *are complete,*
(b) $G = C_4$ *and* $G_1 = G_2 = \overline{K}_2$,
(c) $G = K_4$ *and* $\{G_1, G_2\} = \{\overline{K}_2, 2K_2\}$.

We note in passing that graph spectra can also be used to solve the graph equation $T(G) = L(H)$ when G is regular. Here $T(G)$ denotes the *total graph* of G as defined in Section 1.1. It is well known that the total graph of K_n is isomorphic to $L(K_{n+1})$, a result proved by the use of graph spectra in [Cve3]. A complete solution of the equation $T(G) = L(H)$ is obtained by non-spectral means in [CvSi2].

Proposition 7.7.2 [Doo11]. *The only solutions of the equation $L(G) = G_1 + G_2$ among non-trivial connected graphs are:*

$$G = K_{m,n}, \quad \{G_1, G_2\} = \{K_m, K_n\} \ (m, n = 2, 3, \ldots).$$

Proof. Let q, q_1, q_2 be the least eigenvalues of graphs G, G_1, G_2 respectively. Clearly, $q_1 + q_2 = q \geq -2$. On the other hand, the least eigenvalue of a non-trivial connected graph is always -1 at most, and so $q = -2$, $q_1 = q_2 = -1$. Then G_1 and G_2 are complete graphs, say, K_m and K_n. Since $K_m + K_n = L(K_{m,n})$, the result follows. □

A non-spectral proof of Proposition 7.7.2 appears in [Pal].

Since the complements of line graphs have second largest eigenvalue $\lambda_2 \leq 1$, the following graph equations can be solved immediately.

$$(7.2) \qquad \overline{L(G)} = G_1 \times G_2,$$

$$(7.3) \qquad \overline{L(G)} = G_1 + G_2,$$

$$(7.4) \qquad \overline{L(G)} = G_1 * G_2.$$

For example, the eigenvalues of $G_1 \times G_2$ are all possible products $\lambda_i \mu_j$ where λ_i is an eigenvalue of G_1 and μ_j an eigenvalue of G_2 (cf. Theorem 1.2.32). If G_1 and G_2 are neither trivial nor totally disconnected, then all solutions of equation (7.2) are given by $G = K_{m,n}$, $\{G_1, G_2\} = \{K_m, K_n\}$ for otherwise $G_1 \times G_2$ would have two eigenvalues greater than 1.

Equations (7.3) and (7.4) can be treated similarly. Equations (7.2)–(7.4) have also been solved by non-spectral means [Sim2].

All graphs switching-equivalent to their line graphs have been determined in [CvSi4]. Such graphs can be considered as a solution of a generalized graph equation $L(G) \overset{s}{\sim} G$, in which the equality relation is replaced by switching-equivalence. The solution of this generalized graph equation clearly includes unions of cycles as the solution of the "ordinary" graph equation $L(G) = G$. There are exactly 10 other solutions of the equation $L(G) \overset{s}{\sim} G$: the graphs in question have order at most 7; six are connected and four are disconnected. Finally, the graphs G for which $\overline{L(G)} \overset{s}{\sim} G$ have been determined in [Sim9, Sim10].

7.8 Other topics

Several unrelated topics will be mentioned briefly in this last section.

1. We know from Proposition 4.1.7 that regular exceptional graphs are switching-equivalent to line graphs. However, if we switch a line graph $L(G)$ we may obtain another line graph (possibly isomorphic to the original). This is the case if the switching set U, regarded as a set of edges of G, is such that the adjacencies of edges in U and outside U are unchanged, while each edge from U becomes adjacent to exactly those edges outside U to which it was not adjacent before switching. A switching with respect to such a subset U of $E(G)$ is called a *line-switching* of G (cf. Definition 4.4.6).

We give some examples of line-switching from [Cve9]; the first two are illustrated in Fig. 7.14. The second example here shows that it is sometimes useful to consider root graphs with a number of isolated vertices.

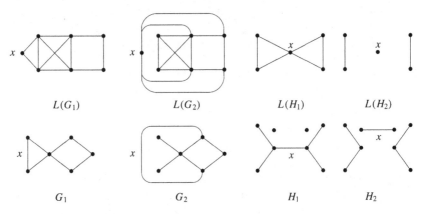

Figure 7.14: Two examples of line-switching.

An interesting effect occurs in line switching due to the fact that non-isomorphic graphs can have isomorphic line graphs. As we have seen in Theorem 2.1.6, the only such case with connected graphs occurs when one is K_3 and the other $K_{1,3}$. Fig. 7.15 shows a line-switching in which a star $K_{1,3}$ is converted into a triangle K_3.

Another example, with two stars converted into triangles, is given in Fig. 7.16. This example is interesting also because a regular graph is switched into a regular graph with a different degree.

2. The study of graphs with least eigenvalue bounded by -2 can be extended to signed graphs. A graph with edge labels from the set $\{-1, 1\}$ is called a *signed graph* (or *sigraph*). A sigraph can also be considered as a pair (X, ϕ)

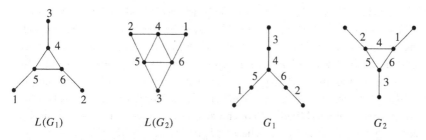

$L(G_1)$ $L(G_2)$ G_1 G_2

Figure 7.15: A star converted to a triangle by line-switching.

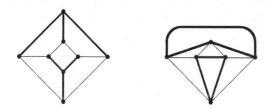

Figure 7.16: Two stars converted to triangles by line-switching.

where X is a finite set (of vertices) and ϕ is a map $X^2 \to \{-1, 0, 1\}$ such that $\phi(x, y) = \phi(y, x)$ and $\phi(x, x) = 0$. The map ϕ defines the adjacency matrix A of a sigraph. If the least eigenvalue of A is greater than or equal to -2, the sigraph can be represented in a root system using a Gram matrix in the same way as for graphs.

The expository paper [ViSi] surveys results and problems concerning the class of signed graphs representable in the root systems D_n ($n \in I\!N$) and E_8, a class which includes the graphs with least eigenvalue -2. In a parallel of the approach for graphs, the paper [Vij4] deals with the classification of sigraphs with least eigenvalue greater than or equal to -2, and the construction of minimal forbidden sub-sigraphs for sigraphs with least eigenvalue less than -2. See also [Zas], [CaST], [BuCST], [ChVi], [Vij2], [Vij3], [Vij5], [SiVi] and [RaSiV].

3. Another generalization involves normal digraphs. A digraph is called *normal* if its adjacency matrix is normal. A *proper normal* digraph is a normal digraph whose adjacency matrix is asymmetric. The paper [Tor5] introduces a *generalized line digraph* (of a digraph) and proves that cycles (with at least three vertices) are the only proper normal generalized line digraphs. For other results on normal digraphs see [CvDSa, Section B.9].

4. Infinite graphs with least eigenvalue -2 have been studied in [Tor1]–[Tor4], while the paper [DePo] investigates properties of the characteristic polynomial of line multidigraphs.

5. Graphs with least eigenvalue -2 and the corresponding proof techniques appear in the study of several other classes of graphs. To a large extent, such research is not described in this book, and we conclude by mentioning some relevant papers. The 187 regular exceptional graphs of Chapter 4 appear, for example, in the study of hypermetric graphs [DeGr, TeDe], regular graphs with four distinct eigenvalues [Dam1, DaSp], non-regular graphs with three distinct eigenvalues [MuKl] and in characterization theorems based on the star complement technique [JaRo]. These 187 graphs, along with other topics related to graphs with least eigenvalue -2, have been used in studying several classes of distance-regular graphs – see [Neu2] and [Ter1]-[Ter5]. In addition, much information on these topics can be found in the expository paper [BuNe] and in the monograph [BrCN].

Appendix

This appendix contains the following graph tables:

Table A1: Some graphs related to graphs with least eigenvalue -2
Table A2: The exceptional graphs with least eigenvalue greater than -2
Table A3: Regular exceptional graphs and their spectra
Table A4: A construction of the 68 connected regular graphs which are not line graphs but cospectral with line graphs
Table A5: One-vertex extensions of exceptional star complements
Table A6: The maximal exceptional graphs
Table A7: The index and vertex degrees of the maximal exceptional graphs.

Each table is accompanied by a description of its structure and content.

Table A1
SOME GRAPHS RELATED TO GRAPHS WITH LEAST
EIGENVALUE −2

Table A1 consists of two parts:
A1.1. Spectra of the Smith graphs and the reduced Smith graphs,
A1.2. Minimal graphs with least eigenvalue less than −2 on 7 and 8 vertices.

A1.1. Spectra of the Smith graphs and the reduced Smith graphs

The Smith graphs are given in Fig. 3.2 and the reduced Smith graphs in Fig. 3.3 of Chapter 3. In both cases, the subscript in the name accorded to a graph denotes the number of vertices. We list the spectra of the Smith graphs and the reduced Smith graphs as given in [CvGu]. Each spectrum includes eigenvalues of the form $2 \cos \frac{\pi}{m} j$ for some m and j. For each graph in the table below we give m and the range of j together with any additional eigenvalues.

Smith graphs

graph	m	j	additional eigenvalues
C_n	n	$2, 4, \ldots, 2n$	
W_n	$n-3$	$1, 2, \ldots, n-4$	$2, 0, 0, -2$
$\mathcal{F}_7$	3	$1, 2, \ldots, 6$	0
$\mathcal{F}_8$	4	$1, 2, 3$	$2, 1, 0, -1, -2$
$\mathcal{F}_9$	5	$1, 2, 3, 4$	$2, 1, 0, -1, -2$

Reduced Smith graphs

graph	m	j	additional eigenvalues
$\mathcal{A}_n$	$n+1$	$1, 2, \ldots, n$	
$\mathcal{D}_n$	$2n-2$	$1, 3, \ldots, 2n-3$	0
$\mathcal{E}_6$	12	$1, 4, 5, 7, 8, 11$	
$\mathcal{E}_7$	18	$1, 5, 7, 9, 11, 13, 17$	
$\mathcal{E}_8$	30	$1, 7, 11, 13, 17, 19, 23, 29$	

A1.2. Minimal graphs with least eigenvalue less than −2 on 7 and 8 vertices

As explained in Section 2.4 there are exactly 1812 minimal graphs with least eigenvalue less than −2. They have between 5 and 10 vertices and some statistics for these graphs are given in Section 2.4. There are 3 graphs on 5 vertices and these are graphs $G^{(1)}$, $G^{(2)}$, $G^{(3)}$ in Fig. 2.4. The 8 graphs on 6 vertices are the graphs $G^{(4)}$, $G^{(5)}$, ..., $G^{(11)}$ in the same figure. Here we describe the 14 graphs on 7 vertices and the 67 graphs on 8 vertices.

The graphs are ordered lexicographically by spectral moments. As usual for a graph on n vertices, the vertices are denoted by $1, 2, \ldots, n$. Each graph G on n vertices ($n = 7, 8$) is represented by a line which contains: the identification number of G; the edges, given as pairs of vertices, or (for graphs of order 8 with nore than 14 edges) the edges of the complement; the maximal vertex degree; the least eigenvalue; and the largest eigenvalue.

The list of graphs on seven vertices includes one further column which contains the identification numbers used in the table of connected graphs on seven vertices in [CvDGT].

Graphs with seven vertices

```
 1.  12 13 16 24 27 35                            3  -2.0529   2.0529     6
 2.  12 13 24 35 46 47 57                         3  -2.1010   2.1010    13
 3.  12 13 15 17 24 26 37                         4  -2.0590   2.4309    37
 4.  12 13 16 17 24 27 35 47                      4  -2.0674   2.7469    97
 5.  12 13 17 24 26 27 35 37                      4  -2.0705   2.7886   101
 6.  12 13 15 17 24 26 27 37                      4  -2.0748   2.8321   104
 7.  12 13 17 24 27 35 37 46 56                   3  -2.1249   2.7616   144
 8.  12 13 15 16 17 24 27 36 47                   5  -2.0772   2.9459   187
 9.  12 13 15 17 24 26 27 46 47 67                4  -2.0818   3.3890   342
10.  12 13 16 17 24 26 27 35 37 46 67             4  -2.0982   3.5821   471

11.  12 13 15 16 17 24 26 27 36 47 67             5  -2.1063   3.6242   475
12.  12 13 15 16 17 24 26 27 46 47 67             5  -2.0886   3.7388   485
13.  12 13 15 16 17 24 26 27 35 37 46 57 67       5  -2.1199   3.9694   682
14.  12 13 15 16 17 24 25 26 27 35 37 46 56 57 67 5  -2.1349   4.5711   809
```

Graphs with eight vertices

1.	12	13	17	24	35	46	78					3	-2.0285	2.0285
2.	12	13	17	24	35	46	48					3	-2.0421	2.0421
3.	12	13	16	24	28	35	48	67				3	-2.0303	2.3163
4.	12	13	17	18	24	35	46	78				4	-2.0321	2.4227
5.	12	13	24	26	28	35	37	48				4	-2.0461	2.4048
6.	12	13	16	18	24	28	35	67				4	-2.0341	2.5019
7.	12	13	16	18	24	27	35	47	68			4	-2.0344	2.5493
8.	12	13	17	24	28	35	46	48	68			3	-2.0496	2.6790
9.	12	13	24	26	27	28	35	47	68			5	-2.0367	2.7361
10.	12	13	18	24	28	35	37	46	48			3	-2.0531	2.7369
11.	12	13	16	17	18	24	28	35	67			5	-2.0393	2.7851
12.	12	13	18	24	26	28	35	37	48			4	-2.0557	2.7799
13.	12	13	17	18	24	28	35	38	46			4	-2.0413	2.8230
14.	12	13	16	18	24	28	35	38	67			4	-2.0423	2.8290
15.	12	13	16	17	24	27	35	37	78			4	-2.0495	2.9057
16.	12	13	24	26	28	35	37	48	58	68		4	-2.0904	2.8729
17.	12	13	17	18	24	35	38	46	68	78		4	-2.0873	2.9136
18.	12	13	15	17	18	24	26	37	46	58		5	-2.0397	2.8051
19.	12	13	18	24	26	27	35	38	47	58		4	-2.0550	2.7521
20.	12	13	16	18	24	27	28	35	38	47		4	-2.0451	2.9647
21.	12	13	16	17	18	24	28	35	38	67		5	-2.0463	3.0366
22.	12	13	15	16	17	18	24	28	36	57		6	-2.0463	3.0366
23.	12	13	16	17	18	24	35	67	68	78		5	-2.0369	3.1978
24.	12	13	17	18	24	28	35	37	46	48	57	4	-2.0651	3.0252
25.	12	13	17	18	24	26	28	35	37	48	57	4	-2.0688	3.0549
26.	12	13	16	18	24	28	35	38	48	58	67	5	-2.0618	3.2877
27.	12	13	17	18	24	28	35	38	46	48	58	5	-2.0663	3.3088
28.	12	13	15	16	17	18	24	36	38	57	68	6	-2.0430	3.3664
29.	12	13	18	24	26	35	37	38	57	58	78	4	-2.0597	3.3610
30.	12	13	18	24	26	27	28	35	47	48	78	5	-2.0493	3.4349
31.	12	13	16	17	18	24	28	35	67	68	78	5	-2.0507	3.4395
32.	12	13	16	17	18	24	27	28	35	38	78	5	-2.0540	3.4852
33.	12	13	15	17	18	24	25	27	28	56	78	5	-2.0611	3.5041
34.	12	13	17	24	26	28	35	37	48	57	58 68	4	-2.1099	3.1099
35.	12	13	16	17	18	24	28	35	38	48	58 67	5	-2.0701	3.4039
36.	12	13	24	26	27	28	35	47	48	58	68 78	5	-2.1047	3.5067
37.	12	13	15	17	18	24	26	28	37	38	46 78	5	-2.0546	3.4943
38.	12	13	14	15	16	17	18	25	28	36	47 58	7	-2.0514	3.5341
39.	12	13	15	16	17	18	24	28	36	38	57 68	6	-2.0564	3.5799
40.	12	13	17	18	24	26	35	37	38	57	58 78	4	-2.0635	3.6983

41.	12	13	17	18	24	27	28	35	37	38	46	78			4	−2.0496	3.7354
42.	12	13	16	17	18	24	27	28	35	37	38	78			5	−2.0561	3.7919
43.	12	13	17	18	24	26	28	35	37	46	48	57	68		4	−2.0749	3.4601
44.	12	13	16	18	24	27	28	35	38	47	48	58	78		6	−2.0767	3.7495
45.	12	13	17	18	24	26	27	28	35	37	38	46	78		5	−2.0550	3.8124
46.	12	13	15	16	17	18	24	36	38	57	58	68	78		6	−2.0630	3.8999
47.	12	13	15	16	17	18	24	27	28	36	38	68	78		6	−2.0680	3.9274
48.	12	13	15	16	17	18	24	25	26	27	28	57	68		6	−2.0791	3.9576
49.	12	13	16	17	18	24	26	28	35	37	46	48	57	68	5	−2.0806	3.8098
50.	12	13	14	15	16	17	18	25	28	36	38	47	58	68	7	−2.0718	4.0118
51.	12	13	17	18	24	27	28	35	37	38	46	48	58	78	6	−2.0888	4.0640
52.	12	13	16	17	18	24	27	28	35	37	38	48	58	78	6	−2.0934	4.0960
53.	12	13	15	16	17	18	24	28	36	37	38	67	68	78	6	−2.0611	4.2651
54.	14	15	16	23	25	34	36	37	38	45	56	57	58		5	−2.0616	4.4507
55.	14	15	23	25	34	36	45	46	47	48	56	57	58		5	−2.0688	4.4842
56.	14	23	25	26	34	35	45	46	47	48	56	57	58		6	−2.0716	4.4951
57.	14	15	16	23	25	34	36	45	47	56	57	67			7	−2.1072	4.3688
58.	14	16	23	25	34	36	45	47	48	56	67	68			5	−2.0698	4.4872
59.	23	24	26	34	35	37	45	46	47	48	56	67			7	−2.0793	4.5685
60.	14	15	23	25	34	45	46	47	48	56	57	58			5	−2.0642	4.7588
61.	14	15	23	25	34	45	46	47	56	57					7	−2.1153	4.9701
62.	14	23	25	34	36	45	46	47	48	56					6	−2.0928	5.0221
63.	23	24	26	34	35	45	46	47	48	56					7	−2.0976	5.0388
64.	14	23	25	26	34	45	46	47	48						6	−2.0806	5.3073
65.	14	23	25	34	45	46	47	48							6	−2.0849	5.5194
66.	23	25	34	36	45	56									7	−2.1142	5.7047
67.	23	25	34	45											7	−2.1231	6.1231

Table A2
THE EXCEPTIONAL GRAPHS WITH LEAST
EIGENVALUE GREATER THAN −2

Table A2, taken from [CvLRS3], contains data on the 573 exceptional graphs with least eigenvalue greater than −2 which appear in Theorem 2.3.20. There are 20 graphs on six vertices, 110 on seven vertices and 443 on eight vertices; in each case, the graphs are ordered lexicographically by spectral moments.

As usual for a graph on n vertices, the vertices are denoted by $1, 2, \ldots, n$. The graphs on 6 vertices are specified by their edges, presented as pairs of vertices. Instead of identification numbers 1–20, the first column contains the names $F_1, F_2, \ldots, F_{20}$ by which these graphs are denoted in [CvRS4]. The next column contains the names $G^{(12)}, \ldots, G^{(31)}$ used in Fig. 2.4. The last three columns contain: the graph identification numbers used in the table of connected graphs on six vertices in [CvPe], the graph names used in [CvDS2], and finally the names of any which appear in Harary's list [Har] of graphs that are forbidden for line graphs. To avoid conflicting notation, the names used in this last category are $H_1, H_2, \ldots, H_9$ instead of $G_1, G_2, \ldots, G_9$ respectively in [Har, Chapter 8]. Each graph G on n vertices ($n = 7, 8$) is represented by a line which contains: the identification number of G; the identification number of the subgraph induced by the vertices $1, 2, \ldots, n - 1$; the vertices to which vertex n is adjacent; the number of edges; the maximal vertex degree; the least eigenvalue; and the largest eigenvalue. The list of graphs on seven vertices includes one further column which contains the identification numbers used in the table of connected graphs on seven vertices in [CvDGT], reproduced from [CvLRS1]. In this book, the graphs of order 8 are labelled $H001$ to $H443$ in order of the graph identification numbers.

Graphs on six vertices

F_1	$G^{(12)}$	12	23	34	36	45						110	G_{18}	
F_2	$G^{(13)}$	12	23	26	34	45	56					105	G_{19}	
F_3	$G^{(14)}$	12	23	34	35	36	45					97	G_{20}	
F_4	$G^{(15)}$	12	23	26	34	36	45	46				80	G_{12}	H_4
F_5	$G^{(16)}$	12	23	24	25	26	34	56				79	G_{23}	
F_6	$G^{(17)}$	12	23	25	26	34	35	56				77	G_{21}	
F_7	$G^{(18)}$	12	15	16	23	26	34	36	45			69	G_{15}	H_7
F_8	$G^{(19)}$	12	23	24	26	34	45	46	56			59	G_{22}	
F_9	$G^{(20)}$	12	23	24	25	26	34	45	56			58	G_{24}	
F_{10}	$G^{(21)}$	12	16	23	25	26	34	35	45	56		44	G_{16}	H_8
F_{11}	$G^{(22)}$	12	14	16	23	24	34	45	46	56		43	G_{27}	
F_{12}	$G^{(23)}$	12	23	25	34	35	36	45	46	56		35	G_{13}	H_5
F_{13}	$G^{(24)}$	12	23	24	25	26	34	45	46	56		33	G_{26}	
F_{14}	$G^{(25)}$	12	15	16	23	26	34	36	45	46	56	28	G_{17}	H_9
F_{15}	$G^{(26)}$	12	13	14	15	16	23	34	36	45	46	22	G_{25}	
F_{16}	$G^{(27)}$	12	23	24	26	34	35	36	45	46	56	20	G_{28}	
F_{17}	$G^{(28)}$	12	13	14	15	16	23	25	34	35	36	46	13	G_{14}	H_6
F_{18}	$G^{(29)}$	12	13	14	16	23	34	35	36	45	46	56	12	G_{29}	
F_{19}	$G^{(30)}$	12	13	14	15	16	23	25	34	35	36	46	56	7	G_{30}	
F_{20}	$G^{(31)}$	12	14	16	23	24	25	26	34	35	36	45	46	56	3	G_{31}

Graphs on seven vertices

1.	1	1	6	3	−1.9696	1.9696	3
2.	1	16	7	3	−1.9449	2.1515	14
3.	2	4	7	3	−1.9122	2.1987	15
4.	1	12	7	3	−1.9672	2.2970	29
5.	3	1	7	4	−1.9653	2.3894	33
6.	1	23	7	4	−1.9624	2.4745	39
7.	2	45	8	3	−1.9354	2.4728	62
8.	2	12	8	4	−1.9354	2.4728	63
9.	1	125	8	3	−1.9323	2.4877	64
10.	2	23	8	4	−1.9210	2.5554	67
11.	3	12	8	4	−1.9620	2.5270	86
12.	4	1	8	3	−1.9586	2.7209	94
13.	3	23	8	5	−1.9555	2.7649	98
14.	6	4	8	4	−1.9555	2.7649	99
15.	1	236	8	4	−1.9542	2.7711	100
16.	6	5	8	4	−1.9434	2.8517	106
17.	2	145	9	3	−1.8662	2.6554	122
18.	2	345	9	3	−1.8944	2.8162	148
19.	2	123	9	4	−1.8944	2.8162	149
20.	2	236	9	4	−1.9134	2.8826	155
21.	4	12	9	4	−1.9537	2.8847	185
22.	6	34	9	4	−1.9497	2.9240	186
23.	3	236	9	5	−1.9483	2.9937	190
24.	1	1236	9	4	−1.9431	3.0245	194
25.	4	23	9	4	−1.9364	3.0536	196
26.	3	346	9	5	−1.9334	3.0569	197
27.	1	1234	9	4	−1.9334	3.0569	198
28.	3	234	9	5	−1.9230	3.0842	201
29.	6	23	9	5	−1.8950	3.1131	205
30.	3	345	9	5	−1.9588	3.1877	215
31.	7	45	10	3	−1.9107	2.9107	261
32.	4	125	10	4	−1.9047	2.9928	262
33.	2	1234	10	4	−1.8620	3.1085	268
34.	3	1236	10	5	−1.9330	3.1843	303
35.	4	123	10	4	−1.9229	3.2100	306
36.	6	345	10	4	−1.8981	3.2370	307
37.	6	126	10	5	−1.9235	3.2530	309
38.	5	123	10	6	−1.9188	3.2554	310
39.	8	45	10	5	−1.9086	3.2755	315
40.	8	26	10	5	−1.9098	3.3132	318

41.	12	1	10	4	−1.9502	3.3571	339
42.	5	234	10	6	−1.9502	3.3571	340
43.	3	3456	10	5	−1.9437	3.4114	343
44.	4	236	10	4	−1.9437	3.4114	344
45.	6	235	10	5	−1.9376	3.4592	348
46.	13	4	10	5	−1.9254	3.4774	349
47.	7	145	11	4	−1.8558	3.1774	388
48.	7	234	11	4	−1.8019	3.2959	390
49.	10	34	11	4	−1.8774	3.3539	427
50.	8	123	11	5	−1.8569	3.3940	428
51.	6	1234	11	5	−1.8943	3.4449	430
52.	7	126	11	4	−1.8812	3.4467	431
53.	9	123	11	6	−1.8248	3.4636	433
54.	2	23456	11	5	−1.8760	3.4926	435
55.	9	236	11	6	−1.8434	3.5366	446
56.	12	12	11	4	−1.9427	3.4219	464
57.	5	1234	11	6	−1.9340	3.5557	468
58.	8	456	11	5	−1.9246	3.5719	469
59.	3	12345	11	5	−1.9208	3.5996	472
60.	5	2345	11	6	−1.9045	3.6147	473
61.	6	1256	11	5	−1.9045	3.6147	474
62.	6	1235	11	5	−1.9059	3.6392	476
63.	8	246	11	5	−1.8738	3.6534	477
64.	16	1	11	4	−1.9460	3.6940	482
65.	6	2356	11	5	−1.9359	3.7524	486
66.	10	123	12	5	−1.8378	3.6254	540
67.	10	345	12	5	−1.9030	3.6519	574
68.	11	234	12	6	−1.9044	3.7161	575
69.	8	1234	12	5	−1.8781	3.7530	578
70.	7	1236	12	4	−1.8890	3.7637	579
71.	9	1234	12	6	−1.8623	3.7759	581
72.	9	1245	12	6	−1.8284	3.8284	588
73.	15	34	12	5	−1.8284	3.8284	589
74.	16	12	12	5	−1.9371	3.7762	599
75.	12	235	12	5	−1.9262	3.8627	601
76.	12	346	12	5	−1.9134	3.8744	602
77.	4	12346	12	5	−1.9080	3.8938	603
78.	6	23456	12	5	−1.9092	3.9118	606
79.	9	2345	12	6	−1.8636	3.923	608
80.	11	2345	13	6	−1.8234	3.9211	663

81.	10	2345	13	5	-1.8658	3.9806	683
82.	11	1234	13	6	-1.8833	4.0059	684
83.	13	1234	13	6	-1.8707	4.0233	685
84.	10	2356	13	5	-1.8019	4.0329	686
85.	8	12456	13	5	-1.8595	4.0561	689
86.	12	2345	13	5	-1.8886	4.1465	701
87.	8	23456	13	5	-1.8772	4.1610	702
88.	9	23456	13	6	-1.8809	4.1747	703
89.	12	3456	13	5	-1.9350	4.2136	707
90.	13	2456	13	6	-1.9266	4.2533	709
91.	7	123456	14	6	-1.8065	4.1736	741
92.	17	125	14	6	-1.8877	4.2121	755
93.	11	12346	14	6	-1.8182	4.2806	757
94.	18	146	14	5	-1.8498	4.2860	758
95.	13	12456	14	6	-1.8741	4.3876	767
96.	16	2346	14	5	-1.9165	4.4636	772
97.	13	23456	14	6	-1.9119	4.4741	773
98.	10	123456	15	6	-1.8019	4.4751	797
99.	15	12345	15	6	-1.7525	4.5114	800
100.	17	1235	15	6	-1.9095	4.5188	807
101.	12	123456	15	6	-1.8799	4.5602	808
102.	15	12346	15	6	-1.8068	4.5898	810
103.	16	23456	15	5	-1.9248	4.7306	812
104.	14	123456	16	6	-1.7016	4.7016	823
105.	17	12345	16	6	-1.8376	4.7520	827
106.	16	123456	16	6	-1.8822	4.8386	832
107.	18	123456	17	6	-1.8116	5.0157	842
108.	19	123456	18	6	-1.8255	5.2434	847
109.	20	23456	18	6	-1.8945	5.2965	849
110.	20	123456	19	6	-1.8882	5.5033	851

Graphs on eight vertices

1.	1	7	7	3	−1.9890	1.9890
2.	1	57	8	3	−1.9738	2.0912
3.	1	56	8	3	−1.9816	2.1648
4.	1	16	8	3	−1.9701	2.2245
5.	1	17	8	3	−1.9886	2.2623
6.	5	7	8	4	−1.9877	2.3920
7.	1	12	8	3	−1.9877	2.3920
8.	1	23	8	4	−1.9863	2.4943
9.	6	7	8	4	−1.9850	2.5606
10.	2	45	9	3	−1.9801	2.3590
11.	3	12	9	4	−1.9661	2.4981
12.	7	7	9	3	−1.9772	2.5466
13.	2	34	9	4	−1.9772	2.5466
14.	1	125	9	3	−1.9760	2.5604
15.	3	23	9	4	−1.9374	2.5919
16.	1	347	9	4	−1.9712	2.6278
17.	5	17	9	4	−1.9871	2.4491
18.	5	12	9	4	−1.9859	2.5697
19.	12	7	9	3	−1.9855	2.7231
20.	4	23	9	4	−1.9855	2.7231
21.	14	7	9	4	−1.9846	2.7668
22.	1	346	9	4	−1.9841	2.7741
23.	5	23	9	5	−1.9841	2.7741
24.	13	7	9	5	−1.9823	2.8194
25.	15	7	9	4	−1.9790	2.8658
26.	2	125	10	3	−1.9476	2.6373
27.	3	145	10	4	−1.9247	2.7393
28.	4	145	10	3	−1.9716	2.6809
29.	5	147	10	4	−1.9697	2.7101
30.	6	145	10	4	−1.9672	2.7210
31.	12	57	10	3	−1.9642	2.7480
32.	1	1457	10	4	−1.9734	2.7810
33.	2	127	10	3	−1.9673	2.8204
34.	6	125	10	4	−1.9698	2.8254
35.	3	126	10	4	−1.9357	2.8278
36.	10	26	10	5	−1.9611	2.8719
37.	2	236	10	4	−1.9719	2.8956
38.	3	345	10	4	−1.9572	2.9028
39.	14	57	10	4	−1.9572	2.9028
40.	16	46	10	4	−1.9654	2.9654

41.	12	17	10	3	−1.9847	2.7633
42.	14	47	10	4	−1.9836	2.8038
43.	4	346	10	4	−1.9830	2.8277
44.	11	23	10	5	−1.9830	2.9139
45.	12	12	10	4	−1.9830	2.9139
46.	14	34	10	4	−1.9817	2.9512
47.	1	3456	10	4	−1.9810	3.0259
48.	6	346	10	5	−1.9810	3.0259
49.	5	346	10	5	−1.9783	3.0587
50.	12	34	10	4	−1.9783	3.0587
51.	1	2345	10	4	−1.9772	3.0625
52.	16	12	10	5	−1.9772	3.0625
53.	5	234	10	5	−1.9747	3.0894
54.	1	1234	10	4	−1.9731	3.0929
55.	6	347	10	5	−1.9666	3.1215
56.	14	23	10	5	−1.9666	3.1215
57.	6	236	10	5	−1.9634	3.1247
58.	16	26	10	5	−1.9439	3.1516
59.	4	127	10	4	−1.9871	3.1112
60.	6	237	10	5	−1.9841	3.2567
61.	7	127	11	4	−1.9434	2.8517
62.	2	1457	11	4	−1.9508	2.9008
63.	8	345	11	4	−1.9623	2.8950
64.	7	123	11	4	−1.9623	2.8950
65.	9	346	11	4	−1.9623	2.9078
66.	31	7	11	3	−1.9704	2.9321
67.	9	236	11	4	−1.9584	3.0031
68.	12	125	11	4	−1.9683	3.0215
69.	10	345	11	4	−1.9277	3.0347
70.	10	126	11	5	−1.9587	3.0662
71.	8	236	11	5	−1.9684	3.0819
72.	2	2345	11	4	−1.9464	3.0898
73.	3	1256	11	4	−1.9369	3.1211
74.	8	123	11	5	−1.9369	3.1211
75.	2	1234	11	4	−1.9547	3.1441
76.	10	236	11	5	−1.9232	3.1964
77.	4	3456	11	4	−1.9795	3.0537
78.	11	346	11	5	−1.9763	3.0932
79.	21	34	11	4	−1.9763	3.1504
80.	4	2345	11	4	−1.9750	3.1598

81.	11	234	11	5	-1.9718	3.1817
82.	6	3456	11	5	-1.9764	3.2053
83.	7	457	11	4	-1.9750	3.2084
84.	8	127	11	5	-1.9750	3.2084
85.	12	345	11	4	-1.9751	3.2133
86.	4	1257	11	4	-1.9735	3.2163
87.	4	1236	11	4	-1.9675	3.2432
88.	14	126	11	5	-1.9751	3.2564
89.	14	345	11	4	-1.9614	3.2620
90.	22	23	11	5	-1.9614	3.2620
91.	13	346	11	6	-1.9720	3.2788
92.	1	12367	11	5	-1.9700	3.2813
93.	5	3457	11	5	-1.9677	3.2837
94.	25	46	11	4	-1.9399	3.2867
95.	37	7	11	5	-1.9652	3.3026
96.	13	236	11	6	-1.9571	3.3073
97.	25	26	11	5	-1.9682	3.3401
98.	6	1234	11	5	-1.9654	3.3422
99.	11	127	11	4	-1.9852	3.1458
100.	41	7	11	4	-1.9831	3.3574
101.	5	3456	11	5	-1.9811	3.4122
102.	13	237	11	6	-1.9811	3.4122
103.	12	236	11	4	-1.9802	3.4171
104.	14	235	11	5	-1.9785	3.4645
105.	44	7	11	4	-1.9774	3.4662
106.	43	7	11	5	-1.9733	3.4842
107.	45	7	11	5	-1.9750	3.5122
108.	17	123	12	4	-1.9220	3.0766
109.	12	1457	12	4	-1.9634	3.1295
110.	21	145	12	4	-1.9597	3.1407
111.	18	126	12	4	-1.9122	3.0644
112.	9	3456	12	4	-1.9500	3.1768
113.	18	236	12	4	-1.9503	3.2242
114.	7	1234	12	4	-1.9505	3.2295
115.	10	1256	12	5	-1.9425	3.2834
116.	9	1457	12	4	-1.9324	3.3086
117.	19	126	12	5	-1.9324	3.3086
118.	7	1457	12	4	-1.9515	3.3183
119.	9	1234	12	4	-1.9335	3.3442
120.	3	12347	12	5	-1.9176	3.3666

121.	10	1234	12	5	−1.9349	3.3808
122.	21	345	12	4	−1.9723	3.2782
123.	22	126	12	5	−1.9724	3.3238
124.	11	1236	12	5	−1.9626	3.3564
125.	12	1237	12	4	−1.9591	3.3726
126.	5	12367	12	5	−1.9660	3.3890
127.	21	123	12	5	−1.9542	3.4120
128.	15	3456	12	5	−1.9543	3.4153
129.	25	345	12	5	−1.9364	3.4320
130.	9	1257	12	4	−1.9346	3.4562
131.	14	1234	12	5	−1.9633	3.4660
132.	15	1234	12	5	−1.9633	3.4689
133.	16	1234	12	5	−1.9552	3.4840
134.	23	346	12	6	−1.9483	3.4857
135.	10	1237	12	5	−1.9483	3.4857
136.	2	12367	12	5	−1.9634	3.4967
137.	12	3467	12	4	−1.9596	3.4984
138.	29	126	12	6	−1.9400	3.5015
139.	29	345	12	5	−1.9113	3.5031
140.	3	23456	12	5	−1.9415	3.5328
141.	26	356	12	6	−1.9560	3.5443
142.	28	235	12	6	−1.9323	3.5762
143.	41	17	12	4	−1.9819	3.3651
144.	21	127	12	5	−1.9819	3.3651
145.	22	347	12	5	−1.9804	3.3849
146.	11	3456	12	5	−1.9795	3.4298
147.	41	12	12	4	−1.9795	3.4298
148.	23	345	12	6	−1.9795	3.4920
149.	21	236	12	5	−1.9786	3.5272
150.	13	3456	12	6	−1.9765	3.5699
151.	22	235	12	5	−1.9765	3.5699
152.	4	12347	12	5	−1.9702	3.5909
153.	25	346	12	5	−1.9722	3.6153
154.	46	12	12	6	−1.9703	3.6166
155.	5	12345	12	5	−1.9703	3.6166
156.	14	1256	12	5	−1.9703	3.6166
157.	13	3457	12	6	−1.9622	3.6326
158.	15	3467	12	5	−1.9580	3.6359
159.	6	23457	12	5	−1.9704	3.6432
160.	14	1235	12	5	−1.9704	3.6432

161.	13	2347	12	6	-1.9658	3.6572
162.	6	12367	12	5	-1.9625	3.6585
163.	25	236	12	5	-1.9455	3.6734
164.	29	237	12	6	-1.9271	3.6746
165.	16	1235	12	5	-1.9465	3.6975
166.	64	7	12	4	-1.9818	3.6943
167.	15	2367	12	5	-1.9775	3.7573
168.	16	2356	12	5	-1.9737	3.8108
169.	18	1256	13	4	-1.9158	3.3606
170.	31	234	13	4	-1.9365	3.3939
171.	18	1234	13	4	-1.9010	3.4670
172.	20	1234	13	5	-1.9244	3.5298
173.	17	1457	13	4	-1.8990	3.5457
174.	31	457	13	4	-1.9664	3.3327
175.	21	1257	13	5	-1.9637	3.4657
176.	31	126	13	4	-1.9559	3.4837
177.	35	456	13	4	-1.9254	3.4774
178.	32	346	13	4	-1.9502	3.5234
179.	36	126	13	5	-1.9291	3.5162
180.	32	236	13	5	-1.9564	3.5748
181.	39	123	13	5	-1.9312	3.5492
182.	24	2345	13	5	-1.9438	3.5691
183.	8	23456	13	5	-1.9569	3.5815
184.	28	1236	13	6	-1.9335	3.6300
185.	35	126	13	5	-1.9522	3.6408
186.	8	12347	13	5	-1.9446	3.6443
187.	9	12367	13	5	-1.9448	3.6464
188.	9	23457	13	5	-1.9360	3.6603
189.	19	2367	13	5	-1.9135	3.6843
190.	38	124	13	7	-1.9460	3.6940
191.	30	1236	13	6	-1.9740	3.5804
192.	22	1256	13	5	-1.9662	3.6612
193.	38	256	13	7	-1.9686	3.6774
194.	22	3457	13	5	-1.9548	3.6783
195.	35	346	13	5	-1.9664	3.6893
196.	22	1235	13	5	-1.9664	3.7297
197.	39	457	13	6	-1.9600	3.7419
198.	37	235	13	6	-1.9665	3.7526
199.	25	1237	13	5	-1.9603	3.7660
200.	13	12367	13	6	-1.9556	3.7671

201.	18	3457	13	4	-1.9492	3.7699
202.	19	1237	13	5	-1.9492	3.7699
203.	40	456	13	5	-1.9637	3.7774
204.	28	3456	13	6	-1.9561	3.7902
205.	15	23457	13	5	-1.9562	3.7918
206.	13	12345	13	6	-1.9603	3.8010
207.	36	235	13	5	-1.9425	3.8028
208.	29	1256	13	6	-1.9249	3.8038
209.	23	2347	13	6	-1.9505	3.8134
210.	20	2367	13	5	-1.9606	3.8212
211.	39	246	13	6	-1.9437	3.8246
212.	26	2347	13	6	-1.9447	3.8455
213.	16	12567	13	5	-1.9447	3.8455
214.	29	1235	13	6	-1.9170	3.8770
215.	64	17	13	4	-1.9804	3.7020
216.	64	12	13	5	-1.9777	3.7836
217.	23	2367	13	6	-1.9753	3.8653
218.	41	235	13	5	-1.9753	3.8653
219.	41	346	13	5	-1.9723	3.8756
220.	75	7	13	5	-1.9705	3.8964
221.	12	23456	13	5	-1.9705	3.8964
222.	12	12346	13	5	-1.9628	3.9083
223.	15	12367	13	5	-1.9707	3.9156
224.	26	3457	13	6	-1.9591	3.9280
225.	28	2347	13	6	-1.9362	3.9563
226.	16	23456	13	5	-1.9635	3.9623
227.	29	2357	13	6	-1.9491	3.9731
228.	30	3457	13	6	-1.9831	4.1090
229.	32	1457	14	4	-1.9050	3.5549
230.	33	1256	14	5	-1.8675	3.6294
231.	17	23456	14	5	-1.8765	3.7191
232.	32	1257	14	5	-1.9141	3.6593
233.	32	1237	14	5	-1.9256	3.7423
234.	19	23456	14	5	-1.8953	3.7521
235.	35	3457	14	5	-1.9270	3.7648
236.	36	1234	14	5	-1.9024	3.7943
237.	19	12567	14	5	-1.9066	3.8148
238.	53	126	14	7	-1.8284	3.8284
239.	41	1237	14	5	-1.9512	3.7306
240.	50	456	14	5	-1.9436	3.7605

241.	34	3456	14	6	-1.9439	3.8027
242.	22	12347	14	5	-1.9571	3.8513
243.	36	1256	14	5	-1.9058	3.8393
244.	23	12345	14	6	-1.9572	3.8863
245.	24	34567	14	5	-1.9456	3.8833
246.	38	1256	14	7	-1.9341	3.8856
247.	39	1234	14	6	-1.9133	3.8941
248.	26	12345	14	6	-1.9360	3.9068
249.	52	236	14	5	-1.9345	3.9186
250.	40	1234	14	6	-1.9463	3.9338
251.	7	234567	14	6	-1.9384	3.9443
252.	19	12347	14	5	-1.9206	3.9464
253.	38	2345	14	7	-1.9217	3.9476
254.	37	1235	14	6	-1.9469	3.9520
255.	10	234567	14	6	-1.9235	3.9642
256.	29	12347	14	6	-1.8951	3.9743
257.	56	127	14	5	-1.9778	3.6329
258.	42	1256	14	7	-1.9742	3.8704
259.	56	346	14	5	-1.9689	3.9053
260.	56	235	14	5	-1.9727	3.9280
261.	21	23456	14	5	-1.9666	3.9595
262.	11	123457	14	6	-1.9666	3.9595
263.	42	2356	14	7	-1.9561	3.9895
264.	35	1237	14	5	-1.9561	3.9895
265.	23	12367	14	6	-1.9667	4.0068
266.	36	3457	14	5	-1.9250	4.0011
267.	37	1267	14	6	-1.9607	4.0162
268.	22	23456	14	5	-1.9512	4.0266
269.	23	34567	14	6	-1.9507	4.0363
270.	38	1237	14	7	-1.9507	4.0363
271.	35	2367	14	5	-1.9437	4.0447
272.	14	234567	14	6	-1.9569	4.0507
273.	45	1256	14	6	-1.9569	4.0507
274.	58	246	14	6	-1.9308	4.0612
275.	24	12347	14	5	-1.9518	4.0690
276.	27	23457	14	5	-1.9338	4.0786
277.	25	12346	14	5	-1.9523	4.0843
278.	29	23456	14	6	-1.9146	4.1099
279.	41	2345	14	5	-1.9664	4.1483
280.	24	12367	14	5	-1.9635	4.1637

281.	44	3467	14	5	-1.9491	4.1729
282.	26	34567	14	6	-1.9638	4.1781
283.	27	12347	14	5	-1.9399	4.1877
284.	29	23567	14	6	-1.9423	4.2146
285.	46	2345	14	6	-1.9517	4.2276
286.	42	2347	14	7	-1.9796	4.1991
287.	44	2367	14	5	-1.9754	4.2559
288.	45	2357	14	6	-1.9725	4.2943
289.	47	1236	15	5	-1.9383	3.9217
290.	66	345	15	5	-1.9403	3.9629
291.	50	1267	15	6	-1.9085	4.0246
292.	32	12346	15	5	-1.9081	4.0522
293.	37	12347	15	6	-1.9301	4.0603
294.	53	1245	15	7	-1.8651	4.1162
295.	58	1234	15	6	-1.9529	4.0586
296.	50	1237	15	6	-1.9135	4.0702
297.	57	2356	15	7	-1.9466	4.1049
298.	51	2356	15	6	-1.9468	4.1309
299.	35	23456	15	5	-1.9237	4.1308
300.	53	2456	15	7	-1.9206	4.1397
301.	37	23456	15	6	-1.9236	4.1548
302.	63	1234	15	6	-1.8794	4.1854
303.	18	234567	15	6	-1.9267	4.1928
304.	36	12357	15	5	-1.9059	4.2012
305.	46	12347	15	6	-1.9286	4.2067
306.	38	23567	15	7	-1.9286	4.2067
307.	33	12347	15	5	-1.8907	4.2081
308.	20	234567	15	6	-1.9404	4.2123
309.	55	2345	15	7	-1.9091	4.2350
310.	40	12456	15	6	-1.9297	4.2403
311.	74	127	15	6	-1.9756	3.9756
312.	56	2345	15	5	-1.9609	4.1965
313.	75	346	15	6	-1.9612	4.2333
314.	21	123467	15	6	-1.9570	4.2347
315.	34	12367	15	6	-1.9570	4.2347
316.	38	12347	15	7	-1.9574	4.2693
317.	41	34567	15	5	-1.9520	4.2705
318.	37	12567	15	6	-1.9465	4.2902
319.	76	456	15	5	-1.9529	4.2958
320.	37	23567	15	6	-1.9371	4.3037

321.	44	12346	15	5	-1.9464	4.3093
322.	25	234567	15	6	-1.9382	4.3159
323.	28	123457	15	6	-1.9231	4.3290
324.	58	4567	15	6	-1.9690	4.3344
325.	30	123457	15	6	-1.9667	4.3661
326.	42	23457	15	7	-1.9607	4.3729
327.	45	12357	15	6	-1.9572	4.4244
328.	46	12456	15	6	-1.9330	4.4317
329.	64	2346	15	5	-1.9726	4.4655
330.	43	34567	15	6	-1.9709	4.4766
331.	96	7	15	5	-1.9665	4.4984
332.	46	23456	15	6	-1.9639	4.5187
333.	68	3456	16	7	-1.9142	4.2612
334.	50	12456	16	6	-1.8691	4.2687
335.	32	234567	16	6	-1.9153	4.2832
336.	53	12567	16	7	-1.8757	4.3107
337.	33	234567	16	6	-1.8803	4.3220
338.	67	1256	16	6	-1.9467	4.2186
339.	68	1456	16	7	-1.9535	4.2741
340.	67	2356	16	6	-1.9262	4.3319
341.	57	12567	16	7	-1.9392	4.3460
342.	50	12347	16	6	-1.9033	4.3591
343.	71	2456	16	7	-1.8898	4.3657
344.	34	123457	16	6	-1.9405	4.3778
345.	51	12347	16	6	-1.8954	4.3849
346.	53	23456	16	7	-1.9051	4.3967
347.	51	12357	16	6	-1.9104	4.4257
348.	53	23457	16	7	-1.8678	4.4331
349.	39	124567	16	6	-1.9146	4.4368
350.	67	3457	16	6	-1.9613	4.3655
351.	68	2347	16	7	-1.9615	4.4231
352.	43	123457	16	6	-1.9474	4.4876
353.	35	123467	16	6	-1.9397	4.4935
354.	57	23457	16	7	-1.9387	4.4989
355.	36	234567	16	6	-1.9265	4.5048
356.	60	23567	16	7	-1.9047	4.5265
357.	61	12357	16	6	-1.9291	4.5308
358.	40	234567	16	6	-1.9101	4.5513
359.	74	2346	16	6	-1.9693	4.5238
360.	75	2357	16	6	-1.9671	4.5505

361.	57	12347	16	7	-1.9671	4.5505
362.	41	123456	16	6	-1.9573	4.5671
363.	44	234567	16	6	-1.9577	4.5915
364.	61	12567	16	6	-1.9352	4.5980
365.	60	23457	16	7	-1.9132	4.6080
366.	62	12357	16	6	-1.9369	4.6214
367.	45	234567	16	6	-1.9380	4.6303
368.	65	23567	16	6	-1.9710	4.7697
369.	31	1234567	17	7	-1.9186	4.4569
370.	81	1256	17	6	-1.8905	4.4808
371.	68	12456	17	7	-1.9312	4.5490
372.	92	235	17	6	-1.9418	4.5598
373.	68	23457	17	7	-1.9321	4.5810
374.	50	234567	17	6	-1.8811	4.5919
375.	51	234567	17	6	-1.9169	4.6160
376.	71	12457	17	7	-1.8912	4.6444
377.	83	2456	17	7	-1.9480	4.6440
378.	69	12347	17	6	-1.9286	4.6732
379.	71	12347	17	7	-1.9291	4.6870
380.	58	124567	17	6	-1.9413	4.6911
381.	61	234567	17	6	-1.9099	4.7053
382.	54	234567	17	6	-1.9154	4.7270
383.	63	124567	17	6	-1.8761	4.7443
384.	72	12457	17	7	-1.8761	4.7443
385.	55	234567	17	7	-1.9172	4.7473
386.	75	34567	17	6	-1.9382	4.7668
387.	70	12367	17	5	-1.9536	4.7742
388.	75	23457	17	6	-1.9479	4.7786
389.	59	123457	17	6	-1.9236	4.7878
390.	65	234567	17	6	-1.9580	4.8745
391.	79	23457	17	7	-1.9367	4.8797
392.	47	1234567	18	7	-1.8959	4.6666
393.	48	1234567	18	7	-1.8241	4.7225
394.	49	1234567	18	7	-1.8924	4.7608
395.	52	1234567	18	7	-1.8916	4.7911
396.	69	124567	18	6	-1.8970	4.7990
397.	73	123457	18	6	-1.8541	4.8541
398.	92	1346	18	7	-1.9544	4.7566
399.	56	1234567	18	7	-1.9489	4.7795
400.	83	12347	18	7	-1.9100	4.8439

401.	68	123467	18	7	-1.9155	4.8636
402.	75	123456	18	6	-1.9324	4.8710
403.	71	234567	18	7	-1.8810	4.8908
404.	74	123456	18	6	-1.9400	4.9060
405.	64	1234567	18	7	-1.9485	4.9204
406.	77	234567	18	6	-1.9090	4.9613
407.	77	123467	18	6	-1.9257	5.0380
408.	78	234567	18	6	-1.9407	5.0472
409.	79	234567	18	7	-1.9285	5.0540
410.	89	34567	18	6	-1.9729	5.1410
411.	90	24567	18	7	-1.9694	5.1706
412.	66.	1234567	19	7	-1.8568	4.9817
413.	67	1234567	19	7	-1.9198	4.9917
414.	70	1234567	19	7	-1.9213	5.0367
415.	83	124567	19	7	-1.8983	5.0506
416.	76	1234567	19	7	-1.9182	5.1125
417.	85	124567	19	6	-1.8826	5.1375
418.	90	124567	19	7	-1.9487	5.2666
419.	87	234567	19	6	-1.9282	5.2702
420.	88	234567	19	7	-1.9417	5.2778
421.	96	23467	19	6	-1.9620	5.3241
422.	100	12346	20	7	-1.9185	5.2182
423.	81	1234567	20	7	-1.8950	5.2475
424.	84	1234567	20	7	-1.8187	5.2840
425.	86	1234567	20	7	-1.8895	5.3486
426.	100	12357	20	7	-1.9624	5.3402
427.	89	1234567	20	7	-1.9495	5.3796
428.	95	124567	20	7	-1.9147	5.4173
429.	97	234567	20	7	-1.9587	5.5336
430.	91	1234567	21	7	-1.8654	5.4269
431.	94	1234567	21	7	-1.8514	5.5016
432.	100	123457	21	7	-1.9342	5.5159
433.	96	1234567	21	7	-1.9176	5.6116
434.	103	234567	21	6	-1.9675	5.7623
435.	98	1234567	22	7	-1.8383	5.6750
436.	101	1234567	22	7	-1.8925	5.7275
437.	103	1234567	22	7	-1.9501	5.8373
438.	104	1234567	23	7	-1.8730	5.8730
439.	106	1234567	23	7	-1.9221	5.9634
440.	107	1234567	24	7	-1.9008	6.1325
441.	109	234567	24	7	-1.9556	6.2151
442.	108	1234567	25	7	-1.9245	6.3347
443.	110	1234567	26	7	-1.9511	6.5605

Table A3
REGULAR EXCEPTIONAL GRAPHS
AND THEIR SPECTRA

This table consists of several tables grouped into three parts:

A3.1. Auxiliary tables,
A3.2. Regular exceptional graphs,
A3.3. Eigenvalues of regular exceptional graphs.

A3.1. Auxiliary tables

These auxiliary tables are necessary for the description of the regular exceptional graphs in part 2. We have two tables:

A3.1.1. Representatives of the line-switching equivalence classes,
A3.1.2. Parameter sets for the second layer.

A3.1.1. REPRESENTATIVES OF THE LINE SWITCHING
EQUIVALENCE CLASSES

The table contains representatives of the line-switching equivalence classes of the even and odd graphs with 8 vertices and m edges ($m = 12, 14, \ldots, 24$); for details see Section 4.4. The graphs are specified by their edges, with each edge given as a pair of vertices. There are 38 equivalence classes, and a regular graph has been chosen as a representative whenever possible. With the exception of $L(K_8)$, all the graphs from Proposition 4.2.3 appear as representatives here. These 16 graphs, and the corresponding line switching equivalence classes, are identified below using the nomenclature of Table A4.

02	06	07	08	09	18	19	20	21	23	26	27	33	34	35	38
Q	S_4	S_2	S_3	S_1	$\overline{S_1}$	$\overline{S_3}$	$\overline{S_4}$	$\overline{S_2}$	$\overline{S_5}$	$\overline{2K_4}$	$K_{6,3}$	$\overline{2C_4}$	$\overline{C_3 \cup C_5}$	$\overline{C_8}$	$CP(4)$

The number of edges is 12

```
01.  1 6   1 7   1 8   2 3   2 4    2 5   3 4   3 5   4 5   6 7
     6 8   7 8 ,
02.  1 2   1 3   1 4   1 6   1 8    2 6   2 7   3 4   3 5   4 5
     5 7   6 7 ,
03.  1 2   1 3   1 4   1 5   1 6    1 7   1 8   2 5   2 6   3 4
     3 6   4 5 ,
04.  1 2   1 3   1 4   1 5   2 3    2 4   2 5   3 4   3 5   4 6
     5 7   6 7 ,
05.  1 6   1 7   1 8   2 5   2 7    2 8   3 4   3 5   3 6   4 5
     4 6   7 8 ,
06.  1 6   1 7   1 8   2 5   2 7    2 8   3 5   3 6   3 8   4 5
     4 6   4 7 ,
```

07. 1 6 1 7 1 8 2 5 2 7   2 8 3 4 3 6 3 8 4 5
    4 6 5 7 ,
08. 1 6 1 7 1 8 2 6 2 7   2 8 3 4 3 5 3 8 4 5
    4 7 5 6 ,
09. 1 6 1 7 1 8 2 5 2 7   2 8 3 4 3 6 3 8 4 5
    4 7 5 6 ,

The number of edges is 14

10. 1 2 1 3 1 4 1 5 1 6   1 7 1 8 2 3 2 4 2 5
    2 6 3 4 3 5 3 6 ,
11. 1 2 1 3 1 4 1 5 2 3   2 4 2 5 3 6 3 7 4 6
    4 7 5 6 5 7 6 7 ,
12. 1 3 1 4 1 5 1 6 1 7   2 3 2 4 2 5 2 6 2 7
    3 4 3 5 3 8 6 7 ,
13. 1 2 1 3 1 4 1 5 2 3   2 4 2 6 3 5 3 7 4 6
    4 7 5 6 5 7 6 7 ,
14. 1 2 1 3 1 4 1 6 1 7   2 3 2 4 2 5 2 6 3 4
    3 5 3 8 5 7 6 7 ,
15. 1 2 1 3 1 4 1 5 1 6   1 7 1 8 2 3 2 4 2 5
    2 6 3 7 4 7 5 6 ,
16. 1 2 1 3 1 4 1 5 1 6   1 7 2 3 2 4 2 5 2 6
    2 7 3 4 3 5 4 5 ,
17. 1 2 1 3 1 4 1 5 1 6   1 7 2 3 2 4 2 5 3 4
    3 5 4 6 5 6 6 7 ,

The number of edges is 16

18. 1 2 1 3 1 4 1 5 2 3   2 4 2 6 3 5 3 7 4 6
    4 8 5 7 5 8 6 7 6 8   7 8 ,
19. 1 2 1 3 1 4 1 5 2 3   2 4 2 5 3 6 3 7 4 6
    4 8 5 7 5 8 6 7 6 8   7 8 ,
20. 1 2 1 3 1 4 1 5 2 3   2 4 2 6 3 4 3 7 4 8
    5 6 5 7 5 8 6 7 6 8   7 8 ,
21. 1 2 1 3 1 4 1 5 2 3   2 4 2 6 3 5 3 7 4 7
    4 8 5 6 5 8 6 7 6 8   7 8 ,
22. 1 2 1 3 1 4 1 5 1 6   1 7 2 3 2 4 2 5 2 6
    2 8 3 4 3 5 3 7 3 8   4 5 ,
23. 1 2 1 3 1 4 1 5 2 3   2 4 2 6 3 7 3 8 4 7
    4 8 5 6 5 7 5 8 6 7   6 8 ,
24. 1 2 1 3 1 4 1 5 1 6   1 7 2 3 2 4 2 5 2 6
    2 7 3 4 3 5 4 6 5 7   6 7 ,
25. 1 2 1 3 1 4 1 5 1 7   2 3 2 4 2 5 2 7 3 4
    3 5 3 6 4 5 4 6 5 8   6 7 ,
26. 1 2 1 3 1 4 1 5 2 6   2 7 2 8 3 6 3 7 3 8
    4 6 4 7 4 8 5 6 5 7   5 8 ,

The number of edges is 18

```
27. 1 2   1 3   1 4   1 5   1 6   1 7   1 8   2 3   2 4   2 5
    2 6   2 7   2 8   3 4   3 5   3 6   3 7   3 8 ,
28. 1 2   1 3   1 4   1 5   1 6   1 7   2 3   2 4   2 5   2 6
    2 8   3 4   3 5   3 7   3 8   4 6   5 7   6 7 ,
29. 1 2   1 3   1 4   1 5   1 6   1 7   2 4   2 5   2 6   2 7
    2 8   3 4   3 5   3 6   3 7   3 8   4 5   6 7 ,
30. 1 3   1 4   1 5   1 6   1 7   2 3   2 4   2 5   2 6   2 7
    3 5   3 6   3 7   4 5   4 6   4 7   5 6   7 8 ,
31. 1 2   1 3   1 4   1 5   1 6   1 7   1 8   2 3   2 4   2 5
    2 6   3 4   3 5   3 6   4 5   4 6   5 7   6 7 ,
32. 1 2   1 3   1 4   1 5   1 6   1 7   2 3   2 4   2 5   2 6
    2 7   3 4   3 5   3 6   3 7   4 5   4 6   4 7 ,
```

The number of edges is 20

```
33. 1 3   1 5   1 6   1 7   1 8   2 4   2 5   2 6   2 7   2 8
    3 5   3 6   3 7   3 8   4 5   4 6   4 7   4 8   5 7   6 8 ,
34. 1 4   1 5   1 6   1 7   1 8   2 4   2 5   2 6   2 7   2 8
    3 4   3 5   3 6   3 7   3 8   4 6   4 7   5 7   5 8   6 8 ,
35. 1 3   1 4   1 5   1 6   1 7   2 4   2 5   2 6   2 7   2 8
    3 5   3 6   3 7   3 8   4 6   4 7   4 8   5 7   5 8   6 8 ,
```

The number of edges is 22

```
36. 1 2   1 3   1 4   1 5   1 6   1 7   1 8   2 3   2 4   2 5
    2 6   2 7   2 8   3 4   3 5   3 6   3 7   3 8   4 6   4 7
    5 6   5 7 ,
37. 1 2   1 3   1 4   1 5   1 6   1 7   2 3   2 4   2 5   2 6
    2 7   3 4   3 5   3 6   3 7   4 5   4 6   4 7   5 6   5 7
    6 8   7 8 ,
```

The number of edges is 24

```
38. 1 2   1 3   1 4   1 5   1 6   1 7   2 3   2 4   2 5   2 6
    2 8   3 4   3 5   3 7   3 8   4 6   4 7   4 8   5 6   5 7
    5 8   6 7   6 8   7 8 ,
```

A3.1.2. PARAMETER SETS FOR THE SECOND LAYER

The six small tables here contain the parameter sets necessary for the construction of regular exceptional graphs in the second layer. For details see Section 4.4.

I. $\quad v = 3,\ n = 9,\ d = 4.$

p	j	r	k	s	ℓ	t	degree sequence	number of graphs F
0	5	0	4	0	3	6	(3,3,3,3,3,3)	2
2	4	0	3	2	2	4	(3,3,2,2,2,2)	4
		1		0		5	(4,2,2,2,2,2)	1
4	3	0	2	4	1	2	(2,2,2,2,1,1)	3
		1		2		3	(3,2,2,1,1,1)	3
		2		0		4	(3,3,1,1,1,1)	1
6	2	0	1	6	0	0	(1,1,1,1,1,1)	1
		1		4		1	(2,1,1,1,1,0)	1
		2		2		2	(2,2,1,1,0,0)	1
		3		0		3	(2,2,2,0,0,0)	1
8	1	2	0	4	−1	0	(1,1,0,0,0,0)	1
								19

II. $\quad v = 4,\ n = 12,\ d = 6.$

p	j	r	k	s	ℓ	t	degree sequence	number of graphs F
0	6	0	5	0	4	6	(4,4,4,4,4,4)	1
2	5	0	4	2	3	4	(4,4,3,3,3,3)	3
		1		0		5	(5,3,3,3,3,3)	1
4	4	0	3	4	2	2	(3,3,3,3,2,2)	4
		1		2		3	(4,3,3,2,2,2)	4
		2		0		4	(4,4,2,2,2,2)	2
6	3	0	2	6	1	0	(2,2,2,2,2,2)	2
		1		4		1	(3,2,2,2,2,1)	3
		2		2		2	(3,3,2,2,1,1)	5
		3		0		3	(3,3,3,1,1,1)	1
8	2	2	1	4	0	0	(2,2,1,1,1,1)	2
		3		2		1	(2,2,2,1,1,0)	2
		4		0		2	(2,2,2,2,0,0)	1
10	1	4	0	2	−1	0	(1,1,1,1,0,0)	1
12	0	6	−1	0	−2	0	(0,0,0,0,0,0)	1
								33

III. $v = 5$, $n = 15$, $d = 8$.

p	j	r	k	s	ℓ	t	degree sequence	number of graphs F
0	7	0	6	0	5	6	(5,5,5,5,5,5)	1
2	6	0	5	2	4	4	(5,5,4,4,4,4)	1
4	5	0	4	4	3	2	(4,4,4,4,3,3)	2
		1		2		3	(5,4,4,3,3,3)	2
		2		0		4	(5,5,3,3,3,3)	1
6	4	0	3	6	2	0	(3,3,3,3,3,3)	2
		1		4		1	(4,3,3,3,3,2)	3
		2		2		2	(4,4,3,3,2,2)	5
		3		0		3	(4,4,4,2,2,2)	1
8	3	2	2	4	1	0	(3,3,2,2,2,2)	4
		3		2		1	(3,3,3,2,2,1)	4
		4		0		2	(3,3,3,3,1,1)	2
10	2	4	1	2	0	0	(2,2,2,2,1,1)	3
		5		0		1	(2,2,2,2,2,0)	1
12	1	6	0	0	−1	0	(1,1,1,1,1,1)	1
								33

IV. $v = 6$, $n = 18$, $d = 10$.

p	j	r	k	s	ℓ	t	degree sequence	number of graphs F
4	6	0	5	4	4	2	(5,5,5,5,4,4)	1
6	5	0	4	6	3	0	(4,4,4,4,4,4)	1
		1		4		1	(5,4,4,4,4,3)	1
		2		2		2	(5,5,4,4,3,3)	1
		3		0		3	(5,5,5,3,3,3)	1
8	4	2	3	4	2	0	(4,4,3,3,3,3)	3
		3		2		1	(4,4,4,3,3,2)	3
		4		0		2	(4,4,4,4,2,2)	1
10	3	4	2	2	1	0	(3,3,3,3,2,2)	4
		5		0		1	(3,3,3,3,3,1)	1
12	2	6	1	0	0	0	(2,2,2,2,2,2)	2
								19

V. $v = 7$, $n = 21$, $d = 12$.

p	j	r	k	s	ℓ	t	degree sequence	number of graphs F
6	6	0	5	6	4	0	(5,5,5,5,5,5)	1
8	5	2	4	4	3	0	(5,5,4,4,4,4)	1
		3		2		1	(5,5,5,4,4,3)	1
		4		0		2	(5,5,5,5,3,3)	0
10	4	4	3	2	2	0	(4,4,4,4,3,3)	2
		5		0		1	(4,4,4,4,4,2)	1
12	3	6	2	0	1	0	(3,3,3,3,3,3)	2
								8

VI. $v = 8$, $n = 24$, $d = 14$.

p	j	r	k	s	ℓ	t	degree sequence	number of graphs F
10	5	4	4	2	3	0	(5,5,5,5,4,4)	1
		5		0		1	(5,5,5,5,5,3)	0
12	4	6	3	0	2	0	(4,4,4,4,4,4)	1
								2

A3.2. Regular exceptional graphs

The regular exceptional graphs are classified within three layers, as defined in Section 4.1:

A3.2.1. First layer (graph nos. 1–163),
A3.2.2. Second layer (graph nos. 164–184),
A3.2.3. Third layer (graph nos. 185–187).

In each layer the graphs are listed according to the number of vertices, and ordered as in [BuCS1].

A3.2.1. FIRST LAYER (graph nos. 1–163)

The graphs 1–5 are defined by a diagram. The graphs 6–160 are defined by specifying line-switching sets in the the odd and even graphs given in 1.1. These line-switching sets are given by the ordinal numbers of the edges listed in 1.1. The graphs 161–163 are just the Chang graphs.

The number of vertices is 10

In this case we have the graphs 1–5; these are the cubic graphs $Z_1, \ldots, Z_5$ shown in Fig. 1. They were constructed in Section 2.4 as the graphs L_2, L_1, L_3, L_5, L_4, respectively.

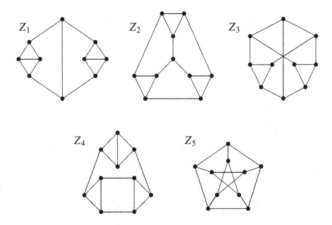

Figure 1: Cubic exceptional graphs

SWITCHING SETS FOR EXCEPTIONAL REGULAR GRAPHS (nos. 6–160)

The number of vertices is 12

```
Class No. 2 :
Graph No. 06. 3 4 7 9 ,

Class No. 3 :
Graph No. 07. 2 3 5 8 ,

Class No. 4 :
Graph No. 08. 1 3 7 8 9 12 ,

Class No. 6 :
Graph No. 09. 1 5 9 10 ,

Class No. 7 :
Graph No. 10. 1 5 9 10 ,
Graph No. 11. 1 6 7 12 ,

Class No. 8 :
Graph No. 12. 1 5 9 10 ,

Class No. 9 :
Graph No. 13. 1 5 9 10 ,
```

The number of vertices is 14

```
Class No. 10 :
Graph No. 14. 1 2 5 9 13 ,

Class No. 11 :
Graph No. 15. 1 2 5 10 11 12 13 ,
Graph No. 16. 1 2 6 8 11 12 13 ,

Class No. 12 :
Graph No. 17. 1 2 6 8 14 ,
Graph No. 18. 2 4 6 10 12 ,

Class No. 13 :
Graph No. 19. 1 2 5 10 11 12 13 ,
Graph No. 20. 1 2 6 8 11 12 14 ,
Graph No. 21. 1 2 6 8 10 13 14 ,

Class No. 14 :
Graph No. 22. 2 3 6 9 13 ,
Graph No. 23. 3 5 6 9 11 ,
```

```
Class No. 15 :
Graph No. 24. 1 3 6 8 14 ,
Graph No. 25. 1 4 5 9 12 ,
Graph No. 26. 4 5 6 8 9 ,
Graph No. 27. 3 5 6 8 10 ,
Graph No. 28. 1 3 5 10 12 ,

Class No. 16 :
Graph No. 29. 1 5 6 8 9 12 13 ,
Graph No. 30. 1 3 5 9 11 12 13 ,

Class No. 17 :
Graph No. 31. 1 2 4 8 10 13 14 ,
Graph No. 32. 1 2 4 8 11 12 14 ,
Graph No. 33. 2 3 4 7 8 13 14 ,
Graph No. 34. 3 4 5 7 8 11 14 ,
```

The number of vertices is 16

```
Class No. 18 :
Graph No. 35. 1 8 11 14 ,
Graph No. 36. 1 2 5 10 11 12 13 14 ,
Graph No. 37. 3 5 13 14 ,

Class No. 19 :
Graph No. 38. 1 8 11 12 ,
Graph No. 39. 1 2 6 8 11 12 13 14 ,
Graph No. 40. 2 6 12 15 ,
Graph No. 41. 1 2 5 10 11 12 13 14 ,
Graph No. 42. 1 2 6 8 10 12 13 16 ,

Class No. 20 :
Graph No. 43. 1 2 6 9 10 11 13 14 ,
Graph No. 44. 1 8 12 15 ,
Graph No. 45. 1 9 10 11 ,

Class No. 21 :
Graph No. 46. 1 8 11 14 ,
Graph No. 47. 1 9 11 12 ,
Graph No. 48. 1 3 5 9 10 12 13 15 ,
Graph No. 49. 3 5 13 14 ,
Graph No. 50. 1 8 10 15 ,
Graph No. 51. 1 3 5 8 10 13 14 15 ,
Graph No. 52. 1 2 6 9 11 12 13 14 ,
Graph No. 53. 1 2 5 10 11 12 13 14 ,
Graph No. 54. 1 3 5 8 11 12 14 16 ,
```

```
Class No. 22 :
Graph No. 55. 2 3 7 9 ,
Graph No. 56. 1 2 7 16 ,
Graph No. 57. 2 3 5 7 9 11 14 16 ,
Graph No. 58. 2 5 6 7 8 11 13 16 ,

Class No. 23 :
Graph No. 59. 1 8 11 12 ,
Graph No. 60. 2 7 10 14 ,

Class No. 24 :
Graph No. 61. 1 3 4 10 11 12 13 16 ,
Graph No. 62. 1 4 6 10 11 12 13 14 ,
Graph No. 63. 1 3 5 9 11 12 13 16 ,
Graph No. 64. 1 3 6 9 10 12 13 16 ,

Class No. 25 :
Graph No. 65. 1 2 8 10 13 16 ,
Graph No. 66. 1 2 7 11 13 16 ,
Graph No. 67. 1 5 8 10 12 13 ,
Graph No. 68. 2 5 7 8 12 13 ,

Class No. 26 :
Graph No. 69. 1 8 12 16 ,
```

The number of vertices is 18

```
Class No. 27 :
Graph No. 70. 2 3 4 8 11 12 18 ,

Class No. 28 :
Graph No. 71. 1 2 9 14 16 ,
Graph No. 72. 1 4 7 14 16 ,
Graph No. 73. 1 2 7 16 17 ,
Graph No. 74. 5 6 7 9 12 ,
Graph No. 75. 2 3 7 9 18 ,
Graph No. 76. 1 3 7 13 18 ,
Graph No. 77. 3 4 7 10 14 ,
Graph No. 78. 2 3 9 10 14 ,
Graph No. 79. 1 3 10 13 14 ,
Graph No. 80. 1 2 8 13 18 ,

Class No. 29 :
Graph No. 81. 1 2 9 15 17 ,
Graph No. 82. 2 5 6 7 8 11 14 15 17 ,
Graph No. 83. 1 3 8 14 15 ,
Graph No. 84. 2 3 8 9 15 ,
Graph No. 85. 2 5 6 7 9 11 13 15 17 ,
```

```
Class No. 30 :
Graph No. 86. 1 2 8 10 12 16 17 ,
Graph No. 87. 1 2 8 9 13 16 17 ,
Graph No. 88. 1 3 7 10 12 16 17 ,
Graph No. 89. 1 3 7 9 13 16 17 ,
Graph No. 90. 1 3 9 10 11 15 16 ,
Graph No. 91. 1 3 8 9 13 15 16 ,
Graph No. 92. 1 3 9 10 12 14 16 ,
Graph No. 93. 1 3 8 10 12 15 16 ,
Graph No. 94. 1 3 9 10 13 14 15 ,
Graph No. 95. 1 5 8 9 13 14 15 ,

Class No. 31 :
Graph No. 96. 1 2 5 8 15 16 17 ,
Graph No. 97. 2 3 5 8 10 16 17 ,
Graph No. 98. 1 2 3 10 14 16 17 ,
Graph No. 99. 2 3 4 8 11 16 17 ,
Graph No. 100. 3 4 5 8 9 14 17 ,
Graph No. 101. 3 5 6 8 10 13 16 ,
Graph No. 102. 2 5 6 8 10 15 16 ,
Graph No. 103. 3 5 6 8 10 14 15 ,
Graph No. 104. 4 5 6 8 9 13 16 ,

Class No. 32 :
Graph No. 105. 4 5 6 7 8 9 12 14 18 ,
Graph No. 106. 2 4 5 8 9 11 12 14 18 ,
Graph No. 107. 2 4 5 8 9 10 12 15 18 ,
```

The number of vertices is 20

```
Class No. 33 :
Graph No. 108. 1 2 7 9 12 17 18 20 ,
Graph No. 109. 1 7 17 20 ,
Graph No. 110. 2 3 9 10 11 12 17 18 ,
Graph No. 111. 2 8 14 17 ,
Graph No. 112. 2 8 13 18 ,

Class No. 34 :
Graph No. 113. 1 7 14 20 ,
Graph No. 114. 1 2 6 9 13 14 19 20 ,
Graph No. 115. 1 2 8 10 13 15 17 18 ,
Graph No. 116. 1 2 6 8 14 15 18 20 ,
Graph No. 117. 1 2 6 9 13 15 18 20 ,
Graph No. 118. 1 2 6 10 13 14 18 20 ,
Graph No. 119. 1 3 6 9 12 15 18 20 ,
Graph No. 120. 1 3 6 9 12 14 19 20 ,
```

```
Class No. 35 :
Graph No. 121. 1 2 7 9 13 15 19 20 ,
Graph No. 122. 1 2 8 9 11 17 18 20 ,
Graph No. 123. 1 2 7 8 13 16 19 20 ,
Graph No. 124. 1 7 16 20 ,
Graph No. 125. 1 4 8 9 11 16 17 19 ,
Graph No. 126. 1 8 16 19 ,
Graph No. 127. 1 8 17 18 ,
Graph No. 128. 1 2 7 8 13 17 18 20 ,
Graph No. 129. 1 6 18 20 ,
Graph No. 130. 1 2 6 8 13 18 19 20 ,
Graph No. 131. 1 2 7 9 14 15 18 20 ,
Graph No. 132. 1 2 8 10 13 15 18 19 ,
Graph No. 133. 2 8 13 19 ,
Graph No. 134. 2 9 12 19 ,
```

The number of vertices is 22

```
Class No. 36 :
Graph No. 135. 1 2 8 19 22 ,
Graph No. 136. 2 4 8 12 19 ,
Graph No. 137. 3 5 7 8 10  12 15 17 19 ,
Graph No. 138. 3 5 8 10 17 ,
Graph No. 139. 3 4 8 11 17 ,
Graph No. 140. 3 6 7 8 10 11 15 17 19 ,
Graph No. 141. 3 4 7 8 10 12 16 17 19 ,
Graph No. 142. 2 3 7 8 10 11 17 19 22 ,
Graph No. 143. 1 3 7 8 10 16 17 19 22 ,
Graph No. 144. 4 6 7 8 9 10 16 17 19 ,
Graph No. 145. 4 6 7 8 9 12 15 16 19 ,
Graph No. 146. 1 2 7 9 10 16 17 19 22 ,
Graph No. 147. 2 4 7 9 10 12 16 17 19 ,
```

```
Class No. 37 :
Graph No. 148. 1 2 7 17 18 19 20 ,
Graph No. 149. 1 2 8 14 18 19 20 ,
Graph No. 150. 1 2 10 14 16 18 20 ,
Graph No. 151. 1 2 10 15 16 17 20 ,
Graph No. 152. 1 2 5 8 10 13 15 16 18 20 21 ,
```

The number of vertices is 24

```
Class No. 38 :
Graph No. 153. 1 2 11 15 16 17 19 20 ,
Graph No. 154. 1 2 3 9 11 14 15 16 18 19 20 22 ,
Graph No. 155. 1 2 8 14 18 19 21 22 ,
```

```
Graph No. 156. 1 2 7 16 17 19 21 24 ,
Graph No. 157. 1 2 8 13 16 21 22 24 ,
Graph No. 158. 1 12 20 23 ,
Graph No. 159. 1 12 19 24 ,
Graph No. 160. 1 2 8 13 16 20 23 24 ,
```

The number of vertices is 28

Graph nos. 161–163. The Chang graphs

A3.2.2. SECOND LAYER (graph nos. 64–184)

Each of graphs 164–183 is defined by a graph on 8 vertices whose edge set is partitioned into two subsets by a separator ∗ ∗ ∗. The line graph should be switched with respect to either subset to obtain the graph in the second layer. Graph no. 184 is just the Schläfli graph.

The number of vertices is 9

```
164.  1 3   2 4   ***   3 4   3 5   3 6   4 5   4 7   6 8   7 8 ,
165.  1 3   2 6   ***   3 4   3 5   3 6   4 5   6 7   6 8   7 8 ,
```

The number of vertices is 12

```
166.  1 3   2 4   ***   3 4   3 5   3 6   3 7   4 5   4 6   4 7
      5 8   6 8   7 8 ,
167.  1 3   2 4   ***   3 4   3 5   3 6   3 7   4 5   4 6   4 8
      5 7   6 8   7 8 ,
168.  1 3   2 8   ***   3 4   3 5   3 6   3 7   4 5   4 8   5 8
      6 7   6 8   7 8 ,
169.  1 3   2 3   ***   3 4   3 5   3 6   3 7   3 8   4 5   4 6
      5 7   6 8   7 8 ,
170.  1 3   1 4   2 3   2 6   ***   3 4   3 5   3 6   3 7   4 5
      4 8   6 7   6 8 ,
```

The number of vertices is 15

```
171.  1 3   2 4   ***   3 4   3 5   3 6   3 7   3 8   4 5   4 6
      4 7   4 8   5 6   5 7   6 8   7 8 ,
172.  1 3   1 4   2 5   2 6   ***   3 4   3 5   3 6   3 7   4 5
      4 6   4 8   5 7   5 8   6 7   6 8 ,
173.  1 3   1 4   2 3   2 5   ***   3 4   3 5   3 6   3 7   3 8
      4 5   4 6   4 7   5 6   5 8   7 8 ,
174.  1 3   1 4   2 3   2 8   ***   3 4   3 5   3 6   3 7   3 8
      4 5   4 6   4 7   5 8   6 8   7 8 ,
175.  1 3   1 4   2 3   2 4   ***   3 4   3 5   3 6   3 7   3 8
      4 5   4 6   4 7   4 8   5 6   7 8 ,
176.  1 3   1 4   1 5   1 6   2 3   2 4   2 5   2 7   ***   3 4
      3 5   3 6   4 5   4 7   5 8   6 7 ,
```

The number of vertices is 18

```
177.   1 3   1 4   2 5   2 6   ***   3 4   3 5   3 6   3 7   3 8
       4 5   4 6   4 7   4 8   5 6   5 7   5 8   6 7   6 8 ,
178.   1 3   1 4   1 5   2 6   2 7   2 8   ***   3 4   3 5   3 6
       3 7   4 5   4 6   4 8   5 7   5 8   6 7   6 8   7 8 ,
179.   1 3   1 4   1 5   2 3   2 4   2 5   ***   3 4   3 5   3 6
       3 7   3 8   4 5   4 6   4 7   4 8   5 6   5 7   5 8 ,
180.   1 3   1 4   1 5   1 6   2 3   2 4   2 7   2 8   ***   3 4
       3 5   3 6   3 7   4 5   4 6   4 8   5 7   6 8   7 8 ,
```

The number of vertices is 21

```
181.   1 3   1 4   1 5   2 6   2 7   2 8   ***   3 4   3 5   3 6
       3 7   3 8   4 5   4 6   4 7   4 8   5 6   5 7   5 8   6 7
       6 8   7 8 ,
182.   1 3   1 4   1 5   1 6   2 3   2 4   2 7   2 8   ***   3 4
       3 5   3 6   3 7   3 8   4 5   4 6   4 7   4 8   5 6   5 7
       6 8   7 8 ,
```

The number of vertices is 24

```
183.   1 3   1 4   1 5   1 6   1 7   2 3   2 4   2 5   2 6   2 8
       ***   3 4   3 5   3 6   3 7   3 8   4 5   4 6   4 7   4 8
       5 6   5 7   5 8   6 7   6 8 ,
```

The number of vertices is 27

184. The Schläfli graph

A3.2.3. THIRD LAYER (graph nos. 185–187)

185. The graph T_1 from Fig. 2

186. The graph T_2 from Fig. 2 (cf. the description of Fig. 4.2 in Section 4.4)

187. The Clebsch graph

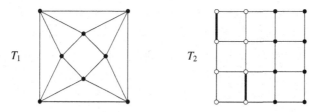

Figure 2: Two graphs in the third layer.

A3.3. Eigenvalues of regular exceptional graphs

Here, n is the number of vertices, C is the class number, G the graph number. An asterisk indicates the existence of a line graph $L(H)$ with the given spectrum; the corresponding graphs H (other than K_8) are identified in Table A3.1.1.

First layer

n	C	G	spectrum								
10	1	3	2	1^3	-1^2	-2^3					
10	2	3	1.879^2	1	-0.347^2	-1.532^2	-2^2				
10	3	3	2.562	1^2	0	-1^2	-1.562	-2^2			
10	4	3	2.414	1.732	0^2	-0.414	-1	-1.732	-2^2		
10	5	3	1^5	-2^4							
*12	2	6	4	2.732^2	0^3	-0.732^2	-2^4				
12	3	7	4	3	1.618^2	0	-0.618^2	-1	-2^4		
12	4	8	4	2.303^2	1^2	0	-1.303^2	-2^4			
*12	6	9	4	2^3	0^3	-2^5					
*12	7	10–11	4	2.732	2	1.414	0^2	-0.732	-1.414	-2^4	
*12	8	12	4	2.562	1.618^2	1	-0.618^2	-1.562	-2^4		
*12	9	13	4	2^2	1.414^2	0	-1.414^2	-2^4			
14	10	14	5	3.562	2	1^3	-0.562	-1	-2^6		
14	11	15–16	5	3	2^2	1^2	-1^2	-2^6			
14	12	17–18	5	3^2	1^2	0^2	-1	-2^6			
14	13	19–21	5	2.802^2	1.445^2	-0.247^2	-1	-2^6			
14	14	22–23	5	3	2.247^2	0.555^2	-0.802^2	-2^6			
14	15	24–28	5	3.343	2.414	1.471	0	-0.414	-0.814	-2^6	
14	16	29–30	5	3	2.562	1^3	0	-1.562	-2^6		
14	17	31–34	5	2.814	2.414	2	1	0.529	-0.414	-1.343	-2^6
*16	18	35–37	6	3.414^2	2	0.586^2	0^2	-2^8			
*16	19	38–42	6	3.562	2.618^2	1	0.382^2	-0.562	-2^8		
*16	20	43–45	6	4	2^3	0^3	-2^8				
*16	21	46–54	6	3.414	2.732	2^2	0.586	0	-0.732	-2^8	
16	22	55–58	6	3.303^2	2	1^2	-0.303^2	-2^8			
*16	23	59–60	6	3.236	2^4	0	-1.236	-2^8			
16	24	61–64	6	3	2.618^2	2	0.382^2	-1	-2^8		
16	25	65–68	6	2.732^2	2^3	-0.732^2	-2^8				
*16	26	69	6	2^6	-2^9						
*18	27	70	7	4^2	1^5	-2^{10}					
18	28	71–80	7	3.532^2	2.347^2	1	0.121^2	-2^{10}			
18	29	81–85	7	4	3^2	1^3	0	-2^{10}			
18	30	86–95	7	3.562	3^2	2	1^2	-0.562	-2^{10}		
18	31	96–104	7	3.732	3	2.414	2	2	0.268	-0.414	-2^{10}
18	32	105–107	7	3^3	2^2	1	-1	-2^{10}			

n		range					
*20	33	108–112	8	4^2	2^4	0	-2^{12}
*20	34	113–120	8	3.618^2	3^2	1.382^2	0	-2^{12}
*20	35	121–134	8	4	3.414^2	2^2	0.586^2	-2^{12}
22	36	135–147	9	4^2	3^3	1^2	-2^{14}
22	37	148–152	9	4	3^5	0	-2^{14}
*24	38	153–160	10	4^4	2^3	-2^{16}
*28		161–163	12	4^7	-2^{20}

Second layer

n	G	spectrum							
9	164	4	1.879^2	-0.347^2	-1.532^2	-2^2
9	165	4	2.562	1	0	-1^2	-1.562	-2^2
12	166	6	2.303^2	1^2	-1.303^2	-2^5
12	167	6	2.732	2	1.414	0	-0.732	-1.414	-2^5
12	168	6	3.236	2	0^3	-1.236	-2^5
12	169	6	3	1.618^2	-0.618^2	-1	-2^5
12	170	6	2.732^2	0^2	-0.732^2	-2^5
15	171	8	3.236	2^3	0	-1.236	-2^8
15	172	8	3	2.618^2	0.382^2	-1	-2^8
15	173	8	3.414	2.732	2	0.586	0	-0.732	-2^8
15	174	8	3.303^2	1^2	-0.303^2	-2^8
15	175	8	4	2^2	0^3	-2^8
15	176	8	2.732^2	2^2	-0.732^2	-2^8
18	177	10	3.562	3^2	2	1	-0.562	-2^{11}
18	178	10	4	3^2	1^2	0	-2^{11}
18	179	10	4^2	1^4	-2^{11}
18	180	10	3.532^2	2.347^2	0.121^2	-2^{11}
21	181	12	4	3^4	0	-2^{14}
21	182	12	4^2	3^2	1^2	-2^{14}
24	183	14	4^4	2^2	-2^{17}
27	184	16	4^6	-2^{20}

Third layer

n	G	spectrum			
8	185	4	1.414^2	0	-1.414^2	-2^2
12	186	7	2^2	1^2	-1	-2^6
16	187	10	2^5	-2^{10}

Table A4.

A CONSTRUCTION OF THE 68 CONNECTED REGULAR GRAPHS WHICH ARE NOT LINE GRAPHS BUT COSPECTRAL WITH LINE GRAPHS

The regular exceptional graphs cospectral with line graphs were constructed in [CvRa1] in such a way that the results can be verified and used without reference to a computer search. In this table we reproduce the relevant data from [CvRa1]. The graphs concerned are obtained by switching graphs $L(H)$ where H is either a regular 3-connected graph on 8 vertices or a semi-regular bipartite graph on $6 + 3$ vertices (cf. Theorem 4.2.9 and Proposition 4.2.3, the remarks at the end of Section 4.2, and Theorem 4.4.17).

All of the relevant regular graphs on 8 vertices are given in Fig. 1 (overleaf) together with a labelling of their edge sets: the case of K_8 was treated in [Hof1] and [Sei4], and exactly three exceptional graphs arise, namely the Chang graphs (graph nos. 161–163 in [BuCS1]). For each of the graphs a table is given. The rows in a table correspond to the inequivalent factorizations of the graph into two regular factors. A row contains the following information (some data are omitted):

- a regular factor (determining the factorization);
- the identification number, from Table A3 (cf. [BuCS1]), of the exceptional graph if such a graph is obtained, or a dash '–' if the graph obtained by switching is isomorphic to the initial line graph;
- in the case that an exceptional graph is obtained, a set of 4 edges forming $K_{1,3}$ (a forbidden subgraph for line graphs);
- the number of 4-cocliques in the graph obtained [5-cocliques do not exist and, in a regular graph, the number of 3-cocliques is determined by the spectrum];
- the number of 4-cliques in the graph obtained [the number of 3-cliques is determined by the spectrum];
- an isomorphism, if any, which maps the graph to the isomorphic mate most highly placed in the table; or, if the graph obtained by switching is again the initial line graph, an isomorphism which maps the original line graph to the graph obtained graph [isomorphisms are given as products of disjoint cycles, with fixed points omitted].

Using the number of 4-cocliques and 4-cliques we can almost always distinguish between non-isomorphic graphs. The remaining cases, indicated in the tables by asterisks, are the graphs $\overline{S}_2$, $\overline{S}_3$, $\overline{C}_8$ and $\overline{C_3 \cup C_5}$. In these cases non-isomorphic exceptional graphs can be distinguished by analysing incidences between vertices and 4-cocliques as follows.

The graph $\overline{S}_2$: the exceptional graph 49 has exactly one vertex (8) belonging to four 4-cocliques, the graph 50 has two such vertices (5, 6) and the graph 46 has none. The graph 51 has two vertices (14, 15) belonging to only one 4-coclique and the graph 52 has no such vertices.

The graph $\overline{S}_3$: every vertex of graph 38 belongs to a 4-coclique and in the graph 42 the vertex 14 does not.

The graph $\overline{C}_8$: in graph 128 there is only one vertex (15) which belongs to exactly two 4-cocliques, while in the graph 131 there are three such vertices (15, 17, 20).

The graph $\overline{C_3 \cup C_5}$: in graph 113 the vertex 5 belongs to six 4-cocliques, and in the graph 116 there is no such vertex.

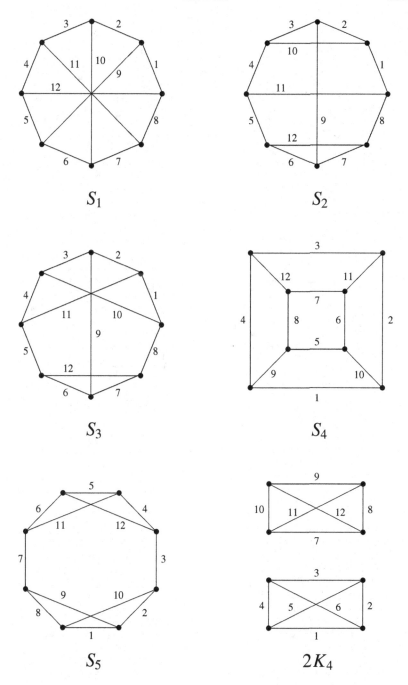

Figure 1: Regular graphs whose line graphs are switched into exceptional graphs.

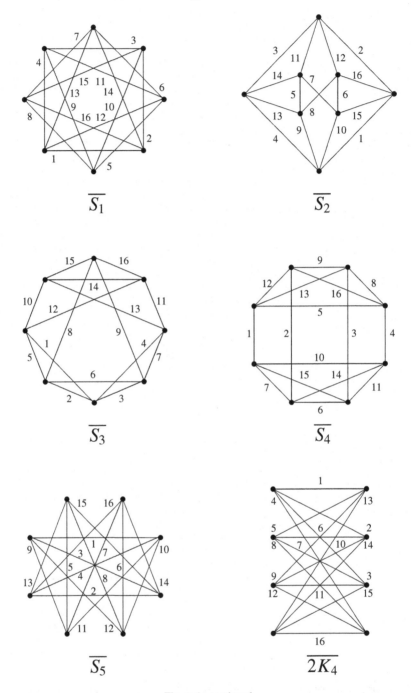

Figure 1: continued.

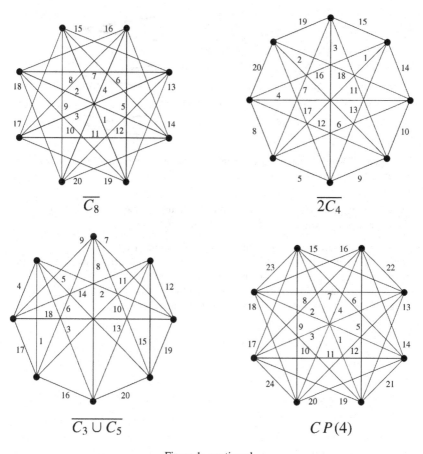

Figure 1: continued.

The two semi-regular bipartite graphs required for our purposes (Q and $L(K_{6,3})$) are given in Fig. 2, together with the factorizations which yield the exceptional graphs 6 and 70 respectively. The details are left to the reader. Other semi-regular bipartite graphs on 9 vertices do not give rise to exceptional graphs.

The tables show how to construct the 68 connected regular graphs that are not line graphs but cospectral with such a graph, and they afford a proof that there are no more such graphs which are switching-equivalent to regular line graphs. Although the tables were produced using a computer, only a modest effort is required to verify the data by hand. For example, 4-cocliques can be enumerated by counting appropriate sets of 4 edges in a given factorization of the initial graph H on 8 vertices: these 4 edges are either independent and belong to the same factor or form a quadrangle with adjacent

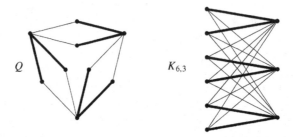

Figure 2: Semi-regular bipartite graphs whose line graphs are switched into exceptional graphs.

edges in different factors. Similarly, to count 4-cliques we have to find the number of sets of 4 edges in H which fulfil one of the following conditions:

- all edges have a common vertex and belong to the same factor;
- three edges belong to one factor and form $K_{1,3}$ or K_3, the fourth edge being adjacent to none of these three and belonging to the other factor;
- the edges are partitioned by the factorization into 2 pairs, each forming $K_{1,2}$, with edges from one pair non-adjacent to edges from the other.

S_1:

1 3 5 7	—				(1 4 3 6 5 8 7 2) (10 12)
9 10 11 12	—				(1 7 5 3) (2 8 6 4)
3 7 9 12	13	2 6 9 10			

S_2 :

9 10 11 12	—				(1 5) (2 6) (3 7) (4 8) (9 11)
1 4 9 12	10	5 8 11 12	2		
2 4 6 8	11	1 4 10 11	3		

S_3:

9 10 11 12	12	1 2 3 10			

S_4:

9 10 11 12	—				(9 11) (10 12)
2 4 5 7	9	1 5 9 10			

S_5:

9 10 11 12	—				(9 12) (10 11)
1 3 5 7	—				(2 4) (3 7) (6 8) (9 11) (10 12)

$2K_4$:

5 6 11 12	—				(5 7 8 11) (6 9 10 12)

$$\overline{S}_1:$$

2 4 6 8	37	2 7 14 15	10	8	
10 11 13 16	–				(10 13) (11 16)
4 6 10 16	35	4 5 9 12	8	8	
1 2 3 4 5 6 7 8	–				(1 3) (2 4) (5 7) (6 8)
2 4 6 7 9 10 12 16	36	2 4 5 7	5		
2 4 6 8 9 12 14 15	37	1 6 9 15	10	8	(4 15) (5 14)
1 3 4 5 6 7 10 16	35	1 2 3 4	8	8	(3 15) (6 12) (9 11) (10 14)

$$\overline{S}_2:$$

1 3 5 6	49	5 7 9 10	10	6*	
1 3 7 8	49	5 7 9 10	10	6	(1 4) (2 11) (3 6) (5 8) (9 13) (10 14) (15 16)
10 11 13 16	50	5 7 9 10	10	6*	
4 5 12 15	46	5 6 7 8	10	6*	
2 8 10 14	47	5 6 7 8	8	6	
9 10 11 12 13 14 15 16	–				(1 3) (2 4) (5 6) (7 8)
1 4 5 6 11 12 13 15	54	3 5 11 13	9		
2 4 5 6 9 12 14 15	51	3 5 11 13	10	8*	
1 2 7 8 10 12 13 14	53	4 5 9 14	7	9	
1 3 5 6 9 12 14 15	51	4 5 9 14	10	8	(1 15) (2 16) (3 11) (4 5) (7 10) (8 13) (9 14)
1 2 3 5 6 8 10 14	47	6 8 9 10	8	6	(4 15) (5 14) (11 12)
1 3 4 5 7 8 12 15	46	1 2 3 4	10	6	(4 13) (6 9) (10 14) (11 15)
2 4 5 8 10 12 14 15	48	5 8 11 12	7	7	
3 4 5 7 9 12 15 16	52	5 7 9 10	10	8*	
2 4 6 7 9 11 13 16	48	6 7 11 12	7	7	(3 4) (5 10) (6 12) (9 11)

$$\overline{S}_3:$$

1 7 8 14	38	2 4 6 7	7	7*	
1 6 13 16	40	7 9 11 16	9		
1 4 6 7 8 10 14 16	39	2 4 6 7	8	8	
1 3 5 6 11 13 15 16	41	1 7 8 14	8	10	
1 2 6 9 10 11 13 16	42	11 13 15 16	7	7*	

$\overline{S}_4$:

1 2 3 4	–				(1 3) (2 4) (13 16) (14 15)
5 6 9 10	–				(1 3) (2 4) (7 8) (11 12) (13 15) (14 16)
6 8 10 12	44	3 4 7 12	12	8	
1 2 8 11	45	7 11 14 15	8	8	
1 2 3 4 7 8 11 12	–				(1 3) (2 4) (7 8) (11 12) (13 16) (14 15)
1 2 3 4 6 8 10 12	44	2 3 6 9	12	8	(4 13)(6 9)(11 15)(12 14)
1 2 6 8 10 11 13 16	43	8 12 13 16	8	12	
1 2 5 6 8 10 11	45	2 3 6 9	8	8	(3 16) (6 13)

$\overline{S}_5$:

1 2 5 6	59	3 6 11 13	12	4	
3 4 5 6	59	4 8 9 14	12	4	(2 16 5 13) (3 5 6 12) (4 8 7 14)
9 10 13 14	59	3 6 11 13	12	4	(2 15 12 5) (3 6 13 16) (4 9 7 10) (8 14)
3 6 11 13	60	4 5 6 14	10	4	
9 10 11 12 13 14 15 16	–				(5 6) (9 13) (10 14) (11 15) (12 16)
3 4 5 6 11 12 13 14	–				(3 6) (4 5) (11 13) (12 14)
3 4 6 8 9 11 13 14	60	4 5 11 13	10	4	(2 4 11 9 5 3) (6 7 8 12 13 10) (15 16)
1 2 7 8 9 10 13 14	59	3 7 9 14	12	4	(2 16 15 6 12 13) (3 5) (4 8 9 7 14 10)
1 2 4 5 6 8 9 14	60	1 2 3 4	10	4	(2 5 16) (4 15 9) (6 10 7) (8 13 12)

$\overline{2K}_4$:

1 6 11 16	69	1 2 9 10	16	0	
1 2 5 6 11 12 15 16	–				(3 7) (4 8) (9 10) (11 16) (12 15) (13 14)
1 2 6 7 11 12 13 16	69	1 2 5 6	16	0	(2 16) (5 14) (9 13) (10 15)

$$\overline{C}_8:$$

1 2 3 4	134	2 8 11 16	23	32	
1 4 7 11	133	5 11 14 19	21	32	
2 4 5 19	127	1 4 6 11	21	34	
4 10 11 15	126	13 15 17 19	20	33	
5 7 9 11	134	4 8 11 20	23	32	(2 15 9 16) (3 4 12 11) (5 6) (7 17 18 8) (13 14) (19 20)
5 7 17 20	124	14 16 18 20	19	34	
15 16 19 20	129	5 11 14 19	23	36	
5 6 7 8 9 10 11 12	–				(1 3) (2 4) (5 9) (6 10) (7 11) (8 12)
13 14 15 16 17 18 19 20	134	3 7 10 19	23	32	(2 9) (3 11) (4 12) (5 6) (7 17) (8 18) (13 14) (19 20)
3 5 10 14 15 17 18 20	121	4 5 10 18	17	36	
5 7 9 11 14 15 18 19	129	2 8 11 16	23	36	(1 18) (3 6 16 20) (4 17 7 14) (5 9 15 11) (10 13) (12 19)
2 4 5 7 9 11 15 19	127	1 4 5 9	21	34	(5 19) (6 15) (12 14) (16 18)
8 10 13 14 15 17 18 20	124	10 12 13 18	19	34	(1 2 20 7 9 10 16 3) (4 6) (5 12 8 18 15 13 11 14)
2 4 13 14 15 17 18 19	127	5 7 14 16	21	34	(1 4 13 9 10 7 17 2) (3 5 8 15 20 19 11 6) (12 16 18 14)
6 7 8 9 13 14 18 19	131	7 9 15 18	18	33*	
1 3 4 7 10 14 17 20	121	1 7 10 15	17	36	(1 2 5 13 9 11 6 3) (4 20 16 19) (7 10 14 15 12 8 17 18)

$\overline{C}_8$ (continued):

2 4 5 10 11 12 15 17	125	4 5 10 18	20	35	
3 4 6 10 13 14 17 18	132	10 12 13 18	17	32	
3 5 9 10 11 15 16 20	128	3 11 12 20	18	33*	
1 2 5 7 8 12 17 20	123	5 11 14 19	21	38	
1 2 3 4 9 11 13 16	124	1 9 12 13	19	34	(1 9) (2 10) (3 7) (4 19) (5 15) (6 17) (8 11) (16 20)
1 2 4 7 8 9 11 13	126	1 7 12 13	20	33	(1 17) (3 7) (4 13 20 12) (5 11 19 10) (6 16 18 15) (9 14)
2 8 9 13 14 15 18 19	128	13 15 17 19	18	33	(2 20) (3 5 4 19) (6 15 14 13) (7 9) (8 16 10 18) (12 17)
4 6 8 10 11 13 15 18	121	10 12 13 18	17	36	(1 5) (3 13) (4 19 16 20) (6 9) (7 8 12 15) (8 17 10 14)
1 2 3 5 12 16 17 20	128	1 5 8 17	18	33	(2 20) (3 19 4 5) (6 13 14 15) (7 9) (8 18 10 16) (12 17)
2 4 5 11 13 15 17 18	122	4 5 12 13	19	36	
2 8 9 10 12 13 14 17	128	2 5 10 14	18	33	(1 9) (2 11) (3 4) (5 18) (6 15) (7 17) (12 20) (13 14) (16 19)
7 8 9 10 13 14 17 20	131	9 11 17 20	18	33	(1 11) (2 10) (3 17) (6 13) (7 15) (12 20) (14 16)
1 2 3 4 5 7 9 11	–				(1 4) (2 3) (5 9) (7 11)
6 8 10 12 14 15 18 19	129	8 10 16 19	23	36	(1 18) (3 20 16 6) (4 14 7 17) (5 11 15 9) (10 13) (12 19)

$\overline{C}_8$ (continued):

3 4 6 10 14 15 18 19	130	4 5 10 18	21	36	
3 5 6 9 10 11 12 16	126	1 5 8 17	20	33	(1 3 14 8 9 7 17 2) (4 5 10 13 20 19 11 12) (6 16 18 15)
1 2 3 4 14 15 18 19	129	1 9 12 13	23	36	(1 18) (3 20) (5 9) (6 16) (10 13) (11 15) (12 19) (14 17)
1 4 7 11 13 14 17 18	130	1 3 13 17	21	36	(2 18) (3 4) (5 10) (6 14) (8 15) (9 12) (13 16) (17 19)
1 4 7 10 11 12 14 17	123	1 7 10 15	21	38	(1 7 18 2 9 3 20 11) (4 15) (6 14 8 13 19 16 10 12)
3 4 5 6 7 9 10 11	133	3 7 10 19	21	32	(2 12) (3 20 5 18 6 17 4 19) (7 15 11 13 8 14 10 16)
5 6 7 8 10 12 17 20	124	6 8 14 17	19	34	(4 6) (5 15) (8 11) (12 13) (14 18) (17 19)

$$\overline{2C_4}:$$

1 2 3 4					(3 4) (5 9) (6 10) (7 11) (8 12) (13 17) (14 18) (15 19) (16 20)
2 4 5 15	109	2 3 11 17	21	36	
5 10 15 20	112	5 7 17 19	25	32	
5 11 14 20	111	1 3 5 15	21	32	
6 8 9 11 14 16 17 19	–				(6 11) (7 13) (8 19) (9 14) (10 20) (16 17)
6 7 9 12 14 15 17 29	110	2 3 6 16	25	40	
2 3 6 8 11 14 16 17	109	9 12 13 16	21	36	(3 18 17 13 7 11 19 14) (4 8 20 12 6 5 16 15) (9 10)
6 8 9 11 14 15 17 20	111	6 8 14 16	21	32	(2 12) (3 20) (4 19 5 18) (6 17) (7 5 11 16) (8 14 10 13)
2 4 5 7 9 14 16 19	108	2 3 9 19	17	36	
5 7 9 12 14 16 18 19	111	5 7 9 11	21	32	(2 18 4) (5 12 19) (7 5 11) (8 10 13)
2 4 5 7 9 15 16 18	108	5 7 9 11	17	36	(4 5) (6 16) (7 11) (15 17)
1 2 3 4 8 9 14 19	112	9 12 17 20	25	32	(2 5 14) (3 7 17) (4 12 8) (6 19 9)
1 2 3 4 6 9 15 20	111	1 4 6 16	21	32	(2 12) (4 18 5 19) (7 16 11 15) (8 13 10 14)
6 7 9 12 13 16 18 19	–				(6 19 8) (7 20 18) (9 14 16) (10 12 13)
6 7 11 12 13 16 17 18	112	6 7 8 19	25	32	(2 18 15 8 6 17) (3 5 12 9 11 16) (10 12 13)
2 3 5 8 12 14 15 18	109	1 3 5 15	21	36	(3 5) (4 14) (6 13) (7 8) (9 10) (11 17) (12 16) (15 20) (18 19)

Appendix

$$\overline{C_3 \cup C_5}:$$

3 6 15 18	113	1 10 13 18	20	32*	
1 5 7 9 12 14 16 20	115	3 4 8 9	22	36	
2 4 6 7 14 15 16 20	114	3 4 8 9	19	33	
2 4 6 9 12 15 16 20	120	2 3 7 8	17		
1 4 7 8 13 15 17 19	117	7 8 19 20	19	31	
1 5 7 8 13 14 17 19	118	7 8 12 13	21		
2 4 6 8 13 15 17 19	119	8 10 13 15	22	34	
1 2 8 9 13 15 17 19	116	8 10 13 15	20	32*	

$$CP(4):$$

5 7 9 11	159	2 5 16 21	44	104	
21 22 23 24	158	1 11 17 21	40	104	
8 13 14 15 18 19 20 23	160	10 13 16 22	36	96	
4 11 13 14 15 18 19 23	157	10 13 16 22	36	100	
2 4 6 8 13 15 18 19	156	13 15 17 19	36	102	
6 8 13 15 16 18 19 20	157	13 15 17 19	36	100	(1 3 13 17 11 22) (2 20 12 4 14 7) (5 23 15 24 6 16) (8 10) (9 21)
8 9 11 13 15 16 18 21	155	6 17 19 21	36	104	
2 4 8 13 15 19 20 23	157	13 15 17 19	36	100	(3 18 7 5) (4 6 8 16) (11 12 14 13) (15 17 23 21) (19 22 24 20)
6 13 14 15 16 18 19 24	158	13 15 17 19	40	104	(3 17) (4 19) (5 21) (6 16) (7 18) (8 23) (10 14) (12 13) (15 20) (22 24)
6 10 14 15 18 19 22 24	–				(2 7) (3 11) (5 9) (6 20 22 13) (8 24 23 10) (12 14 21 15) (16 18 17 19)

$CP(4)$ (continued):

13 14 15 16 17 18 19 20	159	3 5 8 13	44	104	(2 18) (3 9) (5 12) (6 16) (7 24) (10 20) (11 19) (13 23)
1 3 13 14 16 17 18 20	158	10 13 15 23	40	104	(1 5 3 2 19 24) (4 18 11 6 17 13) (7 10 8 20 14 16) (9 23 22 12 21 15)
6 8 13 15 16 18 21 24	153	1 2 21 23	44	112	
1 8 11 13 15 16 18 20	156	14 16 18 20	36	102	(1 14 10) (2 11 9) (3 16 18) (4 22 15 23 19 8) (5 21 7) (20 24)
8 10 12 13 14 15 16 17 20 21 23 24	159	10 18 20 21	44	104	(1 3) (2 15) (4 20) (5 24) (6 14) (7 11) (8 21) (9 12) (13 19) (16 17) (22 23)
2 3 5 8 9 10 13 14 15 20 23 24	–				(2 5 11) (3 9 7) (8 17 21 23 16 12) (10 14 19) (13 20) (15 18 24)
8 10 12 13 14 15 16 17 20 21 23 24	159	13 14 21 22	44	104	(1 3) (2 15) (4 20) (5 24) (6 14) (7 11) (8 21) (9 12) (13 19) (16 17) (22 23)
6 8 10 12 13 14 15 16 19 20 23 24	158	13 15 17 19	40	104	(3 7 22 15) (4 16 5 23) (6 21 8 19) (9 11) (10 13 14 12) (17 18 24 20)
8 9 11 12 13 14 15 16 18 19 21 23	158	13 14 21 22	40	104	(1 4 9 8 2 21 11 16) (3 18 7 22 24 15 20 17) (5 14 6 13 19 10 23 12)
4 7 8 9 11 13 14 15 18 19 21 23	158	13 14 21 22	40	104	(1 5 14 4 11 23 2 19 10 21 9 6) (3 17 7) (8 13 16 12) (15 18) (20 24 22)
5 7 8 9 11 13 14 15 18 19 20 23	159	13 15 17 19	44	104	(1 3 8 17 24 20 22 23 21 14 7 2) (4 19 13 18 5 6 15 12 9 10 11 16)

Appendix

$CP(4)$ (continued):

2 5 8 9 10 11 12 13 14 15 17 18	155	13 15 17 19	36	104	(1 13 11) (2 10 14) (3 18 19) (4 20 8) (5 23 16 24 6 15) (21 22)
2 5 8 9 11 12 13 15 16 21 23 24	156	8 12 22 24	36	102	(1 2) (3 15) (4 18) (5 8) (6 24) (7 23) (9 14) (10 11) (16 19) (17 20) (21 22)
4 5 7 11 13 15 16 17 20 21 23 24	156	13 15 17 19	36	102	(1 14 12 2 11 13) (3 17 7) (4 19 20 23 22 24) (5 6 18) (8 15) (9 10)
2 4 5 7 8 13 15 17 20 21 23 24	155	13 15 17 19	36	104	(1 23 9 5 2 6 12 24) (3 19 17 18 4 8 7 20) (10 15 14 22 13 16 11 21)
1 3 6 7 8 9 13 14 16 18 21 24	154	13 14 21 22	36	108	

Table A5
ONE-VERTEX EXTENSIONS OF EXCEPTIONAL STAR COMPLEMENTS

We have given in Section 5.3 some statistics for one-vertex extensions of the exceptional star complements from Table A2. There are 51 extensions of exceptional star complements on six vertices, 512 extensions of exceptional star complements on seven vertices, and 4206 extensions of exceptional star complements on eight vertices. The one-vertex extensions of $H443$ are described in Section 6.3. Since it is not feasible to list all the extensions here, we restrict the data to the following:

A5.1. Extensions of exceptional star complements on six vertices.
A5.2. Extensions of exceptional star complements on seven vertices for which
an eigenvector corresponding to the eigenvalue -2 has no zero entry.
A5.3. Extensions of the exceptional star complement H440.

In each case, the graphs are ordered lexicographically by spectral moments. As usual for a graph on n vertices, the vertices are denoted by $1, 2, \ldots, n$. Each graph G on n vertices ($n = 7, 8, 9$) is represented by a line which contains: the identification number of G; the identification number from Table A2 of the subgraph induced by the vertices $1, 2, \ldots, n - 1$; the vertices to which vertex n is adjacent; the number of edges; the maximal vertex degree; the largest eigenvalue; the eigenvector of -2 in which the non-zero coordinate with smallest absolute value is ± 1, and the first non-zero coordinate is positive; the symbol * if the eigenvalue -2 is main and # otherwise.

A5.1. Extensions of exceptional star complements on six vertices

First we give the 51 one vertex extensions of the 20 exceptional star complements on six vertices. Below we give a summary.

1.	01	6	06	3	2.0000	1	-2	3	-2	1	-2	1	#
2.	01	14	07	3	2.2143	0	1	-2	2	-1	1	-1	#
3.	01	36	07	4	2.4142	1	-2	3	-2	1	-1	-1	*
4.	01	156	08	3	2.3429	1	-1	1	-1	1	0	-1	#
5.	03	36	08	5	2.7321	1	-2	3	-1	-1	-1	-1	*
6.	01	234	08	4	2.8136	1	-2	2	-2	1	-1	1	#
7.	02	234	09	4	2.8662	1	-2	1	-1	0	1	1	*
8.	01	1346	09	4	2.9537	1	-1	1	0	0	0	-1	#
9.	05	12	09	6	3.0000	1	-3	1	1	1	1	1	*
10.	06	35	09	4	3.0861	1	-2	2	-1	0	1	-1	#
11.	02	2346	10	4	3.1774	0	0	0	1	-1	1	-1	, #
12.	03	1234	10	5	3.2642	1	-1	2	0	-1	-1	-1	*
13.	01	12345	10	5	3.2814	1	-1	2	-1	1	-1	-1	#
14.	05	235	10	6	3.3234	1	-2	0	1	0	1	1	*
15.	03	2345	10	5	3.4321	1	-2	2	-1	-1	-1	1	*

16.	02	12345	11	5	3.3839	0	1	0	0	1	-1	-1	#
17.	03	12346	11	5	3.4909	0	1	-1	1	0	1	-1	*
18.	01	123456	11	6	3.5201	0	1	-1	1	0	1	-1	*
19.	06	2346	11	5	3.5366	0	0	1	0	-1	1	-1	#
20.	03	13456	11	5	3.5366	1	-1	1	0	0	0	-1	#
21.	09	245	11	6	3.6691	1	-2	1	0	0	1	1	*
22.	04	2346	11	4	3.7321	1	-2	1	-2	1	1	1	*
23.	07	1245	12	4	3.5032	0	1	-1	1	0	0	-1	#
24.	07	2346	12	4	3.6458	1	-1	1	-1	0	-1	1	#
25.	11	345	12	6	3.6458	1	-1	1	0	1	-1	-1	#
26.	04	12345	12	5	3.6758	0	1	0	1	0	-1	-1	#
27.	02	123456	12	6	3.6964	1	-1	1	0	0	1	-1	*
28.	06	12345	12	5	3.8039	0	1	-1	1	1	-1	-1	#
29.	09	2346	12	6	3.8478	0	0	0	1	-1	1	-1	#
30.	05	23456	12	6	3.8951	1	-2	1	1	1	1	-1	*
31.	08	2456	12	5	3.9063	1	-2	1	0	-1	1	1	*
32.	07	12345	13	5	3.8256	1	0	1	0	0	-1	-1	#
33.	10	1236	13	5	3.9108	0	0	1	0	-1	1	-1	#
34.	07	12346	13	5	3.9571	1	0	0	1	-1	0	-1	#
35.	08	12345	13	5	3.9832	0	1	0	0	1	-1	-1	#
36.	11	1245	13	6	3.9832	1	0	0	0	1	-1	-1	#
37.	04	123456	13	6	4.0514	1	-1	1	-1	1	1	-1	*
38.	06	123456	13	6	4.0862	1	-1	1	0	0	1	-1	*
39.	10	12345	14	5	4.1807	1	0	0	0	1	-1	-1	#
40.	11	23456	14	6	4.2015	1	-1	0	0	0	-1	1	#
41.	10	12356	14	5	4.2584	1	-1	1	-1	1	0	-1	#
42.	08	123456	14	6	4.3102	1	-1	1	0	0	1	-1	*
43.	12	23456	14	5	4.4495	1	-2	1	-1	1	-1	1	#
44.	14	12346	15	6	4.4317	1	0	0	1	-1	0	-1	#
45.	11	123456	15	6	4.5047	1	0	1	-1	1	0	-1	*
46.	16	12345	15	5	4.5114	0	1	0	0	1	-1	-1	#
47.	18	12345	16	6	4.7102	1	0	0	0	1	-1	-1	#
48.	18	12346	16	6	4.7664	1	-1	0	-1	1	-1	1	#
49.	15	123456	16	6	4.7913	1	-1	0	0	-1	-1	1	*
50.	19	12345	17	6	4.9651	0	0	0	1	1	-1	-1	#
51.	17	123456	17	6	5.0000	1	-1	1	-1	-1	-1	1	*

We summarize the data by giving all non-isomorphic extensions for each of the 20 exceptional star complements. Each line contains the identification number of a star complement from Table A2, the number of non-isomorphic extensions and the identification numbers of extensions as given in the table above.

```
 1.  08    01 02 03 04 06 08 13 18
 2.  06    02 04 07 11 16 27
 3.  07    03 05 07 12 15 17 20
 4.  06    06 07 11 22 26 37
 5.  05    05 09 14 18 30
 6.  11    06 08 10 11 12 13 15 16 19 28 38
 7.  05    07 23 24 32 34
 8.  12    10 11 12 13 16 17 19 26 28 31 35 42
 9.  07    12 14 18 21 27 29 38
10.  07    12 17 23 32 33 39 41

11.  11    13 14 16 18 25 26 27 36 37 40 45
12.  06    15 24 33 34 41 43
13.  09    15 20 21 29 30 31 37 38 42
14.  03    27 32 44
15.  08    21 28 35 36 38 40 42 49
16.  06    22 31 34 41 43 46
17.  04    30 40 45 51
18.  11    31 34 37 38 39 41 42 44 46 47 48
19.  07    41 42 44 45 47 49 50
20.  05    43 48 49 50 51
```

A5.2. Extensions of exceptional star complements on seven vertices for which an eigenvector corresponding to -2 has no zero entry

1.	001	5	07	3	2.0000	2	-3	4	-3	2	-2	-1	-1	*
2.	001	45	08	3	2.3028	2	-3	4	-3	1	-2	-1	1	*
3.	001	34	08	4	2.4812	2	-3	4	-2	1	-2	-1	-1	*
4.	001	167	09	3	2.4812	1	-2	3	-2	1	-2	-1	1	*
5.	001	145	09	3	2.5321	2	-2	2	-1	1	-1	-1	-1	*
6.	004	45	09	3	2.4142	1	-3	4	-3	1	-2	1	1	#
7.	004	34	09	4	2.5869	1	-3	4	-2	1	-2	1	-1	*
8.	012	5	09	3	2.7321	2	-3	2	-3	2	2	-1	-1	#
9.	001	236	09	4	2.7817	2	-3	3	-2	1	-2	-1	1	*
10.	006	34	09	5	2.8136	1	-2	4	-2	1	-2	-1	-1	*
11.	014	5	09	4	2.8608	1	-2	3	-2	-2	2	1	1	*
12.	005	167	10	4	2.6180	1	-2	3	-1	-1	-2	-1	1	*
13.	012	45	10	4	2.8912	2	-3	2	-3	1	2	-1	1	*
14.	004	236	10	4	2.9354	1	-3	3	-2	1	-2	1	1	#
15.	005	236	10	5	3.0000	2	-3	3	-1	-1	-2	-1	1	*
16.	016	34	10	4	3.0000	1	-2	3	-1	-2	2	1	-1	*
17.	001	1236	10	4	3.0627	2	-2	3	-2	1	-1	-1	-1	*
18.	005	345	10	5	3.1889	2	-3	4	-1	-1	-2	-1	-1	*

19.	007	236	11	4	2.9801	1	-2	2	-1	-1	2	1	-1	*
20.	004	1457	11	4	3.0135	1	-2	2	-1	1	-1	1	-1	#
21.	021	45	11	4	3.0000	1	-3	2	-3	1	2	1	1	*
22.	011	236	11	5	3.1071	1	-3	3	-1	-1	-2	1	1	*
23.	005	1236	11	5	3.2116	2	-2	3	-1	-1	-1	-1	-1	*
24.	012	123	11	4	3.2361	2	-2	2	-2	1	1	-1	-1	#
25.	015	346	11	5	3.2814	1	-2	2	-2	1	-2	1	1	#
26.	001	12347	11	5	3.3051	1	-2	2	-2	1	-1	-1	1	*
27.	006	1236	11	5	3.3212	1	-1	3	-2	1	-1	-1	-1	*
28.	011	345	11	5	3.2143	1	-3	4	-1	-1	-2	1	-1	*
29.	013	345	11	6	3.3761	1	-2	4	-1	-1	-2	-1	-1	*
30.	012	346	11	4	3.4142	2	-3	2	-2	1	2	-1	-1	#
31.	006	2367	11	5	3.4663	1	-2	3	-2	1	-2	-1	1	*
32.	016	235	11	5	3.5289	1	-2	2	-1	-2	2	1	1	*
33.	009	2367	12	4	3.3322	1	-1	1	-1	1	-1	-1	1	#
34.	013	1236	12	6	3.4651	1	-1	3	-1	-1	-1	-1	-1	*
35.	021	346	12	4	3.4724	1	-3	2	-2	1	2	1	-1	*
36.	004	12367	12	5	3.5753	1	-2	3	-2	1	-1	1	-1	#
37.	013	2367	12	6	3.6017	1	-2	3	-1	-1	-2	-1	1	*
38.	014	2356	12	5	3.7557	1	-2	3	-2	-1	2	1	-1	*
39.	021	1457	13	4	3.3429	1	-2	1	-1	1	1	1	-1	*
40.	014	12347	13	5	3.6017	1	-1	2	-1	-1	1	1	-1	*
41.	007	23456	13	5	3.6576	1	-2	1	-1	-1	1	1	1	*
42.	011	12367	13	5	3.6488	1	-2	3	-1	-1	-1	1	-1	*
43.	021	1237	13	5	3.6626	1	-2	2	-2	1	1	1	-1	*
44.	023	3456	13	6	3.7310	1	-2	2	-1	-1	-2	1	1	*
45.	005	123457	13	6	3.7683	1	-2	2	-1	-1	-1	-1	1	*
46.	015	34567	13	5	3.7699	1	-2	2	-1	1	-1	1	-1	#
47.	016	12357	13	5	3.8662	1	-1	2	-1	-1	1	1	-1	*
48.	042	256	13	7	3.7321	2	-4	1	1	1	1	1	1	*
49.	022	2356	13	5	3.8325	1	-2	3	-1	-1	2	-1	-1	#
50.	044	346	13	5	3.8951	1	-2	2	-2	1	2	-1	-1	#
51.	014	23456	13	5	3.9261	1	-2	2	-2	-1	1	1	1	*
52.	031	1236	14	4	3.7913	2	-1	2	-1	-1	-1	1	-1	#
53.	012	123467	14	6	4.0353	1	-2	1	-2	1	1	-1	1	#
54.	041	3456	14	5	4.2143	2	-3	2	-1	2	-1	-1	-1	*
55.	030	34567	14	6	4.2564	1	-2	3	-1	-1	-2	-1	1	*
56.	046	2456	14	6	4.3028	1	-2	2	-2	1	1	1	1	*
57.	056	1237	15	5	3.9173	1	-2	2	-1	1	-1	1	-1	#
58.	057	1256	15	7	4.0514	2	-2	1	1	1	1	-1	-1	*
59.	034	34567	15	6	4.1232	1	-1	2	-1	-1	-1	-1	1	*
60.	022	234567	15	6	4.2487	1	-2	2	-1	-1	1	-1	1	#

61.	040	23467	15	6	4.3341	1	-2	1	1	-1	1	1	-1	*
62.	056	3456	15	5	4.2361	1	-3	2	-1	2	-1	1	-1	#
63.	042	12347	15	7	4.3409	2	-3	1	1	1	1	1	-1	*
64.	045	23567	15	6	4.5085	1	-2	2	-1	-1	2	1	-1	*
65.	031	123456	16	6	4.2546	1	-1	1	-1	-1	-1	1	1	#
66.	037	234567	16	6	4.4206	1	-1	1	-1	-1	1	-1	1	#
67.	046	124567	16	6	4.5616	1	-1	1	-1	1	1	1	-1	*
68.	075	3456	16	6	4.5353	1	-2	2	-1	2	-1	-1	-1	*
69.	064	23456	16	5	4.7321	2	-3	1	1	-2	1	-1	1	#
70.	041	1234567	17	7	4.6544	1	-2	1	-1	1	-1	-1	1	*
71.	074	23456	17	6	4.7785	1	-3	1	1	-2	1	1	1	*
72.	064	123456	17	6	4.8448	2	-2	1	1	-1	1	-1	-1	#
73.	100	1346	19	7	5.0000	2	-1	2	-1	-1	-1	-1	-1	*
74.	074	1234567	19	7	5.0536	1	-2	1	1	-1	1	1	-1	*
75.	090	234567	19	7	5.3297	1	-2	2	-1	1	1	1	-1	*
76.	094	134567	20	6	5.3511	1	-1	1	-1	1	-1	1	-1	#
77.	096	234567	20	6	5.5273	1	-2	1	1	-2	1	-1	1	#
78.	109	1234567	25	7	6.3723	2	-1	1	-1	1	-1	1	-1	*

A5.3. Extensions of the exceptional star complement H440

From Table A2 we know that the graph H440 has 8 vertices and the following edges: 12 13 14 16 17 18 23 27 28 34 35 36 37 38 45 46 47 48 56 57 58 67 68 78. We omit the second column from the present format since the identification number is 440 in each case.

1.	4	25	7	6.1538	1	-2	1	-2	0	-1	1	1	1	#
2.	1	25	7	6.1528	2	-1	0	-1	1	-1	0	0	-1	*
3.	3	25	8	6.1581	0	2	-2	1	1	1	-1	-1	1	*
4.	5	25	7	6.1484	1	1	-1	0	2	0	-1	-1	-1	#
5.	14	26	7	6.2160	1	1	-1	1	1	0	-1	-1	-1	#
6.	34	26	8	6.2261	1	0	-1	-1	1	0	0	0	1	*
7.	23	26	8	6.2018	1	-2	0	-1	0	-1	1	1	1	#
8.	35	26	8	6.2148	1	-1	1	-1	1	-1	0	0	-1	*
9.	237	27	8	6.3108	1	0	-1	0	1	0	-1	0	1	*
10.	134	27	8	6.3318	1	-1	1	0	0	-1	0	0	-1	*
11.	123	27	8	6.2974	1	1	0	0	1	0	-1	-1	-1	#
12.	345	27	8	6.3187	0	1	0	1	1	0	-1	-1	-1	#
13.	1346	28	8	6.4754	0	1	0	1	0	1	-1	-1	-1	#
14.	1237	28	8	6.4465	1	-1	1	-1	0	-1	1	0	-1	*
15.	3457	28	8	6.4732	0	1	-1	0	0	1	-1	0	1	*
16.	2357	28	8	6.4284	0	1	0	0	1	0	0	-1	-1	#
17.	13467	29	8	6.6644	0	1	-1	0	1	0	-1	0	1	*
18.	12347	29	8	6.6130	0	1	0	1	0	0	0	-1	-1	#

19.	34567	29	8	6.6464	1	-1	0	-1	0	-1	0	1	1	#
20.	23578	29	8	6.6102	0	1	-1	1	0	1	-1	-1	1	*
21.	123478	30	8	6.8232	0	1	-1	0	1	1	-1	-1	1	*
22.	345678	30	8	6.8594	1	1	-1	0	1	0	-1	-1	1	*
23.	134567	30	8	6.8348	1	0	0	0	1	0	0	-1	-1	#
24.	234578	30	8	6.7999	1	-1	0	-1	0	0	0	0	1	#
25.	1234678	31	8	7.0315	1	-1	0	-1	1	-1	0	0	1	#
26.	1345678	31	8	7.0650	1	-2	1	-1	0	-1	1	1	-1	*
27.	1234578	31	8	7.0032	1	0	0	0	1	-1	0	0	-1	#
28.	12345678	32	8	7.2231	0	2	-1	1	1	1	-1	-1	-1	*

Table A6
THE MAXIMAL EXCEPTIONAL GRAPHS

The tables here are taken from [CvLRS3]. Table A6.1 contains the 430 maximal exceptional graphs of type (a). Here each maximal graph G is determined, in accordance with Theorem 6.2.4, by a graph P on 8 vertices having no dissections. Each row in the table contains: the row ordinal number, the identifier of a maximal exceptional graph G, the number of edges of P, and the edges of P as pairs of vertices.

Tables A6.2 and A6.3 are used to describe the 37 maximal exceptional graphs of type (b) and all their representations by means of dissections of 8-vertex graphs as described in Section 6.2. Table A6.2 contains the 280 graphs P with 8 vertices and at most 14 edges which have dissections: these are the graphs which yield maximal exceptional graphs G in accordance with Theorem 6.2.5. They are ordered by their spectral moments, and the entries for each row of the table are as follows: a number which identifies the graph, the number of edges, the edges as pairs of vertices, an asterisk, the number of dissections, the dissections given as pairs of vertices (dissections of type II) or quadruples of vertices (dissections of type I), and the identification number (name) of G. Table A6.3 contains a row for each of the 37 graphs G. A row contains the identifier of G, the number of vertices of G, the number of graphs P from Theorem 6.2.5 yielding G, and the identification numbers of the graphs P as in Table A6.2.

The six maximal exceptional graphs of type (c) are defined in Section 6.4 by means of representations in the root system E_8.

A6.1. The maximal exceptional graphs of type (a)

1.	G003	8	12 13 24 35 46 57 68 78
2.	G004	8	15 18 23 24 34 56 67 78
3.	G005	4	14 23 56 78
4.	G006	0	
5.	G007	5	14 23 56 67 78
6.	G008	7	12 13 24 35 46 57 68
7.	G009	7	15 23 24 34 56 67 78
8.	G010	7	14 15 23 25 34 67 78
9.	G011	9	12 13 18 24 35 46 57 68 78
10.	G012	3	14 23 56
11.	G013	1	23
12.	G014	7	14 16 23 25 34 56 78
13.	G015	5	14 23 48 56 67
14.	G016	9	12 13 17 24 35 46 58 68 78
15.	G017	7	15 16 23 24 34 56 78
16.	G018	6	14 23 37 45 56 78
17.	G019	10	12 13 16 24 35 37 47 48 58 68
18.	G020	6	16 23 25 34 45 78
19.	G021	8	12 13 18 24 35 46 57 68
20.	G022	6	14 23 45 56 67 78

21.	G023	6	14	23	38	45	56	67			
22.	G024	6	15	23	24	34	67	78			
23.	G025	8	12	13	24	35	46	57	58	68	
24.	G026	4	14	23	56	67					
25.	G027	2	14	23							
26.	G028	5	15	23	24	34	67				
27.	G029	7	14	17	23	25	34	56	67		
28.	G030	9	12	13	16	24	35	48	58	67	68
29.	G031	7	12	13	17	24	35	46	78		
30.	G032	5	14	23	45	56	67				
31.	G033	7	15	23	24	34	47	56	78		
32.	G034	3	14	23	45						
33.	G035	7	12	13	24	28	35	46	57		
34.	G036	7	12	13	18	24	35	46	57		
35.	G037	7	16	23	25	34	45	58	67		
36.	G038	5	14	23	37	45	56				
37.	G039	9	12	13	17	24	35	46	48	58	78
38.	G040	7	16	23	25	34	45	57	78		
39.	G041	7	15	23	24	34	56	67	68		
40.	G042	7	12	13	24	35	46	48	57		
41.	G043	7	14	15	23	25	34	58	67		
42.	G044	5	14	23	56	67	68				
43.	G045	6	14	23	25	34	56	67			
44.	G046	6	16	23	25	34	45	67			
45.	G047	6	15	23	24	34	56	67			
46.	G048	8	12	13	17	24	35	38	46	58	
47.	G049	8	12	13	17	24	35	38	46	78	
48.	G050	8	12	13	17	24	28	35	46	58	
49.	G051	8	12	13	17	24	35	38	46	68	
50.	G052	6	14	23	45	56	58	67			
51.	G053	4	14	23	45	56					
52.	G054	8	12	13	18	24	35	37	46	68	
53.	G055	6	14	15	23	25	34	67			
54.	G056	2	23	34							
55.	G057	8	12	13	24	35	46	48	57	58	
56.	G058	6	14	23	37	45	56	58			
57.	G059	8	12	13	18	24	35	46	48	57	
58.	G060	6	14	23	45	56	67	68			
59.	G061	8	12	13	24	35	37	46	48	58	
60.	G062	4	14	23	36	45					
61.	G063	8	17	23	25	34	36	45	56	78	
62.	G064	8	14	15	23	25	34	46	56	78	

63.	G065	6	14 23 37 38 45 56
64.	G066	7	12 13 17 24 28 35 46
65.	G067	9	12 13 16 24 35 38 48 67 68
66.	G068	3	14 23 34
67.	G069	5	16 23 25 34 45
68.	G070	7	12 13 16 24 28 35 67
69.	G071	7	14 16 23 25 34 45 67
70.	G072	7	14 15 23 25 34 56 67
71.	G073	7	14 15 23 26 34 45 67
72.	G074	7	12 13 17 24 35 38 46
73.	G075	9	12 13 17 24 35 38 46 48 78
74.	G076	7	16 23 25 34 45 48 57
75.	G077	5	14 23 25 34 56
76.	G078	5	14 23 45 56 57
77.	G079	9	12 13 18 24 35 46 48 57 58
78.	G080	7	14 16 23 25 34 56 67
79.	G081	7	15 23 24 34 46 67 68
80.	G082	5	15 23 24 34 56
81.	G083	7	12 13 17 24 35 46 48
82.	G084	7	16 23 25 34 38 45 57
83.	G085	7	12 13 24 35 37 46 48
84.	G086	9	14 15 23 25 26 34 36 45 78
85.	G087	5	14 23 36 45 47
86.	G088	9	14 15 16 23 25 34 36 45 78
87.	G089	6	14 23 25 34 36 67
88.	G090	6	15 23 24 34 46 67
89.	G091	8	14 15 23 25 27 34 56 67
90.	G092	8	14 16 23 25 34 56 57 67
91.	G093	4	15 23 24 34
92.	G094	6	14 16 23 25 34 56
93.	G095	6	14 23 45 48 56 57
94.	G096	10	12 13 17 24 27 35 38 46 48 78
95.	G097	6	15 16 23 24 34 56
96.	G098	8	12 13 16 24 27 35 38 48
97.	G099	8	12 13 16 24 28 35 38 67
98.	G100	6	14 23 25 34 37 56
99.	G101	4	14 23 25 34
100.	G102	6	16 23 25 34 45 57
101.	G103	8	14 16 23 25 34 37 56 67
102.	G104	6	15 23 24 34 47 56
103.	G105	8	12 13 16 24 27 35 38 58
104.	G106	8	12 13 16 24 27 35 38 68
105.	G107	8	14 16 23 25 34 47 56 67

106.	G108	8	12 13 17 24 35 38 46 48
107.	G109	8	12 13 18 24 35 37 46 48
108.	G110	6	14 23 25 34 56 57
109.	G111	8	14 15 23 25 34 38 45 67
110.	G112	8	12 13 17 18 24 35 46 68
111.	G113	4	14 23 45 46
112.	G114	8	12 13 24 35 37 38 46 68
113.	G115	6	15 23 24 34 56 57
114.	G116	10	14 15 16 23 25 26 34 36 45 78
115.	G117	6	14 23 36 38 45 47
116.	G118	3	23 24 34
117.	G119	7	15 23 24 28 34 37 46
118.	G120	7	14 15 23 34 45 56 67
119.	G121	9	14 15 17 23 25 34 46 56 67
120.	G122	9	14 15 23 25 26 34 37 56 67
121.	G123	5	15 23 24 34 46
122.	G124	7	14 16 23 25 34 45 57
123.	G125	7	14 15 23 26 34 37 45
124.	G126	7	14 15 23 26 34 45 57
125.	G127	5	14 23 25 34 36
126.	G128	7	14 15 23 25 34 45 67
127.	G129	9	14 15 23 25 34 37 46 56 67
128.	G130	7	12 13 16 18 24 35 67
129.	G131	9	14 15 23 25 26 34 47 56 67
130.	G132	7	14 15 23 25 27 34 56
131.	G133	9	12 13 16 24 28 35 38 67 68
132.	G134	7	14 16 17 23 25 34 45
133.	G135	3	23 34 35
134.	G136	5	14 15 23 25 34
135.	G137	7	12 13 16 24 27 35 38
136.	G138	7	14 16 23 25 27 34 45
137.	G139	9	12 13 16 17 24 35 38 48 68
138.	G140	5	14 23 25 34 46
139.	G141	7	17 23 25 34 36 45 56
140.	G142	7	14 15 23 25 34 47 56
141.	G143	7	14 15 23 26 27 34 45
142.	G144	9	12 13 15 17 24 26 38 48 58
143.	G145	7	12 13 17 18 24 35 46
144.	G146	7	16 23 25 34 45 57 58
145.	G147	9	12 13 24 26 27 35 38 48 68
146.	G148	7	12 13 24 35 37 38 46
147.	G149	7	15 23 24 34 56 57 58
148.	G150	8	14 15 23 25 26 34 56 67

149.	G151	6	15	23	24	34	37	46				
150.	G152	8	14	15	16	23	25	34	45	67		
151.	G153	8	12	13	16	18	24	27	35	68		
152.	G154	6	14	16	23	25	34	45				
153.	G155	6	14	15	23	26	34	45				
154.	G156	8	14	15	23	25	34	46	56	67		
155.	G157	6	14	23	25	34	37	46				
156.	G158	8	14	15	23	25	26	34	37	56		
157.	G159	8	14	17	23	25	34	36	45	56		
158.	G160	8	14	16	23	25	34	45	56	67		
159.	G161	4	14	23	34	45						
160.	G162	8	12	13	16	24	27	35	37	78		
161.	G163	4	23	25	34	45						
162.	G164	8	12	13	17	24	26	35	37	78		
163.	G165	8	16	23	24	28	34	35	45	57		
164.	G166	6	14	15	23	25	34	56				
165.	G167	8	12	13	16	24	27	28	35	58		
166.	G168	8	14	15	23	25	34	37	46	56		
167.	G169	8	12	13	16	17	24	28	35	48		
168.	G170	8	14	15	23	25	26	34	47	56		
169.	G171	6	14	23	45	56	57	58				
170.	G172	8	12	13	17	18	24	35	46	48		
171.	G173	6	14	23	25	27	34	46				
172.	G174	8	17	23	25	34	36	45	56	58		
173.	G175	8	12	13	24	35	37	38	46	48		
174.	G176	6	14	23	36	45	47	48				
175.	G177	7	14	15	23	34	45	46	67			
176.	G178	7	14	15	17	23	34	45	56			
177.	G179	9	14	15	16	23	25	34	45	57	67	
178.	G180	7	16	23	24	34	35	45	57			
179.	G181	7	17	23	24	34	45	46	56			
180.	G182	9	14	16	23	25	34	45	56	57	67	
181.	G183	5	14	15	23	34	45					
182.	G184	7	14	15	23	25	26	34	56			
183.	G185	9	14	15	16	23	25	34	36	45	67	
184.	G186	7	12	13	16	18	24	27	35			
185.	G187	7	15	23	24	34	37	46	48			
186.	G188	7	14	16	23	25	34	45	56			
187.	G189	7	14	15	23	26	34	36	45			
188.	G190	5	14	23	34	36	45					
189.	G191	5	14	23	25	34	45					
190.	G192	7	14	23	25	34	37	45	56			

191.	G193	7	14 16 23 25 34 45 47
192.	G194	7	14 15 23 34 37 45 56
193.	G195	9	14 15 23 25 34 46 56 57 67
194.	G196	9	12 13 16 24 27 28 35 38 68
195.	G197	9	14 15 23 25 27 34 46 56 57
196.	G198	7	14 15 23 25 34 46 56
197.	G199	7	12 13 16 24 27 28 35
198.	G200	7	14 15 23 25 34 56 57
199.	G201	9	14 15 23 25 34 37 46 56 57
200.	G202	5	14 23 45 46 47
201.	G203	9	12 13 24 35 37 38 46 47 48
202.	G204	7	12 13 24 26 28 35 37
203.	G205	9	18 23 25 34 36 37 45 56 57
204.	G206	8	12 13 16 18 24 26 35 67
205.	G207	10	14 15 17 23 25 26 34 45 56 67
206.	G208	4	23 24 34 45
207.	G209	6	16 23 24 34 35 45
208.	G210	8	14 16 23 25 34 45 46 57
209.	G211	8	14 15 23 25 34 45 56 67
210.	G212	8	14 16 23 25 27 34 35 45
211.	G213	6	14 15 23 34 45 56
212.	G214	8	12 13 16 24 26 28 35 67
213.	G215	8	14 23 25 34 36 45 56 67
214.	G216	8	14 15 23 25 27 34 36 45
215.	G217	6	14 23 25 34 45 56
216.	G218	6	14 15 23 34 36 45
217.	G219	8	14 16 23 25 34 45 56 57
218.	G220	8	14 15 17 23 25 34 36 45
219.	G221	8	17 23 24 26 34 35 45 56
220.	G222	6	14 23 25 34 36 37
221.	G223	6	15 23 24 34 46 47
222.	G224	10	14 15 16 23 25 34 36 45 57 67
223.	G225	8	12 13 16 24 27 28 35 38
224.	G226	6	14 23 25 26 34 45
225.	G227	8	14 16 23 25 27 34 45 46
226.	G228	8	14 17 23 25 26 34 45 46
227.	G229	8	14 15 23 25 34 46 56 57
228.	G230	6	14 23 25 34 46 47
229.	G231	10	18 23 24 26 27 34 35 45 56 57
230.	G232	9	14 15 16 23 25 34 45 56 67
231.	G233	5	14 23 34 35 45
232.	G234	9	12 13 15 17 24 25 28 48 56
233.	G235	7	14 15 23 34 45 47 56

234.	G236	9	14 15 23 25 26 34 45 56 67
235.	G237	7	14 16 23 25 34 35 45
236.	G238	9	14 17 23 25 34 36 45 46 56
237.	G239	7	14 15 23 34 36 45 47
238.	G240	9	14 15 16 23 25 34 36 45 47
239.	G241	7	14 15 16 23 25 34 45
240.	G242	7	14 16 23 25 34 45 46
241.	G243	9	14 15 23 25 34 36 45 56 67
242.	G244	9	14 15 17 23 25 34 36 45 56
243.	G245	7	14 15 23 34 45 56 57
244.	G246	7	14 23 25 34 36 45 56
245.	G247	7	14 15 23 25 34 36 45
246.	G248	7	14 23 25 34 45 56 57
247.	G249	9	14 15 16 23 25 34 36 45 57
248.	G250	5	14 23 34 45 46
249.	G251	7	14 15 23 34 36 37 45
250.	G252	9	14 15 16 23 25 34 36 37 45
251.	G253	9	12 13 16 24 27 28 35 37 38
252.	G254	7	14 23 25 26 34 45 47
253.	G255	9	14 15 23 25 34 46 47 56 57
254.	G256	9	12 13 15 16 17 24 38 48 58
255.	G257	6	14 23 34 35 45 56
256.	G258	8	14 15 23 34 45 46 56 67
257.	G259	8	14 15 23 34 36 45 46 67
258.	G260	10	14 17 23 25 34 36 45 46 56 67
259.	G261	8	14 15 16 23 25 34 45 57
260.	G262	6	14 15 23 34 45 46
261.	G263	8	14 16 23 25 34 35 45 47
262.	G264	8	14 15 23 34 36 45 46 57
263.	G265	8	12 13 16 18 24 27 28 35
264.	G266	8	14 16 23 25 26 34 35 45
265.	G267	6	14 15 23 25 34 45
266.	G268	8	14 15 23 25 34 36 45 47
267.	G269	10	14 15 16 23 25 34 36 37 45 47
268.	G270	8	14 15 16 17 23 25 34 45
269.	G271	8	14 16 23 25 34 36 45 56
270.	G272	8	14 15 23 25 26 34 36 45
271.	G273	6	14 23 34 36 45 47
272.	G274	8	14 15 16 23 25 34 36 45
273.	G275	8	14 23 25 34 36 45 47 56
274.	G276	4	23 34 35 36
275.	G277	6	23 25 34 36 45 56
276.	G278	8	14 23 25 34 36 45 56 57

277.	G279	6	14	23	25	34	45	46				
278.	G280	8	14	15	23	25	34	36	45	57		
279.	G281	10	14	16	23	25	26	27	34	35	45	47
280.	G282	8	14	15	23	25	34	36	37	45		
281.	G283	8	12	13	16	17	18	24	35	48		
282.	G284	8	12	13	24	26	27	28	35	58		
283.	G285	8	12	13	24	26	28	35	37	38		
284.	G286	10	12	13	15	16	17	24	38	48	58	68
285.	G287	9	12	13	16	17	24	27	35	37	78	
286.	G288	5	23	24	34	35	45					
287.	G289	9	14	15	23	25	34	45	46	56	67	
288.	G290	7	14	15	23	34	45	46	56			
289.	G291	7	14	23	34	35	45	56	57			
290.	G292	9	14	23	25	26	34	35	45	56	67	
291.	G293	5	23	24	34	45	46					
292.	G294	7	14	15	23	34	36	45	46			
293.	G295	9	14	15	16	23	25	34	45	47	56	
294.	G296	9	14	15	16	23	25	26	34	36	45	
295.	G297	7	12	13	15	17	24	26	28			
296.	G298	9	12	13	16	18	24	27	28	35	38	
297.	G299	7	14	23	25	26	34	35	45			
298.	G300	7	14	15	23	25	34	45	56			
299.	G301	9	12	13	16	17	18	24	35	38	48	
300.	G302	9	14	23	25	26	34	36	45	56	57	
301.	G303	9	14	23	25	26	34	36	45	56	67	
302.	G304	9	17	23	24	26	34	35	45	46	56	
303.	G305	7	12	13	16	17	18	24	35			
304.	G306	7	15	23	24	34	46	47	48			
305.	G307	9	12	13	18	24	26	28	35	37	38	
306.	G308	7	14	23	25	26	34	45	46			
307.	G309	9	14	15	23	25	34	36	37	45	56	
308.	G310	7	12	13	24	26	27	28	35			
309.	G311	11	12	13	15	16	17	24	38	48	58	68 78
310.	G312	8	12	13	15	17	24	25	28	56		
311.	G313	10	14	15	16	17	23	25	34	45	47	56
312.	G314	6	23	24	34	45	46	56				
313.	G315	6	14	23	25	34	35	45				
314.	G316	8	14	23	25	26	34	35	45	57		
315.	G317	8	14	15	23	34	45	46	47	67		
316.	G318	8	14	15	16	23	25	34	45	56		
317.	G319	10	14	15	16	23	25	26	34	45	56	67
318.	G320	6	14	23	34	35	45	46				
319.	G321	8	14	16	23	25	34	35	45	46		
320.	G322	8	14	15	23	25	34	45	47	56		

321.	G323	8	14 15 23 34 36 37 45 46
322.	G324	8	14 15 23 25 26 34 45 56
323.	G325	8	12 13 16 17 18 24 28 35
324.	G326	8	14 23 25 26 34 36 45 56
325.	G327	10	12 13 16 18 24 27 28 35 37 38
326.	G328	8	14 23 25 26 34 35 45 47
327.	G329	8	14 15 23 25 34 36 45 56
328.	G330	10	14 15 23 25 34 36 37 45 47 56
329.	G331	8	12 13 15 17 18 24 26 48
330.	G332	8	12 13 15 16 17 24 28 48
331.	G333	6	14 23 45 46 47 48
332.	G334	8	14 23 25 26 27 34 45 46
333.	G335	8	12 13 24 26 27 28 35 38
334.	G336	9	14 23 25 34 35 36 45 56 67
335.	G337	9	14 23 25 26 34 35 37 45 56
336.	G338	7	14 23 25 34 35 45 56
337.	G339	9	14 23 25 34 36 45 46 56 67
338.	G340	7	14 23 34 35 37 45 46
339.	G341	9	14 15 16 23 25 26 34 45 56
340.	G342	7	14 23 25 34 35 45 46
341.	G343	9	14 23 25 26 27 34 35 45 56
342.	G344	9	14 23 25 26 34 35 45 46 57
343.	G345	7	14 15 23 34 45 46 47
344.	G346	7	23 24 26 34 35 45 56
345.	G347	9	14 15 23 25 26 34 36 45 56
346.	G348	9	14 15 16 23 25 34 36 45 56
347.	G349	7	12 13 15 17 18 24 26
348.	G350	9	14 23 25 26 27 34 35 45 46
349.	G351	7	14 23 25 34 45 46 47
350.	G352	9	12 13 24 26 27 28 35 37 38
351.	G353	6	23 24 34 35 45 46
352.	G354	8	14 23 25 34 35 37 45 56
353.	G355	10	14 15 23 25 34 36 45 46 56 67
354.	G356	8	14 23 25 34 35 45 47 56
355.	G357	8	14 23 25 26 34 35 45 56
356.	G358	10	17 23 24 26 34 35 45 46 47 56
357.	G359	8	14 15 23 25 34 45 46 56
358.	G360	8	12 13 15 17 18 24 25 56
359.	G361	8	14 23 25 34 36 45 46 56
360.	G362	8	14 23 25 26 34 35 45 46
361.	G363	8	14 15 23 25 34 45 56 57
362.	G364	6	14 23 34 45 46 47

363.	G365	8	23 25 34 36 37 45 56 57
364.	G366	8	14 23 25 26 34 45 46 47
365.	G367	9	14 23 25 34 35 37 45 46 56
366.	G368	7	23 24 34 35 45 46 56
367.	G369	7	23 24 34 45 46 47 56
368.	G370	9	12 13 15 17 18 24 25 28 56
369.	G371	7	14 23 34 35 36 45 46
370.	G372	9	14 16 23 25 34 35 36 45 56
371.	G373	9	14 16 23 25 34 35 45 46 56
372.	G374	11	14 15 23 25 34 36 45 46 56 57 67
373.	G375	9	14 23 25 34 36 45 46 47 56
374.	G376	9	14 15 23 25 34 36 45 46 56
375.	G377	7	14 23 34 35 45 46 47
376.	G378	5	23 34 35 36 37
377.	G379	9	14 23 25 26 34 35 45 46 47
378.	G380	9	23 24 26 27 34 35 45 56 57
379.	G381	9	14 23 25 26 27 34 45 46 47
380.	G382	10	14 23 25 34 35 45 46 56 57 67
381.	G383	8	14 23 25 34 35 36 45 56
382.	G384	8	14 23 25 34 35 45 56 57
383.	G385	8	14 23 25 34 35 45 46 56
384.	G386	10	14 23 25 26 34 35 45 46 56 67
385.	G387	6	23 24 34 45 46 47
386.	G388	10	14 15 16 23 25 26 34 36 45 56
387.	G389	8	12 13 15 17 18 24 26 28
388.	G390	10	14 23 25 26 27 34 35 45 46 56
389.	G391	8	14 23 25 34 35 45 46 47
390.	G392	10	23 24 26 27 34 35 45 56 57 67
391.	G393	10	14 23 25 26 27 34 35 45 46 47
392.	G394	10	12 13 24 26 27 28 35 36 37 38
393.	G395	7	23 24 34 35 36 45 46
394.	G396	9	23 24 34 35 45 46 47 56 67
395.	G397	7	23 24 34 35 45 46 47
396.	G398	9	14 23 25 34 35 45 46 56 57
397.	G399	9	14 16 23 26 34 35 36 45 46
398.	G400	9	14 23 25 34 35 45 46 47 56
399.	G401	9	14 23 25 26 34 35 45 46 56
400.	G402	9	12 13 15 16 17 18 24 38 48
401.	G403	9	12 13 14 15 16 17 28 38 48
402.	G404	8	23 24 34 35 45 46 47 56
403.	G405	10	23 24 27 34 35 45 46 47 56 67
404.	G406	10	14 23 25 26 34 35 45 46 56 57
405.	G407	8	14 23 34 35 36 45 46 47

406.	G408	8	23 24 26 34 35 45 46 56
407.	G409	10	14 15 23 25 26 34 36 45 46 56
408.	G410	8	12 13 15 16 17 18 24 38
409.	G411	10	23 24 26 34 35 37 45 46 56 67
410.	G412	8	12 13 15 16 17 18 24 48
411.	G413	10	23 25 34 36 37 38 45 56 57 58
412.	G414	9	23 24 34 35 45 46 47 56 57
413.	G415	9	23 24 26 34 35 45 46 47 56
414.	G416	9	12 13 15 17 18 24 26 27 28
415.	G417	9	12 13 15 16 17 18 24 28 38
416.	G418	11	23 24 26 34 35 37 45 46 56 57 67
417.	G419	9	12 13 15 16 17 18 24 28 48
418.	G420	7	12 13 15 16 17 18 24
419.	G421	11	12 13 14 15 16 17 28 38 48 58 68
420.	G422	8	23 24 34 35 36 45 46 47
421.	G423	10	23 24 26 34 35 45 46 47 56 67
422.	G424	8	12 13 15 16 17 18 24 28
423.	G426	9	14 23 34 35 36 37 45 46 47
424.	G427	9	12 13 15 16 17 18 24 27 28
425.	G429	12	23 24 26 27 34 35 37 45 46 56 57 67
426.	G430	12	12 13 14 15 16 17 28 38 48 58 68 78
427.	G431	9	23 24 34 35 36 37 45 46 47
428.	G432	10	12 13 15 16 17 18 24 26 27 28
429.	G433	11	12 13 15 16 17 18 24 25 26 27 28
430.	G434	13	12 13 14 15 16 17 18 28 38 48 58 68 78

A6.2. The 280 graphs with 8 vertices and at most 14 edges which have dissections

1.	6	23 34 35 36 37 38	* 1	13	G454
2.	6	23 24 25 34 35 45	* 1	1678	G436
3.	7	12 13 14 15 16 17 18	* 7	21 31 41 51 61 71 81	G473
4.	7	23 24 34 45 46 47 48	* 1	14	G450
5.	7	23 24 25 34 35 45 56	* 1	1678	G438
6.	8	12 13 14 15 16 17 18 28	* 5	31 41 51 61 71	G472
7.	8	23 24 34 45 46 47 48 56	* 1	14	G447
8.	8	23 24 34 35 45 46 47 48	* 1	14	G453

```
 9.   8   14 23 34 35 36 45 46 56 * 1   1278   G441

10.   8   23 24 25 34 35 45 56 57 * 1   1678   G444

11.   8   23 24 25 34 35 45 46 56 * 1   1678   G444

12.   9   12 13 14 15 16 17 18 27 38 * 3   41 51 61   G467

13.   9   12 13 14 15 16 17 18 28 38 * 4   41 51 61 71   G470

14.   9   23 24 34 45 46 47 48 56 78 * 1   14   G439

15.   9   23 24 34 45 46 47 48 56 67 * 1   14   G443

16.   9   23 24 34 35 45 46 47 48 56 * 1   14   G449

17.   9   23 24 34 35 36 45 46 47 48 * 1   14   G456

18.   9   14 23 34 35 36 45 46 56 67 * 1   1278   G437

19.   9   14 23 34 35 36 45 46 47 56 * 1   1278   G443

20.   9   23 24 25 34 35 45 56 57 58 * 2   15 1678   G462

21.   9   14 23 25 34 35 36 45 46 56 * 1   1278   G443

22.   9   23 24 25 34 35 45 46 56 57 * 1   1678   G448

23.   9   23 24 25 34 35 36 45 46 56 * 2   1278 1678   G462

24.  10   12 13 14 15 16 17 18 26 37 48 * 1   51   G452

25.  10   12 13 14 15 16 17 18 27 38 48 * 2   51 61   G464

26.  10   12 13 14 15 16 17 18 27 28 38 * 3   41 51 61   G468

27.  10   12 13 14 15 16 17 18 28 38 48 * 3   51 61 71   G469

28.  10   23 24 34 38 45 46 47 48 56 67 * 1   14   G441

29.  10   23 24 34 45 46 47 48 56 67 78 * 1   14   G440

30.  10   12 13 15 17 24 25 27 56 57 78 * 1   3468   G435

31.  10   12 13 15 17 18 24 25 28 56 58 * 1   3467   G440

32.  10   23 24 34 45 46 47 48 56 67 68 * 1   14   G446
```

33. 10   23 24 34 35 45 46 47 48 56 67 * 1   14   G445

34. 10   12 13 15 17 18 24 26 27 28 78 * 1   3456   G446

35. 10   12 13 15 16 17 18 24 27 28 78 * 1   3456   G453

36. 10   23 24 34 35 45 46 47 48 56 57 * 1   14   G451

37. 10   12 13 14 15 16 17 18 27 28 78 * 5   31 41 51 61 3456   G471

38. 10   23 24 26 34 35 45 46 47 48 56 * 1   14   G452

39. 10   23 24 34 35 36 37 45 46 47 48 * 1   14   G458

40. 10   23 24 25 34 35 45 56 57 58 67 * 1   15   G444

41. 10   14 23 25 34 35 36 45 46 56 67 * 1   1278   G440

42. 10   14 23 25 34 35 36 45 46 47 56 * 1   1278   G447

43. 10   14 23 25 34 35 36 45 46 56 57 * 1   1278   G445

44. 10   14 23 34 35 36 37 45 46 47 56 * 1   1278   G451

45. 10   23 24 25 34 35 45 46 56 57 58 * 2   15 1678   G464

46. 10   14 16 23 25 34 35 36 45 46 56 * 1   1278   G446

47. 10   23 24 25 34 35 37 45 46 56 57 * 1   1678   G451

48. 10   23 24 25 34 35 45 46 47 56 57 * 1   1678   G457

49. 10   14 23 25 26 34 35 36 45 46 56 * 1   1278   G453

50. 10   23 24 25 34 35 36 45 46 56 57 * 2   1278 1678   G464

51. 10   23 24 25 26 34 35 36 45 46 56 * 5   1278 1378 1478 1578 1678   G471

52. 11   12 13 14 15 16 17 18 27 37 48 58 * 1   61   G448

53. 11   12 13 14 15 16 17 18 26 28 37 48 * 1   51   G449

54. 11   12 13 14 15 16 17 18 27 38 48 58 * 1   61   G456

55. 11   12 13 14 15 16 17 18 26 28 37 38 * 2   41 51   G461

56. 11   12 13 14 15 16 17 18 27 28 38 48 * 2   51 61   G465

```
57.  11   12 13 14 15 16 17 18 27 28 37 38 * 3  41 51 61  G467

58.  11   12 13 14 15 16 17 18 28 38 48 58 * 2  61 71  G466

59.  11   23 24 34 35 45 46 47 48 56 67 78 * 1  14  G437

60.  11   12 13 14 15 16 17 18 26 28 37 68 * 2  41 51  G462

61.  11   23 24 26 34 35 45 46 47 48 56 78 * 1  14  G439

62.  11   12 13 15 17 18 24 25 28 48 56 58 * 1  3467  G439

63.  11   12 13 15 17 18 24 25 28 38 56 58 * 1  3467  G437

64.  11   23 24 34 35 45 46 47 48 56 67 68 * 1  14  G443

65.  11   12 13 15 17 18 24 26 27 28 38 78 * 1  3456  G443

66.  11   23 24 34 35 45 46 47 48 56 58 67 * 1  14  G442

67.  11   12 13 15 17 18 24 25 27 28 56 58 * 1  3467  G442

68.  11   23 24 27 34 35 45 46 47 48 56 67 * 1  14  G442

69.  11   12 13 15 16 17 18 24 27 28 48 78 * 1  3456  G450

70.  11   12 13 15 16 17 18 24 27 28 38 78 * 1  3456  G449

71.  11   12 13 15 16 17 18 24 26 27 28 68 * 1  3457  G456

72.  11   23 24 34 35 38 45 46 47 48 56 57 * 1  14  G448

73.  11   12 13 14 15 16 17 18 27 28 38 78 * 4  41 51 61 3456  G470

74.  11   23 24 26 34 35 45 46 47 48 56 67 * 1  14  G449

75.  11   23 24 34 35 45 46 47 48 56 57 58 * 1  14  G456

76.  11   23 24 34 35 36 37 38 45 46 47 48 * 2  13 14  G466

77.  11   23 24 25 34 35 45 56 57 58 67 78 * 1  15  G438

78.  11   23 24 25 34 35 45 48 56 57 58 67 * 1  15  G443

79.  11   23 24 25 34 35 45 46 56 57 58 67 * 1  15  G448

80.  11   14 16 23 25 34 35 36 45 46 56 67 * 1  1278  G443
```

```
 81.   11   14 23 25 34 35 36 45 46 56 57 67 * 1   1278   G442

 82.   11   14 23 25 34 35 36 45 46 47 56 57 * 1   1278   G449

 83.   11   23 24 25 34 35 37 45 46 56 57 58 * 2   15 1678   G461

 84.   11   14 23 25 34 35 36 37 45 46 56 57 * 1   1278   G448

 85.   11   23 24 25 34 35 45 46 47 56 57 58 * 2   15 1678   G465

 86.   11   14 23 25 26 34 35 36 45 46 47 56 * 1   1278   G450

 87.   11   14 23 25 26 34 35 36 45 46 56 67 * 1   1278   G449

 88.   11   14 23 34 35 36 37 45 46 47 56 67 * 2   1258 1278   G461

 89.   11   23 24 25 34 35 36 45 46 56 57 58 * 3   15 1278 1678   G467

 90.   11   23 24 25 34 35 36 45 46 56 57 67 * 1   1278   G456

 91.   11   23 24 25 34 35 36 45 46 47 56 57 * 2   1278 1678   G465

 92.   11   23 24 25 26 34 35 36 45 46 56 67 * 4   1278 1378 1478 1578   G470

 93.   12   12 13 14 15 16 17 18 26 28 37 38 47 * 1   51   G445

 94.   12   12 13 14 15 16 17 18 26 37 38 47 48 * 1   51   G447

 95.   12   12 13 14 15 16 17 18 27 28 37 48 58 * 1   61   G451

 96.   12   12 13 14 15 16 17 18 26 28 37 38 48 * 1   51   G451

 97.   12   12 13 14 15 16 17 18 26 27 37 38 68 * 2   41 51   G460

 98.   12   12 13 14 15 16 17 18 27 37 48 58 78 * 1   61   G457

 99.   12   12 13 14 15 16 17 18 27 28 37 38 48 * 2   51 61   G464

100.   12   12 13 14 15 16 17 18 27 28 38 48 58 * 1   61   G458

101.   12   12 13 14 15 16 17 18 28 38 48 58 68 * 1   71   G459

102.   12   12 13 14 15 16 17 18 26 28 37 47 68 * 1   51   G444

103.   12   23 24 28 34 35 45 46 47 48 56 67 78 * 1   14   G435

104.   12   12 13 14 15 16 17 18 26 28 37 48 68 * 1   51   G453
```

```
105.  12  12 13 15 17 18 24 26 27 28 37 48 78 * 1  3456  G441

106.  12  12 13 15 17 18 24 25 27 28 56 58 68 * 1  3467  G440

107.  12  12 13 15 17 18 24 25 28 48 56 58 68 * 1  3467  G440

108.  12  23 24 27 34 35 45 46 47 48 56 67 78 * 1  14  G440

109.  12  12 13 15 17 18 24 25 28 38 48 56 58 * 1  3467  G440

110.  12  12 13 15 16 17 18 24 27 28 37 48 78 * 1  3456  G447

111.  12  12 13 14 15 16 17 18 26 28 37 38 68 * 2  41 51  G464

112.  12  12 13 15 17 18 24 26 27 28 38 48 78 * 1  3456  G445

113.  12  12 13 15 16 17 18 24 27 28 37 58 78 * 1  3456  G445

114.  12  23 24 26 34 35 45 46 47 48 56 67 78 * 1  14  G440

115.  12  12 13 15 17 18 24 25 28 38 56 58 78 * 1  3467  G441

116.  12  12 13 15 16 17 18 24 27 28 38 48 78 * 1  3456  G453

117.  12  12 13 15 16 17 18 24 26 27 28 38 68 * 1  3457  G451

118.  12  12 13 15 17 18 24 26 27 28 38 58 78 * 1  3456  G444

119.  12  12 13 15 16 17 18 24 27 28 38 58 78 * 1  3456  G451

120.  12  12 13 14 15 16 17 18 27 28 37 48 78 * 3  51 61 3456  G468

121.  12  23 24 26 34 35 38 45 46 47 48 56 67 * 1  14  G447

122.  12  23 24 26 34 35 45 46 47 48 56 58 67 * 1  14  G445

123.  12  12 13 15 16 17 18 24 25 26 27 28 58 * 1  3467  G458

124.  12  12 13 14 15 16 17 18 27 28 38 48 78 * 3  51 61 3456  G469

125.  12  23 24 26 34 35 45 46 47 48 56 67 68 * 1  14  G453

126.  12  23 24 26 34 35 37 45 46 47 48 56 67 * 1  14  G453

127.  12  23 24 25 34 35 45 46 56 57 58 67 78 * 1  15  G441

128.  12  23 24 25 34 35 45 46 56 57 58 67 68 * 1  15  G444
```

```
129.  12   23 24 25 34 35 38 45 46 56 57 58 67 * 1   15  G445

130.  12   12 13 15 17 18 24 26 27 28 37 38 78 * 1   3456  G447

131.  12   14 16 23 25 34 35 36 45 46 56 57 67 * 1   1278  G445

132.  12   12 13 15 17 18 24 25 27 28 56 58 78 * 1   3467  G445

133.  12   12 13 15 16 17 18 24 27 28 47 48 78 * 1   3456  G454

134.  12   23 24 25 34 35 45 46 48 56 57 58 67 * 1   15  G451

135.  12   12 13 15 16 17 18 24 27 28 37 38 78 * 1   3456  G452

136.  12   23 24 25 28 34 35 37 45 46 56 57 58 * 2   15 1678  G460

137.  12   14 16 23 26 34 35 36 37 45 46 47 67 * 1   1258  G452

138.  12   12 13 15 17 18 24 25 27 28 56 57 58 * 2   3467 3468  G460

139.  12   14 16 23 25 34 35 36 45 46 47 56 67 * 1   1278  G444

140.  12   12 13 15 16 17 18 24 26 27 28 68 78 * 2   3456 3457  G464

141.  12   23 24 26 34 35 45 46 47 48 56 57 67 * 1   14  G451

142.  12   12 13 14 15 16 17 18 27 28 37 38 78 * 5   41 51 61 2456 3456  G472

143.  12   23 24 25 34 35 37 45 46 48 56 57 58 * 2   15 1678  G464

144.  12   23 24 25 34 35 45 46 47 48 56 57 58 * 3   14 15 1678  G469

145.  12   23 24 25 34 35 45 56 57 58 67 68 78 * 1   15  G436

146.  12   23 24 25 34 35 36 45 46 56 57 58 78 * 1   15  G447

147.  12   23 24 25 34 35 45 46 47 56 57 58 67 * 1   15  G457

148.  12   14 23 25 26 34 35 36 45 46 47 56 67 * 1   1278  G453

149.  12   14 23 25 34 35 36 45 46 47 56 57 67 * 1   1278  G451

150.  12   23 24 25 34 35 36 45 46 56 57 58 67 * 2   15 1278  G464

151.  12   14 23 25 26 34 35 36 45 46 56 57 67 * 1   1278  G451

152.  12   14 23 25 34 35 36 37 45 46 47 56 57 * 2   1268 1278  G464
```

153. 12   23 24 25 34 35 36 45 46 47 56 57 58 * 3   15 1278 1678   G468

154. 12   14 23 34 35 36 37 45 46 47 56 57 67 * 3   1258 1268 1278   G468

155. 12   23 24 25 26 34 35 36 45 46 56 67 68 * 5   16 1278 1378 1478 1578   G472

156. 12   23 24 25 34 35 36 45 46 47 56 57 67 * 1   1278   G458

157. 12   23 24 25 34 35 36 37 45 46 47 56 57 * 3   1268 1278 1678   G469

158. 12   23 24 25 26 34 35 36 45 46 56 57 67 * 3   1278 1378 1478   G469

159. 13   12 13 14 15 16 17 18 25 27 36 37 58 68 * 1   41   G437

160. 13   12 13 14 15 16 17 18 26 27 37 38 48 68 * 1   51   G442

161. 13   12 13 14 15 16 17 18 26 28 37 38 47 48 * 1   51   G443

162. 13   12 13 14 15 16 17 18 27 28 37 38 47 58 * 1   61   G449

163. 13   12 13 14 15 16 17 18 26 27 28 37 38 48 * 1   51   G448

164. 13   12 13 14 15 16 17 18 27 28 37 38 48 58 * 1   61   G456

165. 13   12 13 14 15 16 17 18 27 28 37 38 47 48 * 2   51 61   G462

166. 13   12 13 14 15 16 17 18 26 28 37 38 47 68 * 1   51   G443

167. 13   12 13 14 15 16 17 18 26 28 37 47 68 78 * 1   51   G448

168. 13   12 13 14 15 16 17 18 26 28 37 38 47 78 * 1   51   G449

169. 13   12 13 15 17 18 24 26 27 28 37 48 58 78 * 1   3456   G437

170. 13   12 13 15 17 18 24 25 27 28 38 48 56 58 * 1   3467   G439

171. 13   12 13 15 16 17 18 24 27 28 37 48 58 78 * 1   3456   G443

172. 13   12 13 15 17 18 24 25 28 38 48 56 58 68 * 1   3467   G442

173. 13   12 13 15 17 18 24 26 27 28 37 48 68 78 * 1   3456   G438

174. 13   12 13 15 16 17 18 24 26 27 28 38 48 68 * 1   3457   G449

175. 13   12 13 15 16 17 18 24 27 28 38 47 58 78 * 1   3456   G443

176. 13   12 13 14 15 16 17 18 26 28 37 38 48 68 * 1   51   G456

```
177.  13   12 13 15 17 18 24 25 28 38 48 56 58 78 * 1   3467  G443
178.  13   12 13 14 15 16 17 18 26 27 28 37 48 78 * 2   51 3456  G461
179.  13   12 13 15 17 18 24 26 27 28 38 48 58 78 * 1   3456  G448
180.  13   23 24 27 34 35 45 46 47 48 56 58 67 78 * 1   14  G439
181.  13   12 13 15 16 17 18 24 27 28 38 48 58 78 * 1   3456  G456
182.  13   12 13 14 15 16 17 18 26 27 28 37 38 68 * 2   41 51  G461
183.  13   23 24 26 34 35 45 46 47 48 56 58 67 78 * 1   14  G437
184.  13   12 13 14 15 16 17 18 27 28 37 48 58 78 * 2   61 3456  G465
185.  13   12 13 15 16 17 18 24 27 28 38 58 68 78 * 1   3456  G456
186.  13   23 24 26 34 35 37 45 46 47 48 56 67 78 * 1   14  G443
187.  13   12 13 14 15 16 17 18 27 28 38 48 58 78 * 2   61 3456  G466
188.  13   23 24 26 34 35 37 45 46 47 48 56 67 68 * 1   14  G450
189.  13   12 13 14 15 16 17 18 25 27 36 38 57 68 * 1   41  G438
190.  13   12 13 14 15 16 17 18 26 37 38 47 48 78 * 1   51  G450
191.  13   23 24 27 34 35 45 46 47 48 56 67 68 78 * 1   14  G437
192.  13   12 13 15 17 18 24 26 27 28 37 38 48 78 * 1   3456  G443
193.  13   12 13 15 17 18 24 25 27 28 56 58 68 78 * 1   3467  G442
194.  13   12 13 15 16 17 18 24 27 28 38 47 48 78 * 1   3456  G450
195.  13   12 13 14 15 16 17 18 26 28 37 38 68 78 * 2   41 51  G465
196.  13   12 13 15 16 17 18 24 27 28 37 38 48 78 * 1   3456  G449
197.  13   12 13 15 16 17 18 24 26 27 28 37 38 78 * 1   3456  G449
198.  13   12 13 15 16 17 18 24 26 27 28 37 68 78 * 1   3456  G448
199.  13   23 24 26 34 35 45 46 47 48 56 67 68 78 * 1   14  G443
200.  13   12 13 15 17 18 24 26 27 28 37 38 58 78 * 1   3456  G443
```

201.   13   23 24 26 34 35 38 45 46 47 48 56 57 67 * 1   14   G443

202.   13   12 13 15 17 18 24 25 27 28 38 56 58 78 * 1   3467   G443

203.   13   23 24 26 34 35 45 46 47 48 56 57 67 78 * 1   14   G442

204.   13   12 13 14 15 16 17 18 26 27 28 37 38 78 * 3   41 51 3456   G467

205.   13   12 13 15 16 17 18 24 27 28 37 38 58 78 * 1   3456   G449

206.   13   12 13 15 16 17 18 24 26 27 28 48 68 78 * 2   3456 3457   G462

207.   13   12 13 15 16 17 18 24 26 27 28 38 68 78 * 2   3456 3457   G461

208.   13   23 24 26 34 35 45 46 47 48 56 57 67 68 * 1   14   G449

209.   13   12 13 15 16 17 18 24 25 26 27 28 58 68 * 2   3457 3467   G465

210.   13   12 13 14 15 16 17 18 27 28 37 38 48 78 * 4   51 61 2456 3456   G470

211.   13   23 24 25 34 35 45 46 56 57 58 67 68 78 * 1   15   G438

212.   13   23 24 25 34 35 36 45 46 56 57 58 67 78 * 1   15   G443

213.   13   23 24 25 34 35 45 46 47 56 57 58 67 78 * 1   15   G448

214.   13   23 24 25 34 35 36 45 46 47 56 57 58 78 * 1   15   G449

215.   13   23 24 25 28 34 35 36 45 46 56 57 58 67 * 1   15   G449

216.   13   14 16 23 26 34 35 36 37 45 46 47 57 67 * 1   1258   G449

217.   13   14 16 23 25 34 35 36 45 46 47 56 57 67 * 1   1278   G448

218.   13   23 24 25 34 35 36 45 46 56 57 58 67 68 * 2   15 1278   G462

219.   13   23 24 25 34 35 36 45 46 48 56 57 58 67 * 2   15 1278   G461

220.   13   23 24 25 34 35 45 46 47 48 56 57 58 67 * 2   14 15   G465

221.   13   23 24 25 34 35 36 38 45 46 47 56 57 58 * 3   15 1278 1678   G467

222.   13   23 24 25 27 34 35 36 45 46 56 57 58 67 * 1   15   G456

223.   13   23 24 25 34 35 36 45 46 47 48 56 57 58 * 4   14 15 1278 1678   G470

224.   13   12 13 15 17 18 24 25 27 28 56 57 58 78 * 2   3467 3468   G461

225. 13   12 13 15 16 17 18 24 26 27 28 67 68 78 * 3   3456 3457 3458 G467

226. 13   14 23 25 26 34 35 36 45 46 47 56 57 67 * 1   1278 G456

227. 13   12 13 14 15 16 17 18 26 27 28 67 68 78 * 7   31 41 51 2345 3456 3457 3458 G473

228. 13   23 24 25 34 35 36 45 46 47 56 57 58 67 * 2   15 1278 G465

229. 13   14 23 25 26 34 35 36 37 45 46 56 57 67 * 1   1278 G456

230. 13   23 24 25 34 35 36 37 45 46 47 56 57 58 * 4   15 1268 1278 1678 G470

231. 13   23 24 25 26 34 35 36 45 46 56 67 68 78 * 1   16 G450

232. 13   14 23 25 34 35 36 37 45 46 47 56 57 67 * 2   1268 1278 G465

233. 13   23 24 25 26 34 35 36 45 46 56 57 67 68 * 4   16 1278 1378 1478 G470

234. 13   23 24 25 26 34 35 36 45 46 47 56 57 67 * 2   1278 1378 G466

235. 14   12 13 14 15 16 17 18 26 27 37 38 47 48 68 * 1   51 G440

236. 14   12 13 14 15 16 17 18 25 27 36 37 58 68 78 * 1   41 G441

237. 14   12 13 14 15 16 17 18 26 27 28 37 38 47 48 * 1   51 G444

238. 14   12 13 14 15 16 17 18 27 28 37 38 47 48 58 * 1   61 G453

239. 14   12 13 14 15 16 17 18 25 27 28 36 37 58 68 * 1   41 G440

240. 14   12 13 15 17 18 24 26 27 28 37 47 58 68 78 * 1   3456 G435

241. 14   12 13 15 17 18 24 26 27 28 37 48 57 68 78 * 1   3456 G436

242. 14   12 13 14 15 16 17 18 26 28 37 38 47 48 68 * 1   51 G446

243. 14   12 13 14 15 16 17 18 25 27 28 36 37 38 58 * 1   41 G445

244. 14   12 13 14 15 16 17 18 25 27 36 37 38 58 78 * 1   41 G445

245. 14   12 13 15 17 18 24 25 28 38 48 56 58 68 78 * 1   3467 G451

246. 14   12 13 14 15 16 17 18 26 27 28 37 38 48 68 * 1   51 G451

247. 14   12 13 15 17 18 24 26 27 28 38 48 58 68 78 * 1   3456 G457

248. 14   12 13 15 16 17 18 24 27 28 38 48 58 68 78 * 1   3456 G458

249. 14   12 13 14 15 16 17 18 27 28 38 48 58 68 78 * 1   3456   G459

250. 14   12 13 14 15 16 17 18 25 27 28 36 38 57 68 * 1   41   G441

251. 14   12 13 14 15 16 17 18 25 27 28 36 37 58 78 * 1   41   G447

252. 14   12 13 14 15 16 17 18 26 28 37 38 47 68 78 * 1   51   G451

253. 14   12 13 15 17 18 24 26 27 28 37 47 48 58 78 * 1   3456   G440

254. 14   12 13 15 16 17 18 24 27 28 37 47 48 58 78 * 1   3456   G447

255. 14   12 13 15 16 17 18 24 26 27 28 37 38 48 78 * 1   3456   G445

256. 14   12 13 15 17 18 24 26 27 28 37 38 47 58 78 * 1   3456   G441

257. 14   12 13 14 15 16 17 18 26 28 37 38 47 48 78 * 1   51   G453

258. 14   12 13 14 15 16 17 18 25 27 28 36 37 38 78 * 1   41   G452

259. 14   12 13 15 16 17 18 24 26 27 28 37 48 68 78 * 1   3456   G444

260. 14   12 13 15 17 18 24 26 27 28 37 38 48 58 78 * 1   3456   G445

261. 14   12 13 15 16 17 18 24 27 28 38 47 48 58 78 * 1   3456   G453

262. 14   12 13 14 15 16 17 18 26 27 28 38 48 67 68 * 2   51 3457   G464

263. 14   12 13 14 15 16 17 18 26 27 28 37 38 48 78 * 2   51 3456   G464

264. 14   12 13 14 15 16 17 18 26 28 37 38 48 68 78 * 1   51   G458

265. 14   12 13 14 15 16 17 18 27 28 37 38 47 58 78 * 3   61 2456 3456   G468

266. 14   12 13 14 15 16 17 18 26 27 28 36 37 38 68 * 2   41 51   G464

267. 14   12 13 14 15 16 17 18 27 28 37 38 48 58 78 * 3   61 2456 3456   G469

268. 14   12 13 15 17 18 24 26 27 28 37 38 47 48 78 * 1   3456   G446

269. 14   12 13 15 16 17 18 24 27 28 37 38 47 48 78 * 1   3456   G453

270. 14   12 13 15 16 17 18 24 26 27 28 37 38 68 78 * 1   3456   G451

271. 14   12 13 15 16 17 18 24 25 26 27 28 57 58 68 * 1   3467   G457

272. 14   12 13 14 15 16 17 18 26 27 28 37 38 68 78 * 3   41 51 3456   G468

273. 14 12 13 15 17 18 24 26 27 28 37 38 57 58 78 * 1 3456 G446

274. 14 12 13 14 15 16 17 18 27 28 37 38 47 48 78 * 5 51 61 2356 2456 3456 G471

275. 14 12 13 14 15 16 17 18 25 27 28 36 57 58 78 * 1 41 G454

276. 14 12 13 15 17 18 24 25 27 28 38 56 57 58 78 * 1 3467 G451

277. 14 12 13 15 16 17 18 24 26 27 28 48 67 68 78 * 2 3456 3457 G464

278. 14 12 13 15 16 17 18 24 26 27 28 38 67 68 78 * 2 3456 3457 G464

279. 14 12 13 15 16 17 18 24 25 26 27 28 57 58 78 * 3 3456 3467 3468 G468

280. 14 12 13 14 15 16 17 18 26 27 28 38 67 68 78 * 5 41 51 2345 3456 3457 G472

A6.3. The maximal exceptional graphs of type (b) and their representations by graphs of order 8.

```
G435  30   3    30 103 240
G436  30   3     2 145 241
G437  30   7    18  59  63 159 169 183 191
G438  30   5     5  77 173 189 211
G439  30   5    14  61  62 170 180
G440  30  11    29  31  41 106 107 108 109 114 235 239 253

G441  30   8     9  28 105 115 127 236 250 256
G442  30   8    66  67  68  81 160 172 193 203
G443  30  19    15  19  21  64  65  78  80 161 166 171 175 177 186 192 199
      200 201 202 212
G444  30   9    10  11  40 102 118 128 139 237 259
G445  30  13    33  43  93 112 113 122 129 131 132 243 244 255 260
G446  30   6    32  34  46 242 268 273
G447  30   9     7  42  94 110 121 130 146 251 254
G448  30  11    22  52  72  79  84 163 167 179 198 213 217
G449  30  16    16  53  70  74  82  87 162 168 174 196 197 205 208 214
      215 216
G450  30   7     4  69  86 188 190 194 231

G451  30  16    36  44  47  95  96 117 119 134 141 149 151 245 246 252
      270 276
G452  30   5    24  38 135 137 258
G453  30  12     8  35  49 104 116 125 126 148 238 257 261 269
G454  30   3     1 133 275
G456  30  12    17  54  71  75  90 164 176 181 185 222 226 229
```

```
G457  30   5    48   98  147  247  271
G458  30   6    39  100  123  156  248  264
G459  30   2   101  249
G460  31   3    97  136  138

G461  31   8    55   83   88  178  182  207  219  224
G462  31   6    20   23   60  165  206  218
G464  31  14    25   45   50   99  111  140  143  150  152  262  263  266  277  278
G465  31   9    56   85   91  184  195  209  220  228  232
G466  31   4    58   76  187  234
G467  32   6    12   57   89  204  221  225
G468  32   7    26  120  153  154  265  272  279
G469  32   6    27  124  144  157  158  267
G470  33   7    13   73   92  210  223  230  233

G471  34   3    37   51  274
G472  34   4     6  142  155  280
G473  36   2     3  227
```

Table A7
THE INDEX AND VERTEX DEGREES OF THE MAXIMAL EXCEPTIONAL GRAPHS

This table, taken from [CvLRS3], contains the largest eigenvalue (index) and the degree sequence for each maximal exceptional graph G. Often these parameters are common to several maximal exceptional graphs. In each row of the table, identification numbers ijk for graphs $Gijk$ are followed by the number of vertices, the number of edges, the largest eigenvalue and the degree sequence (where a^b indicates b vertices of degree a). Here graphs with the same index are in fact cospectral.

```
      001  22 140 14.0000   07^08 16^14

      002  28 224 17.0000   10^07 16^14 22^07

003 - 006  29 196 14.0000   13^28 28^01

007 - 013  29 199 14.2915   12^12 14^15 16^01 28^01
014 - 017  29 199 14.2915   10^01 12^09 14^18 28^01

018 - 027  29 200 14.3852   11^04 13^16 15^08 28^01

028 - 029  29 203 14.6569   12^14 14^07 16^07 28^01
030 - 033  29 203 14.6569   10^01 12^11 14^10 16^06 28^01
034 - 040  29 203 14.6569   10^02 12^08 14^13 16^05 28^01
041 - 043  29 203 14.6569   10^03 12^05 14^16 16^04 28^01
      044  29 203 14.6569   10^04 12^02 14^19 16^03 28^01

045 - 056  29 204 14.7446   11^06 13^10 15^10 17^02 28^01
057 - 063  29 204 14.7446   09^01 11^04 13^10 15^12 17^01 28^01
064 - 065  29 204 14.7446   09^02 11^02 13^10 15^14 28^01

066 - 073  29 207 15.0000   10^02 12^09 14^08 16^08 18^01 28^01
074 - 082  29 207 15.0000   10^03 12^06 14^11 16^07 18^01 28^01
083 - 084  29 207 15.0000   08^01 10^01 12^07 14^10 16^09 28^01
085 - 088  29 207 15.0000   08^01 10^02 12^04 14^13 16^08 28^01

089 - 093  29 208 15.0828   11^06 13^10 15^06 17^06 28^01
094 - 097  29 208 15.0828   11^09 15^18 19^01 28^01
098 - 107  29 208 15.0828   09^01 11^04 13^10 15^08 17^05 28^01
108 - 113  29 208 15.0828   09^02 11^02 13^10 15^10 17^04 28^01
114 - 115  29 208 15.0828   09^03 13^10 15^12 17^03 28^01
116 - 117  29 208 15.0828   07^01 11^06 15^21 28^01

      118  29 211 15.3246   12^15 16^10 18^03 28^01
      119  29 211 15.3246   10^01 12^12 14^03 16^09 18^03 28^01
120 - 123  29 211 15.3246   10^02 12^09 14^06 16^08 18^03 28^01
```

```
124 - 131   29 211 15.3246   10^03 12^06 14^09 16^07 18^03 28^01
132 - 135   29 211 15.3246   10^04 12^03 14^12 16^06 18^03 28^01
      136   29 211 15.3246   10^05 14^15 16^05 18^03 28^01
      137   29 211 15.3246   08^01 12^10 14^05 16^10 18^02 28^01
138 - 141   29 211 15.3246   08^01 10^01 12^07 14^08 16^09 18^02 28^01
142 - 145   29 211 15.3246   08^01 10^02 12^04 14^11 16^08 18^02 28^01
      146   29 211 15.3246   08^01 10^03 12^01 14^14 16^07 18^02 28^01
      147   29 211 15.3246   08^02 12^05 14^10 16^10 18^01 28^01
      148   29 211 15.3246   08^02 10^01 12^02 14^13 16^09 18^01 28^01
      149   29 211 15.3246   08^03 14^15 16^10 28^01

150 - 153   29 212 15.4031   11^07 13^06 15^08 17^06 19^01 28^01
154 - 161   29 212 15.4031   09^01 11^05 13^06 15^10 17^05 19^01 28^01
162 - 165   29 212 15.4031   09^02 13^16 17^10 28^01
166 - 170   29 212 15.4031   09^02 11^03 13^06 15^12 17^04 19^01 28^01
      171   29 212 15.4031   09^03 11^01 13^06 15^14 17^03 19^01 28^01
172 - 173   29 212 15.4031   07^01 11^04 13^06 15^11 17^06 28^01
174 - 175   29 212 15.4031   07^01 09^01 11^02 13^06 15^13 17^05 28^01
      176   29 212 15.4031   07^01 09^02 13^06 15^15 17^04 28^01

177 - 180   29 215 15.6332   10^02 12^08 14^07 16^05 18^06 28^01
181 - 183   29 215 15.6332   10^03 12^05 14^10 16^04 18^06 28^01
184 - 190   29 215 15.6332   10^04 12^04 14^06 16^12 18^01 20^01 28^01
191 - 195   29 215 15.6332   08^01 10^01 12^06 14^09 16^06 18^05 28^01
196 - 197   29 215 15.6332   08^01 10^02 12^03 14^12 16^05 18^05 28^01
198 - 199   29 215 15.6332   08^01 10^02 12^05 14^05 16^14 20^01 28^01
200 - 201   29 215 15.6332   08^02 12^04 14^11 16^07 18^04 28^01
      202   29 215 15.6332   08^02 10^01 12^01 14^14 16^06 18^04 28^01
203 - 204   29 215 15.6332   06^01 10^02 12^03 14^07 16^14 18^01 28^01
      205   29 215 15.6332   06^01 08^01 12^04 14^06 16^16 28^01

206 - 209   29 216 15.7082   11^06 13^08 15^04 17^08 19^02 28^01
210 - 214   29 216 15.7082   09^01 11^04 13^08 15^06 17^07 19^02 28^01
215 - 224   29 216 15.7082   09^02 11^02 13^08 15^08 17^06 19^02 28^01
225 - 228   29 216 15.7082   07^01 11^03 13^08 15^07 17^08 19^01 28^01
229 - 230   29 216 15.7082   07^01 09^01 11^01 13^08 15^09 17^07 19^01 28^01
      231   29 216 15.7082   07^02 13^08 15^10 17^08 28^01

232 - 234   29 219 15.9282   10^02 12^08 14^04 16^08 18^05 20^01 28^01
235 - 238   29 219 15.9282   10^03 12^05 14^07 16^07 18^05 20^01 28^01
239 - 241   29 219 15.9282   10^04 12^02 14^10 16^06 18^05 20^01 28^01
      242   29 219 15.9282   08^01 12^09 14^03 16^10 18^04 20^01 28^01
243 - 245   29 219 15.9282   08^01 10^01 12^06 14^06 16^09 18^04 20^01 28^01
246 - 250   29 219 15.9282   08^01 10^02 12^03 14^09 16^08 18^04 20^01 28^01
251 - 252   29 219 15.9282   08^02 12^04 14^08 16^10 18^03 20^01 28^01
      253   29 219 15.9282   06^01 12^07 14^05 16^10 18^05 28^01
      254   29 219 15.9282   06^01 10^01 12^04 14^08 16^09 18^05 28^01
```

```
      255   29 219 15.9282    06^01 10^02 12^01 14^11 16^08 18^05 28^01
      256   29 219 15.9282    06^01 08^01 12^02 14^10 16^10 18^04 28^01

257 - 260   29 220 16.0000    11^06 13^06 15^06 17^06 19^04 28^01
261 - 264   29 220 16.0000    09^01 11^04 13^06 15^08 17^05 19^04 28^01
265 - 266   29 220 16.0000    09^01 11^06 13^01 15^10 17^09 21^01 28^01
267 - 270   29 220 16.0000    09^02 11^02 13^06 15^10 17^04 19^04 28^01
271 - 274   29 220 16.0000    09^02 11^04 13^01 15^12 17^08 21^01 28^01
275 - 276   29 220 16.0000    09^03 13^06 15^12 17^03 19^04 28^01
277 - 281   29 220 16.0000    07^01 11^03 13^06 15^09 17^06 19^03 28^01
282 - 283   29 220 16.0000    07^01 09^01 11^01 13^06 15^11 17^05 19^03 28^01
      284   29 220 16.0000    07^02 13^06 15^12 17^06 19^02 28^01
      285   29 220 16.0000    05^01 11^04 13^01 15^12 17^10 28^01
      286   29 220 16.0000    05^01 09^01 11^02 13^01 15^14 17^09 28^01

287 - 288   29 223 16.2111    10^01 12^08 14^08 18^10 20^01 28^01
289 - 291   29 223 16.2111    10^02 12^07 14^04 16^08 18^05 20^02 28^01
292 - 295   29 223 16.2111    10^03 12^04 14^07 16^07 18^05 20^02 28^01
296 - 297   29 223 16.2111    10^06 16^21 22^01 28^01
298 - 299   29 223 16.2111    08^01 12^06 14^10 16^01 18^09 20^01 28^01
300 - 301   29 223 16.2111    08^01 10^01 12^05 14^06 16^09 18^04 20^02 28^01
302 - 304   29 223 16.2111    08^01 10^02 12^02 14^09 16^08 18^04 20^02 28^01
      305   29 223 16.2111    08^02 12^03 14^08 16^10 18^03 20^02 28^01
      306   29 223 16.2111    08^02 10^01 14^11 16^09 18^03 20^02 28^01
      307   29 223 16.2111    06^01 12^04 14^12 16^01 18^10 28^01
      308   29 223 16.2111    06^01 12^06 14^05 16^10 18^05 20^01 28^01
      309   29 223 16.2111    06^01 10^01 12^03 14^08 16^09 18^05 20^01 28^01
      310   29 223 16.2111    06^01 08^01 12^01 14^10 16^10 18^04 20^01 28^01
      311   29 223 16.2111    04^01 10^03 16^24 28^01

      312   29 224 16.2801    11^06 13^05 15^04 17^10 19^02 21^01 28^01
313 - 314   29 224 16.2801    11^07 15^14 19^07 28^01
315 - 317   29 224 16.2801    09^01 11^02 13^10 15^04 17^05 19^06 28^01
318 - 321   29 224 16.2801    09^01 11^04 13^05 15^06 17^09 19^02 21^01 28^01
322 - 325   29 224 16.2801    09^02 11^02 13^05 15^08 17^08 19^02 21^01 28^01
      326   29 224 16.2801    09^03 13^05 15^10 17^07 19^02 21^01 28^01
327 - 328   29 224 16.2801    07^01 11^01 13^10 15^05 17^06 19^05 28^01
      329   29 224 16.2801    07^01 11^03 13^05 15^07 17^10 19^01 21^01 28^01
      330   29 224 16.2801    07^01 11^04 15^17 19^06 28^01
      331   29 224 16.2801    07^01 09^01 11^01 13^05 15^09 17^09 19^01 21^01
                              28^01
      332   29 224 16.2801    07^02 13^05 15^10 17^10 21^01 28^01
      333   29 224 16.2801    07^02 11^01 15^20 19^05 28^01
      334   29 224 16.2801    05^01 11^02 13^05 15^08 17^10 19^02 28^01
      335   29 224 16.2801    05^01 09^01 13^05 15^10 17^09 19^02 28^01
```

```
      336   29 227 16.4833   10^01 12^08 14^04 16^06 18^06 20^03 28^01
337 - 338   29 227 16.4833   10^02 12^05 14^07 16^05 18^06 20^03 28^01
      339   29 227 16.4833   10^03 12^02 14^10 16^04 18^06 20^03 28^01
340 - 341   29 227 16.4833   10^04 12^02 14^04 16^11 18^06 22^01 28^01
342 - 343   29 227 16.4833   08^01 12^06 14^06 16^07 18^05 20^03 28^01
344 - 345   29 227 16.4833   08^01 10^01 12^03 14^09 16^06 18^05 20^03 28^01
      346   29 227 16.4833   08^01 10^02 14^12 16^05 18^05 20^03 28^01
347 - 348   29 227 16.4833   08^01 10^02 12^03 14^03 16^13 18^05 22^01 28^01
      349   29 227 16.4833   08^02 12^04 14^02 16^15 18^04 22^01 28^01
      350   29 227 16.4833   06^01 12^04 14^08 16^07 18^06 20^02 28^01
      351   29 227 16.4833   06^01 10^01 12^01 14^11 16^06 18^06 20^02 28^01
      352   29 227 16.4833   04^01 10^01 12^02 14^04 16^14 18^06 28^01

353 - 354   29 228 16.5498   11^05 13^05 15^06 17^06 19^05 21^01 28^01
355 - 358   29 228 16.5498   09^01 11^03 13^05 15^08 17^05 19^05 21^01 28^01
359 - 360   29 228 16.5498   09^02 13^10 17^14 21^02 28^01
      361   29 228 16.5498   09^02 11^01 13^05 15^10 17^04 19^05 21^01 28^01
362 - 363   29 228 16.5498   07^01 11^02 13^05 15^09 17^06 19^04 21^01 28^01
      364   29 228 16.5498   07^01 09^01 13^05 15^11 17^05 19^04 21^01 28^01
      365   29 228 16.5498   05^01 13^10 17^16 21^01 28^01
      366   29 228 16.5498   05^01 11^01 13^05 15^10 17^06 19^05 28^01

367 - 369   29 231 16.7460   10^02 12^04 14^06 16^08 18^03 20^05 28^01
370 - 371   29 231 16.7460   10^02 12^05 14^04 16^07 18^08 20^01 22^01 28^01
372 - 373   29 231 16.7460   10^03 12^02 14^07 16^06 18^08 20^01 22^01 28^01
      374   29 231 16.7460   08^01 12^05 14^05 16^10 18^02 20^05 28^01
      375   29 231 16.7460   08^01 10^01 12^02 14^08 16^09 18^02 20^05 28^01
376 - 377   29 231 16.7460   08^01 10^01 12^03 14^06 16^08 18^07 20^01 22^01
                             28^01
      378   29 231 16.7460   08^02 14^10 16^10 18^01 20^05 28^01
      379   29 231 16.7460   06^01 12^03 14^07 16^10 18^03 20^04 28^01
      380   29 231 16.7460   06^01 12^04 14^05 16^09 18^08 22^01 28^01
      381   29 231 16.7460   04^01 12^02 14^07 16^09 18^08 20^01 28^01

382 - 383   29 232 16.8102   11^04 13^06 15^04 17^08 19^04 21^02 28^01
384 - 385   29 232 16.8102   09^01 11^02 13^06 15^06 17^07 19^04 21^02 28^01
386 - 387   29 232 16.8102   09^02 13^06 15^08 17^06 19^04 21^02 28^01
388 - 389   29 232 16.8102   09^02 11^03 15^07 17^14 19^01 23^01 28^01
390 - 391   29 232 16.8102   07^01 11^01 13^06 15^07 17^08 19^03 21^02 28^01
      392   29 232 16.8102   07^01 11^04 15^06 17^16 23^01 28^01
      393   29 232 16.8102   05^01 13^06 15^08 17^08 19^04 21^01 28^01
      394   29 232 16.8102   03^01 11^02 15^08 17^16 19^01 28^01

      395   29 235 17.0000   12^09 16^09 18^06 20^03 22^01 28^01
      396   29 235 17.0000   10^01 12^06 14^03 16^08 18^06 20^03 22^01 28^01
397 - 398   29 235 17.0000   10^02 12^03 14^06 16^07 18^06 20^03 22^01 28^01
      399   29 235 17.0000   10^03 14^09 16^06 18^06 20^03 22^01 28^01
      400   29 235 17.0000   08^01 12^04 14^05 16^09 18^05 20^03 22^01 28^01
```

```
401   29 235 17.0000    08^01 10^01 12^01 14^08 16^08 18^05 20^03 22^01
                         28^01
402   29 235 17.0000    06^01 12^02 14^07 16^09 18^06 20^02 22^01 28^01
403   29 235 17.0000    04^01 14^09 16^09 18^06 20^03 28^01

404   29 236 17.0623    11^04 13^03 15^08 17^06 19^04 21^03 28^01
405   29 236 17.0623    11^05 13^02 15^05 17^10 19^05 23^01 28^01
406   29 236 17.0623    09^01 11^02 13^03 15^10 17^05 19^04 21^03 28^01
407   29 236 17.0623    09^01 11^03 13^02 15^07 17^09 19^05 23^01 28^01
408   29 236 17.0623    09^02 13^03 15^12 17^04 19^04 21^03 28^01
409   29 236 17.0623    09^02 11^01 13^02 15^09 17^08 19^05 23^01 28^01
410   29 236 17.0623    07^01 11^01 13^03 15^11 17^06 19^03 21^03 28^01
411   29 236 17.0623    07^01 11^02 13^02 15^08 17^10 19^04 23^01 28^01
412   29 236 17.0623    05^01 13^03 15^12 17^06 19^04 21^02 28^01
413   29 236 17.0623    03^01 13^02 15^10 17^10 19^05 28^01

414   29 239 17.2462    10^01 12^04 14^06 16^04 18^10 20^01 22^02 28^01
415   29 239 17.2462    10^02 12^02 14^04 16^13 20^06 22^01 28^01
416   29 239 17.2462    10^04 14^01 16^12 18^10 24^01 28^01
417   29 239 17.2462    08^01 12^02 14^08 16^05 18^09 20^01 22^02 28^01
418   29 239 17.2462    08^01 10^02 12^01 16^14 18^09 24^01 28^01
419   29 239 17.2462    06^01 14^10 16^05 18^10 22^02 28^01
420   29 239 17.2462    06^01 12^01 14^05 16^15 20^05 22^01 28^01
421   29 239 17.2462    02^01 12^01 16^16 18^10 28^01

422   29 240 17.3066    11^03 13^05 15^03 17^10 19^05 21^01 23^01 28^01
423   29 240 17.3066    09^01 11^01 13^05 15^05 17^09 19^05 21^01 23^01
                         28^01
424   29 240 17.3066    07^01 13^05 15^06 17^10 19^04 21^01 23^01 28^01

425   29 239 17.3899    10^02 12^08 18^16 24^02 26^01

426   29 243 17.4853    10^02 12^02 14^03 16^08 18^10 20^02 24^01 28^01
427   29 243 17.4853    08^01 12^03 14^02 16^10 18^09 20^02 24^01 28^01

428   29 244 17.5440    09^03 17^24 25^01 28^01
429   29 244 17.5440    01^01 17^27 28^01

430   29 243 17.5887    10^04 14^06 18^16 22^01 26^02

431   29 247 17.7178    12^04 14^06 18^16 22^01 24^01 28^01

432   29 248 17.7750    09^01 11^02 15^04 17^14 19^06 25^01 28^01

433   29 255 18.1652    10^02 16^08 18^16 20^01 26^01 28^01

434   29 271 19.0000    18^27 28^02
```

```
435 - 436  30 244 17.2111   12^01 13^16 20^12 28^01

437 - 438  30 247 17.3859   12^07 14^09 16^01 19^06 21^06 28^01
      439  30 247 17.3859   10^01 12^04 14^12 19^06 21^06 28^01

440 - 441  30 248 17.4434   11^02 12^01 13^08 15^06 18^02 20^08 22^02 28^01

      442  30 251 17.6139   12^07 14^05 16^05 17^01 19^05 21^05 23^01 28^01
      443  30 251 17.6139   10^01 12^04 14^08 16^04 17^01 19^05 21^05 23^01
                            28^01

444 - 445  30 252 17.6700   11^02 12^01 13^06 15^06 17^02 18^04 20^04 22^04
                            28^01
      446  30 252 17.6700   11^04 12^01 15^12 16^01 20^10 24^01 28^01
      447  30 252 17.6700   09^01 12^01 13^06 15^08 17^01 18^04 20^04 22^04
                            28^01

      448  30 255 17.8365   12^06 14^04 16^06 17^02 18^01 19^04 21^04 23^02
                            28^01
      449  30 255 17.8365   10^01 12^03 14^07 16^05 17^02 18^01 19^04 21^04
                            23^02 28^01
      450  30 255 17.8365   08^01 12^01 14^09 16^06 17^02 19^04 21^04 23^02
                            28^01

      451  30 256 17.8913   11^02 12^01 13^04 15^06 16^01 17^04 18^02 20^06
                            22^02 24^01 28^01
      452  30 256 17.8913   11^03 12^01 15^12 18^06 19^01 22^06 28^01
      453  30 256 17.8913   09^01 12^01 13^04 15^08 16^01 17^03 18^02 20^06
                            22^02 24^01 28^01
      454  30 256 17.8913   07^01 12^01 15^15 18^06 22^06 28^01

      455  30 256 17.9589   11^04 13^04 14^02 16^01 17^08 20^08 24^01 26^02

      456  30 259 18.0539   10^01 12^03 14^03 15^01 16^09 18^01 19^05 21^05
                            25^01 28^01

      457  30 260 18.1075   12^01 13^08 16^02 17^08 20^08 24^02 28^01

      458  30 264 18.3190   11^02 12^01 14^01 15^06 17^08 18^01 20^08 22^01
                            26^01 28^01

      459  30 276 18.9282   12^02 17^16 20^10 28^02

      460  31 269 18.2621   13^12 18^12 23^06 28^01
```

461 31 272 18.4136 12^03 13^02 14^06 16^01 17^06 19^06 22^03 24^03
                     28^01
462 31 272 18.4136 10^01 13^02 14^09 17^06 19^06 22^03 24^03 28^01

463 31 272 18.4462 12^06 15^06 16^04 19^08 22^04 26^03

464 31 273 18.4637 11^01 13^06 15^05 16^02 18^08 20^02 21^01 23^04
                     25^01 28^01

465 31 276 18.6125 12^02 13^02 14^04 15^02 16^04 17^02 19^08 22^04
                     24^01 26^01 28^01

466 31 284 19.0000 13^04 16^10 19^10 22^05 28^02

467 32 290 18.9833 12^02 14^07 17^12 20^03 22^03 25^04 28^01

468 32 291 19.0292 13^04 14^03 15^02 16^06 18^06 21^06 24^02 26^02
                     28^01

469 32 295 19.2111 14^06 15^06 18^09 21^06 24^03 28^02

470 33 309 19.5498 14^03 15^08 17^08 20^06 23^04 26^02 28^02

471 34 326 20.0000 16^20 22^10 28^04
472 34 326 20.0000 13^01 16^15 19^10 25^05 28^03

473 36 364 21.0000 18^28 28^08

Bibliography

[AiDo]   Aigner M., Dowling T. A., A geometric characterization of the line graph of a symmetric balanced incomplete block design, *Studia Sci. Math. Hungar.* 7(1972), 137–145.

[AkHa]   Akiyama J., Harary F.: A graph and its complement with specified properties VII: a survey. in *The theory and applications of graphs*, Proc. Fourth Internat. Conf. on Theory and Appl. of Graphs, (Eds. Chartrand G., Alavi Y., Goldsmith D. L, Lesniak-Foster L., Lick D. R), Wiley (New York) 1981, pp.1–12.

[BaCLS]   Balińska K., Cvetković D., Lepović M., Simić S., There are exactly 150 connected integral graphs up to 10 vertices, *Univ. Beograd, Publ. Elektrotehn. Fak., Ser. Mat.* 10(1999), 95–105.

[BaCRSS]   Balińska K. T., Cvetković D., Radosavljević Z., Simić S. K., Stevanović D., A survey on integral graphs, *Univ. Beograd, Publ. Elektrotehn. Fak., Ser. Mat.* 13(2002), 42–65.

[BaKZ]   Balińska K., Kupczyk M., Zwierzyński K. *Methods of generating integral graphs*, Computer Science Center Report No. 457, Technical University of Poznań, 1997.

[BaKSZ1]   Balińska K. T., Kupczyk M., Simić S. K., Zwierzyński K. T., *On generating all integral graphs on 11 vertices*, Computer Science Center Report No. 469, Technical University of Poznań, 1999/2000.

[BaKSZ2]   Balińska K. T., Kupczyk M., Simić S. K., Zwierzyński K. T., *On generating all integral graphs on 12 vertices*, Computer Science Center Report No. 482, Technical University of Poznań, 2001.

[BaSi1]   Balińska K. T., Simić S. K., Some remarks on integral graphs with maximum degree four, *Novi Sad J. Math.* 31(2001), 19–25.

[BaSi2]   Balińska K. T., Simić S. K., The nonregular, bipartite, integral graphs with maximum degree four, *Discrete Math.* 236(2001), 13–24.

[BaSi3]   Balińska K. T., Simić S. K., Which nonregular, bipartite, integral graphs with maximum degree four do not have ±1 as eigenvalues? *Discrete Math.*, to appear.

[BaLa1]   Bapat R. B., Lal A. K., Path-positive graphs, *Linear Algebra Appl.* 149(1991), 125–149.

[BaLa2]   Bapat R. B., Lal A. K., Path positivity and infinite Coxeter groups, *Linear Algebra Appl.* 196(1994), 19–35.

[Bei1]   Beineke L. W., Characterization of derived graphs, *J. Combin. Theory* 9(1970), 129–135.

[Bei2]   Beineke L. W.: Derived graphs with derived complements, in: *Recent Trends in Graph Theory*, Springer Verlag (Berlin), Lecture Notes in Mathematics Vol. 186 (1971), pp.15–24.

[Bel1]   Bell F. K., Characterizing line graphs by star complements, *Linear Algebra Appl.* 296(1999), 15–25.

[Bel2]   Bell F. K., Line graphs of bipartite graphs with Hamiltonian paths, *J. Graph Theory* 43(2003), 137–149.

[BeCRS]   Bell F. K., Cvetković D., Rowlinson P., Simić S. K., Some additions to the theory of star partitions of graphs, *Discussiones Math. – Graph Theory*, 19(1999), 119–134.

[BLMS]   Bell F. K., Li Marzi E. M., Simić S. K., Some new results on graphs with least eigenvalue not less than -2, to appear.

[BeRo]   Bell F. K., Rowlinson P., On the multiplicities of graph eigenvalues, *Bull. London Math. Soc.* 35(2003), 401–408.

[BeSi]   Bell F. K., Simić S. K., On graphs whose star complement for -2 is a path or cycle, *Linear Algebra Appl.* 377(2004), 249–265.

[BeHa]   Benzaken C., Hammer P. L., Linear separation of dominating sets in graphs, *Annals Discrete Math.* 3(1978), 1–10.

[Big]   Biggs N. L., *Algebraic Graph Theory* (second edition), Cambridge University Press, Cambridge, 1993.

[Bos]   Bose R. C., Strongly regular graphs, partial geometries and partially balanced designs, *Pacific J. Math.* 13(1963), 389–419.

[Bra]   Branković Lj., Usability of secure statistical databases, Ph.D Thesis, University of Newcastle (Australia), 1998.

[BrMS]   Branković Lj., Miller M., Siráň J., Graphs, (0, 1)-matrices and usability of statistical databases, *Congressus Numerantium* 120(1996), 169–182.

[BrCv]   Branković Lj., Cvetković D., The eigenspace of the eigenvalue -2 in generalized line graphs and a problem in the security of statistical databases, *Univ. Beograd, Publ. Elektrotehn. Fak., Ser. Mat.*, to appear.

[BrMe]   Bridges W. G., Mena R. A., Multiplicative cones – a family of three-eigenvalue graphs, *Aequationes Math.* 22(1981), 208–214.

[BrCN]   Brouwer A. E., Cohen A. M., Neumaier A., *Distance Regular Graphs*, Springer-Verlag (Berlin), 1989.

[BrLi]   Brouwer A. E., van Lint J. H., Strongly regular graphs and partial geometries, in: *Enumeration and Design* (Eds. Jackson, D. M., Vanstone S. A.), Academic Press (Toronto), 1984, pp.85–122.

[BuCST]   Bussemaker F C., Cameron P. J., Seidel J. J., Tsaranov S. V., Tables of signed graphs. Eindhoven University of Technology, Report 91-WSK-01, Eindhoven, 1991.

[BuCv]   Bussemaker F. C., Cvetković D., There are exactly 13 connected, cubic, integral graphs, *Univ. Beograd, Publ. Elektrotehn. Fak., Ser. Mat. Fiz.*, Nos. 544–576(1976), 43–48.

[BuCS1]   Bussemaker F. C., Cvetković D., Seidel J. J., Graphs related to exceptional root systems, T. H. Report 76-WSK-05, Univ. Eindhoven, 1976.

[BuCS2]   Bussemaker F. C., Cvetković D., Seidel J. J., Graphs related to exceptional root systems, in: *Combinatorics I, II*, Proc. 5th Hungarian Coll. on Combinatorics, Keszthely 1976 (Eds. Hajnal A., Sos V. T.), North-Holland (Amsterdam), 1978, pp.185–191.

[BuNe]   Bussemaker F. C., Neumaier A., Exceptional graphs with smallest eigenvalue -2 and related problems, *Mathematics of Computation* 59(1992), 583–608.

[Cam]   Cameron P. J., A note on generalized line graphs, *J. Graph Theory* 4(1980), 243–245.

[CaGSS]   Cameron P. J., Goethals J. M., Seidel J. J., Shult E. E., Line graphs, root systems, and elliptic geometry, *J. Algebra* 43(1976), 305–327.

[CaLi]   Cameron P. J., van Lint J. H., *Designs, Graphs, Codes and their Links*, Cambridge University Press, 1991.

[CaST]   Cameron P. J., Seidel J. J., Tsaranov S. V., Signed graphs, root lattices and Coxeter groups, *J. Algebra* 164(1994), 173–209.

[Card]   Cardoso D. M., Convex quadratic programming approach to the maximum matching problem, *J. Global Opt.* 21(2001), 91–106.

[Car]   Carter R. W., *Simple Groups of Lie Type*, Wiley, 1972.

[CaHo]   Cao D., Hong Y., Graphs characterized by the second eigenvalue, *J. Graph Theory* 17(1993), 325–331.

[Cha1]   Chang L. C., The uniqueness and non-uniquennes of the triangular association scheme, *Sci. Record* 3(1959), 604–613.

[Cha2]   Chang L. C, Association schemes of partially balanced block designs with parameters $v = 28$, $n_1 = 12$, $n_2 = 15$, $p_{11}^2 = 4$, *Sci. Record* 4(1960), 12–18.

[ChVi]   Chawathe P. D., Vijayakumar G. R., A characterization of signed graphs represented by root system D_∞, *Europ. J. Combinatorics* 11(1990), 523–533.

[Clarke4] Clarke F. H., A graph polynomial and its application. *Discrete Math.* 3(1972), 305–315.

[Conn]   Connor W. S., The uniqueness of the triangular association scheme, *Ann. Math. Statist.* 29(1958), 262–266.

[Cons]   Constantine G., Lower bound on the spectra of symmetric matrices with non-negative entries, *Linear Algebra Appl.* 65(1985), 171–178.

[Cox]   Coxeter H. S. M., Discrete groups genereted by reflections, *Ann. Math.* 35(1934), 588–621.

[Cve1]   Cvetković D., Bipartiteness and the spectrum of a graph (Serbian), in: *Uvodjenje mladih u naučni rad VI*, Matematička Biblioteka 41, Janić R.R., Zavod za izdavanje udžbenika SR Srbije, Beograd, 1969, pp.193–194.

[Cve2]   Cvetković D., Graphs and their spectra (Thesis), *Univ. Beograd, Publ. Elektrotehn. Fak. Ser. Mat. Fiz.* No. 354–No. 356 (1971), 1–50.

[Cve3]   Cvetković D., Spectrum of the total graph of a graph, *Publ. Inst. Math. (Beograd)* 16(30)(1973), 49–52.

[Cve4]   Cvetković D., Spectra of graphs formed by some unary operations, *Publ. Inst. Math. (Beograd)* 19(33)(1975), 37–41.

[Cve5]   Cvetković D., Cubic integral graphs, *Univ. Beograd, Publ. Elektrotehn. Fak. Ser. Mat. Fiz.* No. 498–No. 541 (1975), 107–113.

[Cve6]   Cvetković D., The main part of the spectrum, divisors and switching of graphs, *Publ. Inst. Math. (Beograd)* 23(37)(1978), 31–38.

[Cve7]   Cvetković D., Some possible directions in further investigation of graph spectra, in: *Algebraic Methods in Graph Theory*, Vols. I, II, (Eds. Lovász L., Sós V. T.), North Holland (Amsterdam) 1981, pp.47–67.

[Cve8]   Cvetković D., On graphs whose second largest eigenvalue does not exceed 1, *Publ. Inst. Math (Beograd)* 31(45)(1982), 15–20.

[Cve9]   Cvetković D., Spectral characterizations of line graphs. Variations on the theme, *Publ. Inst. Math. (Beograd)* 34(48)(1984), 31–35.

[Cve10]   Cvetković D., Star partitions and the graph isomorphism problem, *Linear and Multilinear Algebra* 39(1995), 109–132.

[Cve11]   Cvetković D., Characterizing properties of some graph invariants related to electron charges in the Hückel molecular orbital theory, in: *Discrete Mathematical Chemistry (DIMACS Workshop, March 1998)*, American Mathematical Society (Providence, RI), *DIMACS Series in Discrete Mathematics and Theoretical Computer Science* 51(2000), 79–84.

[Cve12]   Cvetković D., On the reconstruction of the characteristic polynomial of a graph, *Discrete Math.* 212(2000), 45–52.

[Cve13]   Cvetković D., Graphs with least eigenvalue −2; a historical survey and recent developments in maximal exceptional graphs, *Linear Algebra Appl.* 356(2002), 189–210.

[CvDo1]   Cvetković D., Doob M., On spectral characterizations and embeddings of graphs, *Linear Algebra Appl.* 27(1979), 17–26.

[CvDo2]   Cvetković D., Doob M., Root systems, forbidden subgraphs and spectral characterizations of line graphs, in: *Graph Theory*, Proc. Fourth Yugoslav Seminar on Graph Theory, Novi Sad, April 15–16. 1983, (Eds. Cvetković D., Gutman I., Pisanski T., Tošić R.), Univ. Novi Sad Inst. Math., Novi Sad, 1984, pp.69–99.

[CvDG]   Cvetković D., Doob M., Gutman I., On graphs whose eigenvalues do not exceed $\sqrt{2 + \sqrt{5}}$, *Ars Combinatoria* 14(1982), 225–239.

[CvDGT]   Cvetković D., Doob M., Gutman I., Torgašev A., *Recent Results in the Theory of Graph Spectra*, North-Holland (Amsterdam), 1988.

[CvDSa]   Cvetković D., Doob M., Sachs H., *Spectra of Graphs*, 3rd edition, Johann Ambrosius Barth Verlag (Heidelberg), 1995.

[CvDS1]   Cvetković D., Doob M., Simić S., Some results on generalized line graphs, *Comptes Rendus Math. Rep. Acad. Sci. Canada* 2(1980), 147–150.

[CvDS2]   Cvetković D., Doob M., Simić S., Generalized line graphs, *J. Graph Theory* 5(1981), 385–399.

[CvGu]   Cvetković D., Gutman I., On the spectral structure of graphs having the maximal eigenvalue not greater than two, *Publ. Inst. Math. (Beograd)* 18(32)(1975), 39–45.

[CvGT]   Cvetković D., Gutman I., Trinajstić N., Conjugated molecules having integral graph spectra, *Chem. Phys. Letters* 29(1974), 65–68.

[CvLe]   Cvetković D., Lepović M., Seeking counterexamples to the reconstruction conjecture for the characteristic polynomial of a graph and a positive result,

Bull. Acad. Serbe Sci. Arts, Cl. Sci. Math. Natur., Sci. Math., 116(1998), No. 23, 91–100.

[CvLRS1] Cvetković D., Lepović M., Rowlinson P., Simić S., A database of star complements of graphs, *Univ. Beograd, Publ. Elektrotehn. Fak., Ser. Mat.* 9(1998), 103–112.

[CvLRS2] Cvetković D., Lepović M., Rowlinson P., Simić S., The maximal exceptional graphs, *J. Combin. Theory Ser. B* 86(2002), 347–363.

[CvLRS3] Cvetković D., Lepović M., Rowlinson P., Simić S., Computer investigations of the maximal exceptional graphs, Technical Report CSM-160, Department of Computing Science and Mathematics, University of Stirling, 2001.

[CvPe] Cvetković D., Petrić M., A table of connected graphs on six vertices, *Discrete Math.* 50(1984), 37–49.

[CvRa1] Cvetković D., Radosavljević Z., A Construction of the 68 connected regular graphs, non-isomorphic but cospectral to line graphs, in: *Graph Theory*, Proc. Fourth Yugoslav Seminar on Graph Theory, Novi Sad, 15–16 April 1983, (Eds. Cvetković D., Gutman I., Pisanski T., Tošić R.), Univ. Novi Sad Inst. Math., Novi Sad, 1984, pp.101–123.

[CvRa2] Cvetković D., Radosavljević Z., A table of regular graphs with at most 10 vertices, in: *Graph Theory*, Proc. VI Yugoslav Seminar on Graph Theory, Dubrovnik, 18–19 April 1985, (Eds. Tošić R., Acketa D., Petrović V.), Univ. Novi Sad Inst. Math., Novi Sad, 1986, pp.71–105.

[CvRo1] Cvetković D., Rowlinson P., Some properties of graph angles, *Scientia (Valparaiso) Ser. A* 1(1988), 41–51.

[CvRo2] Cvetković D., Rowlinson P., Some results in the theory of graph spectra, in: *Selected Topics in Algebraic Graph Theory* (eds. Beineke L. W, Wilson R. J.), to appear.

[CvRS1] Cvetković D., Rowlinson P., Simić S. K., A study of eigenspaces of graphs, *Linear Algebra Appl.* 182(1993), 45–66.

[CvRS2] Cvetković D., Rowlinson P., Simić S. K., *Eigenspaces of Graphs*, Cambridge University Press (Cambridge), 1997.

[CvRS3] Cvetković D., Rowlinson P., Simić S. K., Some characterizations of graphs by star complements, *Linear Algebra Appl.*, 301(1999), 81–97.

[CvRS4] Cvetković D., Rowlinson P., Simić S. K., Graphs with least eigenvalue −2: the star complement technique, *J. Algebraic Combinatorics* 14(2001), 5–16.

[CvRS5] Cvetković D., Rowlinson P., Simić S. K., Constructions of the maximal exceptional graphs with largest degree less than 28, Department of Computing Science and Mathematics, University of Stirling, Technical Report CSM-156, 2000.

[CvRS6] Cvetković D., Rowlinson P., Simić S. K., The maximal exceptional graphs with largest degree less than 28, *Bull. Acad. Serbe Sci. Arts, Cl. Sci. Math. Natur., Sci. Math.* 22(2001), No. 26, 115–131.

[CvRS7] Cvetković D., Rowlinson P., Simić S., Graphs with least eigenvalue -2; a new proof of the 31 forbidden subgraphs theorem, *Designs, Codes and Cryptography*, to appear.

[CvSi1]   Cvetković D., Simić S., Some remarks on the complement of a line graph, *Publ. Inst. Math. (Beograd)* 17(31)(1974), 37–44.

[CvSi2]   Cvetković D., Simić S., Graph equations for line graphs and total graphs, *Discrete Math.* 13(1975), 315–320.

[CvSi3]   Cvetković D., Simić S., Graph equations, in: *Beiträge zur Graphentheorie und deren Anwendungen*, vorgetragen auf dem Internat. Koll. Oberhof (DDR), 10–16 April 1977, pp.40–56.

[CvSi4]   Cvetković D., Simić S., Graphs which are switching equivalent to their line graphs, *Publ. Inst. Math. (Beograd)* 23(37)(1978), 39–51.

[CvSi5]   Cvetković D., Simić S., A bibliography of graph equations, *J. Graph Theory* 3(1979), 311–324.

[CvSi6]   Cvetković D., Simić S., Graph-theoretical results obtained by the support of the expert system "Graph", *Bull. Acad. Serbe Sci. Arts, Cl. Sci. Math. Natur., Sci. Math.* 107(1994), No. 19, 19–41.

[CvSi7]   Cvetković D., Simić S., On the graphs whose second largest eigenvalue does not exceed $(\sqrt{5} - 1)/2$, *Discrete Math.* 138(1995), 213–227.

[CvSi8]   Cvetković D., Simić S., The second largest eigenvalue of a graph – a survey, Int. Conf. on Algebra, Logic & Discrete Math., Niš, 14–16 April 1995 (Eds. Bogdanović S., Ćirić M., Perović Ž.), *FILOMAT* 9(1995) No. 3, 449–472.

[CvSi9]   Cvetković D., Simić S., Minimal graphs whose second largest eigenvalue is not less than $(\sqrt{5} - 1)/2$, *Bull. Acad. Serbe Sci. Arts, Cl. Sci. Math. Natur., Sci. Math.* 121(2000), No. 25, 47–70.

[CvSS]   Cvetković D., Simić S. and Stevanović D., 4-regular integral graphs, *Univ Beograd, Publ. Elektrotehn. Fak., Ser. Mat.* 9(1998), 89–102.

[CvSt]   Cvetković D., Stevanović D., Graphs with least eigenvalue at least $-\sqrt{3}$, *Publ. Inst. Math. (Beograd)*, 73(87) (2003), 249–265.

[Dam1]   van Dam E. R., Regular graphs with four eigenvalues, *Linear Algebra Appl.* 228(1995), 139–162.

[Dam2]   van Dam E. R., Nonregular graphs with three eigenvalues, *J. Combin. Theory Ser. B*, 73(1998), 101–118.

[DaSp]   van Dam E. R., Spence E., Small regular graphs with four eigenvalues, *Discrete Math.* 189(1998), 233–257.

[DeGr]   Deza M., Grishukhin P., Hypermetric graphs, *Quart. J. Math. Oxford* 44(1993), 399–433.

[DePo]   Dedo E., Porcu L., Algebraic properties of line multidigraphs of a multidigraph, *J. Combin. Inform. System Sci.* 12(1987), 113–118.

[Doo1]   Doob M., On characterizing a line graph by the spectrum of its adjacency matrix, Ph. D. Thesis, The City University of New York, 1969.

[Doo2]   Doob M., A geometrical interpretation of the least eigenvalue of a line graph., in: *Proc. Second Conference on Comb. Math. and Appl.*, Univ. North Carolina, Chapel Hill NC, 1970, pp.126–135.

[Doo3]   Doob M., Graphs with a small number of distinct eigenvalues, *Ann. New York Acad. Sci.* 175(1970), 104–110.

[Doo4]   Doob M., On characterizing certain graphs with four eigenvalues by their spectra, *Linear Algebra Appl.* 3(1970), 461–482.

[Doo5]   Doob M., Graphs with a small number of distinct eigenvalues, *Ann. New York Acad. Sciences* 175(1970), 104–110.

[Doo6]   Doob M., On the spectral characterization of the line graph of a BIBD., in: *Proc. Second Louisiana Conf. on Comb., Graph Theory and Computing, 8–11 March 1971* (Eds. Mullin R. C., Reid K. B., Roselle D. P., Thomas R. D.), Louisiana State University, Baton Rouge LA, 1971, pp.225–234.

[Doo7]   Doob M., On the spectral characterization of the line graph of a BIBD, II., in: *Proc. Manitoba Conference on Numerical Mathematics*, University of Manitoba, 1971, pp.117–126.

[Doo8]   Doob M., An interrelation between line graphs, eigenvalues, and matroids, *J. Combin. Theory Ser. B* 15(1973), 40–50.

[Doo9]   Doob M., A spectral characterization of the line graph of a BIBD with $\lambda = 1$, *Linear Algebra Appl.* 12(1975), 11–20.

[Doo10]   Doob M., A note on eigenvalues of a line graph., in: *Proc. Conf. on Algebraic Aspects of Combinatorics, University of Toronto, January 1975* (Eds. Corneil D., Mendelsohn E.), *Congressus Numerantium XIII*, Utilitas Math. (Winnipeg, Manitoba), 1975, pp.209–211.

[Doo11]   Doob M., A note on prime graphs, *Utilitas Math.* 9(1976), 297–299.

[Doo12]   Doob M., A surprising property of the least eigenvalue of a graph, *Linear Algebra Appl.* 46(1982), 1–7.

[Doo13]   Doob M., The limit points of eigenvalues of graphs, *Linear Algebra Appl.* 114/115(1989), 659–662.

[DoCv]   Doob M., Cvetković D., On spectral characterizations and embedding of graphs, *Linear Algebra Appl.* 27(1979), 17–26.

[DoLa]   Dowling T. A., Lascar R., A geometric characterization of the line graph of a projective plane, *J. Combinatorial Theory* 3(1967), 402–410.

[Ell]   Ellingham M. N., Basic subgraphs and graph spectra, *Australasian J. Combinatorics* 8(1993), 247–265.

[Err]   Erriksson K., Convergence of Mozes's game of numbers, *Linear Algebra Appl.* 166(1992), 151–165.

[Fin]   Finck H.-J., Vollstandiges Produkt, chromatische Zahl und characteristisches Polynom regularer Graphen II, *Wiss. Z., T. H. Ilmenau* 11(1965), 81–87.

[Gan]   Gantmacher F.R., *The Theory of Matrices*, Chelsea (New York), 1959.

[GoRo]   Godsil C., Royle G., *Algebraic Graph Theory*, Springer (New York), 2001.

[GoHJ]   Goodman F. M., de la Harpe P., Jones V. F. R., *Coxeter Graphs and Towers of Algebras*, Springer-Verlag (New York), 1989.

[GuCv]   Gutman I., Cvetković D., The reconstruction problem for characteristic polynomials of graphs, *Univ. Beograd, Publ. Elektrotehn. Fak., Ser. Mat. Fiz.*, Nos. 498–541(1975), 45–48.

[Hae]   Haemers W., A generalization of the Higman-Sims technique, *Proc. Kon. Ned. Akad. Wet. A* 81(4)(1978), 445–447.

[Hall]   Hall M., Jr., *The Theory of Groups*, Macmillan (New York), 1959.

[Har]   Harary F., *Graph Theory*, Addison-Wesley (Reading, Mass.), 1969.

[HaSc1]   Harary F., Schwenk A. J, Which graphs have integral spectra?, in: *Graphs and Combinatorics, Proc. Capit. Conf. Graph Theory and Combinatorics, George Washington Univ., June 1973* (Eds. Bari R. and Harary F.), Springer-Verlag, 1974, pp. 45–51.

[HaSc2]   Harary F., Schwenk A. J., The spectral approach to determining the number of walks in a graph, *Pacific J. Math.* 80(1979), 443–449.

[HaWe]   de la Harpe P., Wenzl H., Operations sur le rayons spectraux de matrices symmetriques entières positives. *Comptes Rendues Acad. Sci. Paris* 305(1987), 733–736.

[Hay]   Haynsworth E. V., Applications of a theorem on partitioned matrices, *J. Res. Nat. Bureau Stand.* 62(1959), 73–78.

[Haz]   Hazama F., On the kernels of the incidence matrices of graphs, *Discrete Math.* 254(2002), 165–174.

[Hemm]   Hemminger, R. L., On reconstructing a graph, *Proc. Amer. Math. Soc.* 20(1969), 185–187.

[HeBe]   Hemminger R. L., Beineke L. W., Line graphs and line digraphs, in: *Selected Topics in Graph Theory* (Eds. Beineke L. W., Wilson R. J.), Academic Press (New York), 1978, pp.127–167.

[HeHi]   Hestenes M., Higman D. G., Rank 3 groups and strongly regular graphs, in: *Computers in Linear Algebra and Number Theory*, SIAM-AMS Proc., Vol. IV, Providence RI, 1971, 141–159.

[Hof1]   Hoffman A. J., On the exceptional case in a characterization of the arcs of a complete graph, *IBM J. Res. Develop.* 4(1960), 487–496.

[Hof2]   Hoffman A. J., On the uniqueness of the triangular association scheme, *Ann. Math. Stat.* 31(1960), 492–497.

[Hof3]   Hoffman A. J., On the polynomial of a graph, *Amer. Math. Monthly* 70(1963), 30–36.

[Hof4]   Hoffman A. J., On the line graph of the complete bipartite graph, *Ann. Math. Stat.* 35(1964), 883–885.

[Hof5]   Hoffman A. J., On the line graph of a projective plane, *Proc. Amer. Math. Soc.* 16(1965), 297–302.

[Hof6]   Hoffman A. J., Some recent results on spectral properties of graphs, in: *Beiträge zur Graphentheorie* (Eds. Sachs H., Voss H-J., Walther H.), Teubner (Leipzig), 1968, pp.75–80.

[Hof7]   Hoffman A. J., The change in the least eigenvalue of the adjacency matrix of a graph under embedding, *SIAM J. Appl. Math.* 17(1969), 664–671.

[Hof8]   Hoffman A. J., $-1 - \sqrt{2}$, in: *Combinatorial Structures and their Applications*, Proc. Calgary International Conf. Combinatorial Structures and their Applications, June 1969 (Eds. Guy R., Hanani H., Sauer N., Schönheim J.), Gordon and Breach (New York), 1970, pp.173–176.

[Hof9]   Hoffman A. J., On spectrally bounded graphs, in: *A Survey of Combinatorial Theory* (Eds. Srivastava N. *et al*), North Holland, 1973, pp.277–283.

[Hof10]   Hoffman A. J., Eigenvalues and the partitioning of the edges of a graph, *Linear Algebra Appl.* 5(1972), 137–148.

[Hof11]   Hoffman A. J., On graphs whose least eigenvalues exceeds $-1 - \sqrt{2}$, *Linear Algebra Appl.* 16(1977), 153–165.

[Hof12]   Hoffman A. J., On limit points of the least eigenvalue of a graph, *Ars Combinatoria* 3(1977), 3–14.

[HoJa]   Hoffman A. J., Jamil B. A., On the line graph of the complete tripartite graph, *Linear and Multilinear Algebra* 5(1977), 19–25.

[HoMc]   Hoffman A. J., McAndrew M. H., The polynomial of a directed graph, *Proc. Amer. Math. Soc.* 16(1965), 303–309.

[HoRa1]   Hoffman A. J., Ray-Chaudhuri D. K., On the line graph of a finite affine plane, *Canad. J. Math.* 17(1965), 687–694.

[HoRa2]   Hoffman A. J., Ray-Chaudhuri D. K., On the line graph of a symmetric balanced incomplete block design, *Trans. Amer. Math. Soc.* 116(1965), 238–252.

[HoRa3]   Hoffman A. J., Ray-Chaudhuri D. K., On a spectral characterization of regular line graphs, unpublished manuscript.

[Hon]   Hong Y., On the least eigenvalue of a graph, *System Sci. Math. Sci.* 6(1993), 269–275.

[How1]   Howes L., On subdominantly bounded graphs, Thesis, City Univ. of New York, 1970.

[How2]   Howes L., On subdominantly bounded graphs – summary of results, in: *Recent Trends in Graph Theory*, Proc. First New York City Graph Theory Conf., 11–13 June 1970 (Eds. Capobianco M., Frechen J. B., Krolik M.), Springer-Verlag (Berlin), 1971, pp.181–183.

[Hum]   Humphreys J. E., *Reflection groups and Coxeter groups*, Cambridge University Press, Cambridge, 1990.

[HuPi]   Hughes D. R., Piper F. C., *Design Theory*, Cambridge University Press (Cambridge), 1985.

[JaRo]   Jackson P. S., Rowlinson P., On graphs with complete bipartite star complements, *Linear Algebra Appl.* 298(2000), 9–20.

[Kra]   Krausz J., Demonstration nouvelle d'une théorème de Whitney sur les reseaux, *Mat. Fiz. Lapok* 50(1943), 75–89.

[KuRS]   Kumar V., Rao S. B., Singhi N. M., Graphs with eigenvalues at least −2, *Linear Algebra Appl.* 46(1982), 27–42.

[LaSo]   Lai H.-J., Šoltés Ľ., Line graphs and forbidden induced subgraphs, *J. Combin. Theory Ser. B* 82(2001), 38–55.

[Leh]   Lehot P. G. H., An optimal algorithm to detect a line graph and output its root graph, *J. Assoc. Comput. Machinery* 21(1974), 569–575.

[LeSe]   Lemmens P. W. H., Seidel J. J., Equiangular lines, *J. Algebra* 24(1973), 494–512.

[Li]   Li J., Ph.D. Thesis, The University of Manitoba, Winnipeg, 1994.

[Lov]   Lovász, L. A note on the line reconstruction problem, *J. Combin. Theory Ser. B* 13(1972), 309–310.

[LuPS]   Lubotzky A., Phillips R., Sarnak P., Ramanujan graphs, *Combinatorica* 8(1988), 261–277.

[MaMi]   Marcus M., Minc H., *A Survey of Matrix Theory and Matrix Inequalities*, Allyn and Bacon (Boston, Mass.), 1964.

[Max]   Maxwell G., Hyperbolic Trees, *J. Algebra* 54(1978), 46–49.

[MiHu]   Milnor J., Husemoller D., Symmetric bilinear forms, Springer Verlag (New York), 1973.

[Moo]   Moon J. W., On the line graph of the complete bigraph, *Ann. Math. Stat.* 34(1963), 664–667.

[MuKl]   Muzychuk M., Klin M., On graphs with three eigenvalues, *Discrete Math.* 189(1998), 191–207.

[Neu1]   Neumaier A., The second largest eigenvalue of a tree, *Linear Algebra Appl.* 46(1982), 9–25.

[Neu2]   Neumaier A., Characterization of a class of distance-regular graphs, *J. Reine Angew. Math.* 357(1985), 182–192.

[NeSe]   Neumaier A., Seidel J. J., Discrete hyperbolic geometry, *Combinatorica* 3(1983), 219–237.

[Nuff]   van Nuffelen C., On the rank of the incidence matrix of a graph, *Cahiers Centre Étud. Rech. Opér. (Bruxelles)* 15(1973), 363–365.

[Pal]   Palmer E. M., Prime line graphs, *Nanta Math.* 6(1973), 75–76.

[PeSa]   Petersdorf M., Sachs H., Über Spektrum, Automorphismengruppe und Teiler eines Graphen, *Wiss. Z., T. H. Ilmenau*, 15(1969), 123–128.

[Pet1]   Petrović M., On graphs whose spectral spread does not exceed 4, *Publ. Inst. Math. (Beograd)* 34(48)(1983), 169–174.

[Pet2]   Petrović M., On graphs with exactly one eigenvalue less than -1, *J. Combin. Theory Ser. B* 52(1991), 102–112.

[Pet3]   Petrović M., On graphs whose second largest eigenvalue does not exceed $\sqrt{2} - 1$. *Univ. Beograd, Publ. Elektrotehn. Fak., Ser. Mat.* 4(1993), 70–75.

[PeMi1]   Petrović M., Milekić B., On the second largest eigenvalue of line graphs, *J. Graph Theory* 27(1998), 61–66.

[PeMi2]   Petrović M., Milekić B., On the second largest eigenvalue of generalized line graphs, *Publ. Inst. Math. (Beograd)* 68(82)(2000), 37–45.

[PeRa]   Petrović M., Radosavljević Z., *Spectrally Constrained Graphs*, Faculty of Science, University of Kragujevac, 2001.

[Pra]   Prasolov V. V., *Problems and Theorems in Linear Algebra*, Amer. Math. Soc. (Providence, RI), 1994.

[Pri]   Prisner E., Line graphs and generalizations – a survey, *Congr. Numer.* 116(1996), 193–230.

[Rad]   Radosavljević Z., Inequivalent regular factors of regular graphs on 8 vertices, *Publ. Inst. Math. (Beograd)* 29(43)(1981), 171–190.

[RaSi1]   Radosavljević Z., Simić S., Computer aided search for all graphs such that both graph and its complement have its spectrum bounded from below by -2, *Ars Combinatoria A* 24(1987), 21–27.

[RaSi2]   Radosavljević Z., Simić S., There are just thirteen connected nonregular nonbipartite integral graphs having maximum vertex degree four, in: *Graph Theory*, Proc. IV Yugoslav Seminar on Graph Theory, Dubrovnik, 18–19 April 1985 (Eds. Tošić R., Acketa D., Petrović V.), University of Novi Sad Institute of Mathematics, Novi Sad, 1987, pp.183–187.

[RaSi3]   Radosavljević Z., Simić S., Which bicyclic graphs are reflexive? *Univ. Beograd, Publ. Elektrotehn. Fak. Ser. Mat.* 7(1996), 90–104.

[RaSST]   Radosavljević Z., Simić S., Syslo M., Topp J., A note on generalized line graphs, *Publ. Inst. Math. (Beograd)* 34(48)(1983), 193–198.

[RaST]   Radosavljević Z., Simić S., Tuza Zs., Complementary pairs of graphs orientable to line digraphs, *J. Comb. Math. Comb. Comp.* 13(1993), 65–75.

[RaRa]   Rao S. B., Rao A. R., A characterization of the line graph of a BIBD with $\lambda = 1$, *Sankhyā A* 31(1969), 369–370.

[RaSV1]   Rao S. B., Singhi N. M., Vijayan K. S., The minimal forbidden subgraphs for generalized line-graphs, in: *Combinatorics and Graph Theory*,

Proc. Symp. Indian Statistical Institute, Calcutta, 25–29 February 1980 (Ed. Rao S.B.), Lecture Notes in Math. 885, Springer-Verlag (Berlin), 1981, pp.459–472.

[RaSV2] Rao S. B., Singhi N. M., Vijayan K. S., Spectral characterization of the line graph of K_ℓ^n, in: *Combinatorics and Graph Theory*, Proc. Symp. Indian Statistical Institute, Calcutta, 25–29 February 1980 (Ed. Rao S.B.), Lecture Notes in Math. 885, Springer-Verlag (Berlin), 1981, pp.473–480.

[Ray] Ray-Chaudhuri D. K, Characterization of line graphs, *J. Combin. Theory* 3(1967), 201–214.

[RaSiV] Ray-Chaudhuri D. K., Singhi N. M., Vijayakumar G. R., Signed graphs having least eigenvalue around -2, *J. Combin. Inform. System Sci.* 17(1992), 148–165.

[RoWi] van Rooij A. C. M., Wilf H. S., The interchange graphs of a finite graph, *Acta Math. Acad. Sci. Hungarica* 16(1965), 263–270.

[Rou] Roussopoulos N. D., A max$\{m, n\}$ algorithm for determining the graph H from its line graph, *Information Processing Letters* 2(1982), 28–30.

[Row1] Rowlinson P., Eutactic stars and graph spectra, in: *Combinatorial and Graph-Thoeretical Problems in Linear Algebra* (Eds. Brualdi R. A., Friedland S. and Klee V.), Springer-Verlag (New York), 1993, pp.153–164.

[Row2] Rowlinson P., On graphs with multiple eigenvalues, *Linear Algebra Appl.* 283(1998), 75–85.

[Row3] Rowlinson P., Star sets and star complements in finite graphs: a spectral construction technique, in: *Discrete Mathematical Chemistry*, American Mathematical Society (Providence, RI), Proc. DIMACS Workshop, March 1998, *DIMACS Series in Discrete Mathematics and Theoretical Computer Science* 51(2000), 323–332.

[Row4] Rowlinson P., Star complements and the maximal exceptional graphs, *Publ. Inst. Math. (Beograd)*, to appear.

[Row5] Rowlinson P., Star complements in finite graphs: a survey, *Rendiconti Sem. Mat. Messina* 8(2002), 145–162.

[RoJa] Rowlinson P., Jackson P. S., Star complements and switching in graphs, *Linear Algebra Appl.* 356(2002), 145–156.

[Rys] Ryser H. J., Geometries and incidence matrices, *Amer. Math. Monthly* 62(1955), No. 7, part II, 25–31.

[Sac1] Sachs H., On a theorem connecting the factors of a regular graph with the eigenvectors of its line graph, in: *Combinatorics I, II*, Proc. 5th Hungarian Coll. on Combinatorics, Keszthely 1976 (Eds. Hajnal A., Sós V. T.), North-Holland (Amsterdam), 1978, pp.947–957.

[Sac2] Sachs H., Über Teiler, Faktoren und charakteristische Polynome von Graphen, Teil I, *Wiss. Z., T. H. Ilmenau* 12(1966), 7–12.

[Sac3] Sachs H., Über Teiler, Faktoren und charakteristische Polynome von Graphen, Teil II, *Wiss. Z., T. H. Ilmenau* 13(1967), 405–412.

[Schm] Schmeichel, E. F., A note on the reconstruction conjecture, *Bull. Austral. Math. Soc.* 12(1975), 27–30.

[Sch1] A. J. Schwenk, Exactly thirteen connected cubic graphs have integral spectra, in: *Theory and Applications of Graphs*, Proc. Internat. Conf. Western

Michigan Univ., Kalamazoo, Mich., May 1976 (Eds. Alavi Y., Lick D.), Springer-Verlag, 1978, pp.516–533.

[Sch2]   Schwenk A. J., Spectral reconstruction problems, *Ann. N. Y. Acad. Sci.* 328(1978), 183–189.

[Sci]   Sciriha I., Polynomial reconstruction: old and new techniques, *Rendiconti Sem. Mat. Messina* 8(2002), 163–179.

[Sei1]   Seidel J. J., Strongly regular graphs with $(-1, 1, 0)$ adjacency matrix having eigenvalue 3, *Linear Algebra Appl.* 1(1968), 281–298.

[Sei2]   Seidel J. J., A survey of two-graphs, in: *Proc. Coll. Theorie Combinatorie*, Acc. Naz. Lincei, Roma, 1976, pp.481–511.

[Sei3]   Seidel J. J., On two-graphs and Shult's characterization of symplectic and orthogonal geometries over $GF(2)$, T. H. Report 73-WSK-02, Univ. Eindhoven, 1973.

[Sei4]   Seidel J. J., Graphs and two-graphs, in: Proc. 5th South Eastern Conf. on Combinatorics, Graph Theory and Computing, Boca Raton (Fla), 1974, *Congr. Num.* X, Utilitas Math., Winnipeg, Man., 1974, pp.125–143.

[Sei5]   Seidel J. J., Eutactic stars, in: *Combinatorics* (Eds. Hajnal A., Sós V. T.), North-Holland (Amsterdam), 1978, pp.983–989.

[Sei6]   Seidel J. J., Strongly regular graphs, in: *Surveys in Combinatorics* (Ed. Bollobás B.), Cambridge University Press, 1979, pp.157–180.

[Sei7]   Seidel J. J., Graphs and their spectra, in: *Combinatorics and Graph Theory*, Banach Center Publ. Vol. 25, PWN Polish Scientific Publ., Warsaw, 1989, pp.147–162.

[Sei8]   Seidel J. J., A note on path-zero graphs, *Discrete Math.* 106–107(1992), 435–438.

[Seid]   Seiden E., On a geometric method of construction of partially balanced designs with two associate classes, *Ann. Math. Stat.* 32(1961), 1177–1180.

[Shr1]   Shrikhande S. S., On a characterization of the triangular association scheme, *Ann. Math. Stat.* 30(1959), 39–47.

[Shr2]   Shrikhande S. S., On the uniqueness of the L_2 association scheme, *Ann. Math. Stat.* 30(1959), 781–798.

[ShBh]   Shrikhande S. S., Bhagwandas, Duals of incomplete block designs, *J. Indian Stat. Assoc.* 3(1965), 30–37.

[Sim1]   Simić S. K., *Graph Equations* (Serbian), Master's Thesis, University of Belgrade, 1977.

[Sim2]   Simić S. K., On the decomposition of the line (total) graphs with respect to some binary operations, *Publ. Inst. Math (Beograd)* 24(38)(1978), 163–172.

[Sim3]   Simić S. K., *Contributions to the investigations of graph operations* (Serbian), Ph.D Thesis, University of Belgrade, 1979.

[Sim4]   Simić S. K., An algorithm to recognize a generalized line graph and its root graph, *Publ. Inst. Math. (Beograd)* 49(63)(1990), 21–26.

[Sim5]   Simić S. K., A note on reconstructing the characteristic polynomial of a graph, in: Proc. 4th Czechoslovakian Symposium on Combinatorics, Graphs and Complexity, (Eds. Nešetřil J., FiedlerM.), Elsevier Science Publishers, 1992, pp.315–319.

[Sim6]   Simić S. K., Some notes on graphs whose second largest eigenvalue is less than ($\sqrt{5} - 1)/2$, *Linear and Multilinear Algebra* 39(1995), 59–71.

[Sim7]   Simić S. K, Complementary pairs of graphs with the second largest eigenvalue not exceeding ($\sqrt{5} - 1)/2$, *Publ. Inst. Math. (Beograd)* 57(71)(1995), 179–188.

[Sim8]   Simić S., Arbitrarily large graphs whose second largest eigenvalue is less than ($\sqrt{5} - 1)/2$, *Rendiconti Sem. Mat. Messina* 8(2002), 181–205.

[Sim9]   Simić S., Graphs which are switching equivalent to their complementary line graphs I, *Publ. Inst. Math. (Beograd)* 27(41)(1980), 229–235.

[Sim10]   Simić S. K., Graphs which are switching equivalent to their complementary line graphs II, *Publ. Inst. Math. (Beograd)* 31(45)(1982), 183–194.

[SiRa]   Simić S. K., Radosavljević Z., The nonregular, nonbipartite, integral graphs with maximum degree four, *J. Combin. Inform. System Sci.* 20(1995), 9–26.

[SiVi]   Singhi N. M., Vijayakumar G. R., Signed graphs with least eigenvalue < -2, *Europ. J. Combinatorics* 13(1992), 219–220.

[Smi]   Smith J. H., Some properties of the spectrum of a graph, in: *Combinatorial Structures and Their Applications* (Eds. Guy R., Hanani H., Sauer N., Schönheim J.), Gordon and Breach (New York), 1970, pp.403–406.

[Sol]   Šoltés Ľ., Forbidden induced subgraphs for line graphs, *Discrete Math.* 132(1994), 391–394.

[Ste1]   Stevanović D., 4-Regular integral graphs avoiding ±3 in the spectrum, to appear.

[Ste2]   Stevanović D., Nonexistence of some 4-regular integral graphs, *Univ. Beograd, Publ. Elektrotehn. Fak., Ser. Mat.* 10(1999), 81–86.

[Ter1]   Terwilliger P., Distance regular graphs with girth 3 or 4, I, *J. Combin. Theory, Ser. B* 39(1985), 265–281.

[Ter2]   Terwilliger P., A class of distance regular graphs that are Q-polynomial, *J. Combin. Theory, Ser. B* 40(1986), 213–223.

[Ter3]   Terwilliger P., A new feasibility condition for distance regular graphs, *Discrete Math.* 61(1986), 311–315.

[Ter4]   Terwilliger P., Root systems and the Johnson and Hamming graphs, *Europ. J. Combinatorics* 8(1987), 73–102.

[Ter5]   Terwilliger P., Root system graphs, *Linear Algebra Appl.* 94(1987), 157–163.

[TeDe]   Terwilliger P., Deza M., The classification of finite connected hypermetric spaces, *Graphs Comb.* 3(1987), 293–298.

[Tor1]   Torgašev A., The spectrum of line graphs of some infinite graphs, *Publ. Inst. Math. (Beograd)* 31(45)(1982), 209–222.

[Tor2]   Torgašev A., On infinite graphs whose spectrum is greater than -2, *Bull. Acad. Serbe Sci. Arts, Cl. Sci. Math. Nat., Sci. Math.* 13(1984), 21–35.

[Tor3]   Torgašev A., A note on infinite generalized line graphs, in: *Graph Theory*, Proc. 4th Yugoslav Seminar on Graph Theory, Novi Sad, 15–16 April 1983, (Eds. Cvetković D., Gutman I., Pisanski T., Tošić R.), Univ. Novi Sad Inst. Math., Novi Sad, 1984, pp.291–297.

[Tor4]   Torgašev A., Infinite graphs with the least limiting eigenvalue greater than -2, *Linear Algebra Appl.* 82(1986), 133–141.

[Tor5]   Torgašev A., On the proper normal generalized line digraphs, *Bull. Acad. Serbe Sci. Arts, Cl. Sci. Math. Nat., Sci. Math.* 104(1992), No. 18, 1–6.

[Tut1]   Tutte W. T., All the king's horses: a guide to reconstruction, in: *Graph Theory and Related Topics*, Proc. Conf. University of Waterloo, Waterloo, Ont. 1977, (Eds. Bondy J. A., Murty U. S. R.), Academic Press (New York), 1979, pp.15–33.

[Tut2]   Tutte W. T., Lectures on matroids, *J. Res. Nat. Bur. Standards Sect. B* 69(1965), 1–47.

[Vij1]   Vijayakumar G. R., A characterization of generalized line graphs and classification of graphs with eigenvalues at least -2, *J. Combin. Inform. System Sci.* 9(1984), 182–192.

[Vij2]   Vijayakumar G. R., Signed graphs represented by D_∞, *Europ. J. Combinatorics* 8(1987), 103–112.

[Vij3]   Vijayakumar G. R., Signed graphs represented by root system E_8, in: *Combinatorial Mathematics and Applications* (Calcutta, 1988), Sankya, Ser. A 54(1992), 511–517.

[Vij4]   Vijayakumar G. R., Algebraic equivalence of signed graphs with all eigenvalues ≥ -2, *Ars Combinatoria* 35(1993), 173–191.

[Vij5]   Vijayakumar G. R., Representation of signed graphs by root system E_8, *Graphs Comb.* 10(1994), 383–388.

[ViSi]   Vijayakumar G. R., Singhi N. M., Some recent results on signed graphs with least eigenvalues ≥ -2, in: *Coding Theory and Design Theory, Part II: Design Theory*, Proc. Workshop IMAI Program Appl. Comb., Minneapolis, MN (USA) 1987–88, IMA Vol. Math. Appl. 21(1990), pp. 213–218.

[Whi]   Whitney H., Congruent graphs and the connectivity of graphs, *Amer. J. Math.* 54(1932), 150–168.

[Wol]   Wolk E. S., A note on the comparability graph of a tree, *Proc. Amer. Math. Soc.* 16(1965), 17–20.

[WoNe]   Woo R., Neumaier A., On graphs whose smallest eigenvalue is at least $-1 - \sqrt{2}$, *Linear Algebra Appl.* 226–228(1995), 577–591.

[Yong]   Yong X., On the distribution of eigenvalues of a simple undirected graph, *Linear Algebra Appl.* 295(1999), 73–80.

[Zas]   Zaslavsky T., The geometry of root systems and signed graphs, *Amer. Math. Monthly* 88(1981), No.1, 88–105.

Index